The Race between Education and Technology

The Race between Education and Technology

CLAUDIA GOLDIN

LAWRENCE F. KATZ

The Belknap Press of
Harvard University Press
Cambridge, Massachusetts, and London, England   2008

Library of Congress Cataloging-in-Publication Data
Goldin, Claudia Dale.
   The race between education and technology / Claudia Goldin, Lawrence F. Katz.
       p.   cm.
   Includes bibliographical references and index.
   ISBN 978-0-674-02867-8 (alk. paper)
     1. Education—Economic aspects—United States.   2. Economic development—Effect
of education on—United States.   3. Education—Effect of technological innovations on—
United States.   4. Human capital—United States.   I. Katz, Lawrence F.   II. Title.
LC66.G65   2008
     338.4'737—dc22   2007045158

Contents

The Race between Education and Technology

Introduction

At the dawn of the twentieth century the United States became the richest nation in the world. Its people had a higher average standard of living than those in Britain, the previous leader. America was poised to ascend further. The gap between it and other front-runners would widen and the standard of living of its residents would continue to grow, even when its doors were open to the world's poor. American economic supremacy would be maintained to the end of the century, and beyond. In economic terms, the twentieth century fully merits the title "The American Century."

The twentieth century could also be titled the "Human Capital Century." By the end of the twentieth century all nations, even the poorest, provided elementary schooling and beyond to most of their citizens. At the start of the century and even by its midpoint many nations, including relatively rich ones, educated only those who could personally afford to attend school. The United States was different. Its educational system had always been less elite than those of European countries. By 1900, if not before, it had begun to educate its masses at the secondary level not just in primary schools, at which it had remarkable success in the nineteenth century.

That the twentieth century was both the American Century *and* the Human Capital Century is no historical accident. Economic growth in the more modern period requires educated workers, managers, entrepreneurs, and citizens. Modern technologies must be invented, innovated, put in place, and maintained. They must have capable workers at the helm. Rapid technological advance, measured in various ways, has characterized the twentieth century. Because the American people were the most educated in the world, they were in the best position to invent, be entrepreneurial, and produce goods and services using advanced technologies.

The connection between the American Century and the Human Capital Century concerns the role of education in economic growth and individual productivity. A greater level of education results in higher labor productivity. Moreover, a greater level of education in the entire nation tends to foster a higher rate of aggregate growth. The nation that invested the most in education, and did much of that investment during the century in which education would critically matter, was the nation that had the highest level of per capita income.

We do not mean to imply that economic growth is a simple matter of investing in education. If it were, then any poor nation could invest in education, wait a few years, and reap enormous economic returns. But given a set of important preconditions, such as the type of government and the security of property rights, the notion that the American Century and the Human Capital Century occurred together follows directly from the relationships among growth, technology, and education. Invest in education, get higher levels of technology and productivity, and attain a rapid rate of economic growth and a higher standard of living. However, the benefits from economic growth might be unequally distributed and a high average standard of living might not translate into betterment for all.

If these statements regarding the role of education in technological change and economic growth are correct, then rapid technological change would also increase the demand for more educated workers at all levels. With increased demand for their services, the earnings of the more educated would rise relative to the less educated. In the absence of an increased supply of educated workers, the gap between the earnings of those with more and less education would expand. If there are various educational groups in society, ranging from the lowest to the most highly educated, and if the groups were fixed in relative proportions, then technological advances would unambiguously increase economic inequality since the relative income gap between the less and the more educated would rise. Yet if, in addition to technological progress, the quantity and possibly the quality of education increases, then inequality could decrease.

The American Century with its great technological progress and economic growth could well have been an era of ever-expanding inequality. Economic growth could have entailed considerably more income for some, with scant increases, if any, for others. Instead, the first three-quarters of the American Century was an era of long-term economic growth and *declining* inequality. For much of the twentieth century the gains from economic growth became *more* equally distributed. But by the end of the 1970s, an abrupt and substantial rise in economic inequality ensued. In addition, average real wage growth slowed. In the last three decades of the century there were times when most Americans gained, although those at the top gained considerably more. However, there were also times when the real incomes of those in the lowest third of the distribution stagnated.

The economic well-being of Americans increased monumentally and almost continuously throughout the twentieth century despite various setbacks, such as several small recessions and the Great Depression of the 1930s. Income per capita in 2000 was five to six times its level in 1900, using standard measures of income and of the price level. Adjustments to the quality of goods and services would serve to increase the figure, perhaps substantially. The rate of increase in the income of the entire nation, known as Gross Domestic Product (GDP), was rapid and remarkably constant across the century at around 3.2 percent average annually. On a per capita basis, the measure increased somewhat more after the 1940s. From 1900 to 1929 real income per capita increased by about 1.7 percent on an average annual basis. It increased to 1.9 percent after 1950. Thus, there was a slight acceleration in economic growth in per capita terms across the century.

In sharp contrast to economic growth, which was relatively continuous, economic inequality was highly discontinuous. The twentieth century contains two distinct inequality components. Inequality initially declined, in several stages, from 1900 to about the third quarter of the century. Inequality then rose, often spectacularly, to the end of the century. By most measures, economic inequality is now as high as it was prior to its great decline. That is, inequality today is as high as it was during the Great Depression and probably for some time before.

One of the key links between these two parts of the economic system—technological change and inequality—is educational progress. Educational attainment, as measured by the completed schooling levels of successive cohorts, was exceptionally rapid and continuous for the first three-quarters of the twentieth century. But educational advance slowed considerably for young adults beginning in the 1970s and for the overall labor force by the early 1980s. For cohorts born from the 1870s to about 1950, every decade was accompanied by an increase of about 0.8 years of education. During that 80-year period the vast majority of parents had children whose educational attainment greatly exceeded theirs. Educational change between the generations then came to an abrupt standstill. An important part of the American dream, that children will do better than their parents, was threatened, and this danger was even greater than the educational data would suggest. The reason is that the decrease in inequality and then the increase in inequality during the American Century are mimicked in another important economic indicator—productivity change.

Productivity change in the United States, as measured by the increase in output per worker hour, had been rapid during much of the twentieth century but it slowed during the latter part. The slowdown, it appears, ended in the late 1990s, but not soon enough. National income was considerably lower than it would have been had productivity change kept pace. In fact, the only reason that real income per capita could sustain its rapid clip in the face of a slowing of labor productivity was that the labor force expanded more rapidly than did the population. Americans were running harder just to maintain their previous rate of economic growth.

At the beginning of the twentieth century America was confident, even exuberant. There were, to be sure, industries such as steel and chemicals that still lagged behind their European competitors. But a vast sea of manufactured goods flowed from American ports. In industries such as book publishing, carriages, business machines, agricultural equipment, and industrial machinery Americans were portrayed as invaders. We were, as well, superior producers of raw and semiprocessed goods, such as grains, flour, meat, leather, and a host of nonreproducible resources including petroleum. In the first two decades of the twentieth century America emerged as the world's leading producer of manufactured goods, including the automobile—the symbol of the modern age.

America's economic competitors watched over their shoulders to see what Americans were doing and what they could emulate. The British, aware that they were losing their competitive edge, were frantically searching for "the secret of American success," as one account, *The American Invaders*, noted. Chief among the reasons offered for American supremacy was "their better education." Americans were winning the "battle" of economic competition with their "brains, enterprise and energy . . . their longer worker hours, their willingness to receive new ideas, their better plant, and perhaps most of all . . . their freedom from hampering traditions."[1] Some of these "hampering traditions" concerned education.

Today, at the start of the twenty-first century, the United States is somewhat less exuberant than it was a hundred years earlier. It had once demonstrated to the world the importance of universal education. The nations of Europe and Asia eventually followed America's lead, and some, in recent years, have begun to exceed U.S. high school and college graduation rates among younger cohorts. On standardized reading, math, and science exams the United States has lagged considerably, as demonstrated by the Third [also Trends in] International Math and Science Study (TIMSS) and Program for International Assessment (PISA).

The U.S. educational system from almost its inception was built on a set of "virtues" that contain many elements of American egalitarianism. The existence of slavery and the absence of equal access to education for most free African Americans during slavery and for some time after must qualify our use of the term *egalitarianism*. By the mid-nineteenth century schooling was, for most European-descent children, publicly funded, open, forgiving, gender neutral in most respects, secular, and publicly provided by a multitude of competing school districts.[2] In the chapters that follow we will explain in detail what we mean by each of these virtues and why they were meritorious at many moments in U.S. history. The important point here is that these virtues once furthered education at all levels but that they appear, to some, to be failing us today.

Rising inequality, lagging productivity for a prolonged period, and a rather non-stellar educational report card have led many to question the qualities that once made America the envy of all and a beacon for the world's people. Americans have never been complacent about the quality of their children's schooling, and the recent past has brought an onslaught of proposed and enacted reforms. Many of these reforms alter the qualities of American schooling that had been the virtues of the past. Vouchers, charter schools, public funding for church-based schools, and high-stakes testing with real consequences are some of the enacted reforms. Whether the virtues have run their course and whether the reforms will have praiseworthy outcomes is yet to be determined.

More important is that we have developed a form of collective amnesia about our past accomplishments. We may well be doing something wrong now that we once did right and there may be ways of altering our institutions to create an even more productive and equitable society. But an obsession with current problems has caused us to forget the special and spectacular history of American education and has led us, as well, to overlook the fact that higher education in America is still the finest in the world.

Our recent experience with rising inequality has also led to several misunderstandings about the role of technological change in the economy. Advancing technology does *not* inevitably produce an increase in the relative demand for skilled and educated workers. Grand technological changes during much of the nineteenth century probably did not increase the relative demand for skill; however, during most of the twentieth century technological change did increase the relative demand for skill and therefore was skill biased.

Rapid technological change does not always increase economic inequality, even when it is skill biased. Similarly, rising inequality in the latter part of the twentieth century does not imply that the rate of technological change accelerated the relative demand for educated and skilled workers. Economic inequality can decrease even with rapidly increasing demands for educated workers. Likewise, soaring inequality need not be due to acceleration in the relative demand for educated workers. In both instances, the supply of educated workers could be varying, increasing rapidly at some times and slowing down at others. This scenario is precisely what happened. One must not overlook the crucial other half of the inequality equation: the supply side.

The supply of educated Americans increased greatly and almost unceasingly from 1900 to around 1980. The enormous increase in educational attainment in the early part of the twentieth century came primarily from a grass-roots movement that propelled the building and staffing of public high schools. It was not due to a top-down mandate or pressure from the federal government, nor did it a result from powerful local interest groups or arise because of legal compulsion. Later in the century, after high schools had spread and attendance in them had grown, the expansion of state colleges and universities led to further increases in education.

But after around 1980 the supply of educated Americans slowed considerably. The sluggish growth in the educated workforce in the last quarter century has been mainly due to a slowing down in the educational attainment of those schooled in the United States, rather than to an increase in the foreign-born component of the workforce.

This book concerns a remarkable century of economic growth, technological change, advancing education, and even a narrowing of inequality during many of its years. It is about a unique set of enabling institutions that allowed the United States to have mass education and a level of schooling that far exceeded that of other rich nations until late in the twentieth century. It is also about why rapid technological advance in the twentieth century did not produce ever-increasing inequality and why the fruits of economic growth were often more equally distributed, at least until the last three decades.

The book is also concerned with what many see as the current malaise. In the years since the end of the 1970s, economic inequality has widened to levels as great as they were in the early twentieth century. We address the educational response and why, after generations of enormous advance, increases in years of schooling have stalled. The slowdown in the growth of educational attainment has been most extreme and disturbing for those at the bottom of the income distribution, particularly for racial and ethnic minorities. On the positive side, however, educational advances for women relative to men have been substantial. In fact, gender differences in both education and earnings relative to those for comparable men have gone against the general rising inequality tide of the past three decades.

The three main topics of this book—technological change, education, and inequality—are intricately related in a kind of "race." During the first three-quarters of the twentieth century, the rising supply of educated workers outstripped the increased demand caused by technological advances. Higher real incomes were accompanied by lower inequality. But during the last two decades of the century the reverse was the case and there was sharply rising inequality. Put another way, in the first half of the century, education raced ahead of technology, but later in the century, technology raced ahead of educational gains.[3] The skill bias of technology did not change much across the century, nor did its rate of change. Rather, the sharp rise in inequality was largely due to an educational slowdown.

The virtues of the past may not function as well in the present, and part of the recent rise in inequality may be a consequence. We are not advocating a particular way of fixing the system, but some aspects about what is wrong are fairly obvious. We end the book with a discussion of these issues.

I

ECONOMIC GROWTH AND DISTRIBUTION

1

The Human Capital Century

Investment in physical capital became vital to a nation's economic growth with the onset of the Industrial Revolution in the nineteenth century. But the path to ongoing economic success for nations and individuals eventually became investment in human capital. Human capital became supreme in the twentieth century and America led the way. At the start of that century, Americans had embraced the novel idea that the "wealth of nations" would be embodied in its human capital stock; it would take even the richer nations of Europe about five decades or more to catch on to that notion.

For most Americans in the early twentieth century access to schooling, at least through high school, was largely unlimited by personal station and residence. Education was publicly provided and funded and was free of direct charge, except at the highest levels. Even the most rural Americans had the privilege of sending their children to public secondary schools, although African Americans, especially in the South, were often excluded from various levels of schooling, especially above the common school.[1] Americans had a strong tradition of educating their youth at public charge and the expansion of education beyond the common school and elementary grades continued a commitment rooted in basic democratic and egalitarian principles. These "virtues" were many and together they created a commitment to equality of opportunity.[2]

Rather than being praised throughout the world, American educational institutions in the early twentieth century were roundly criticized. "There are those who are critical of the American high school because its doors are open to pupils of all types," said one notable commentator.[3] Because they allowed youths of all abilities to use public resources, European observers termed the American educational system "wasteful." In contrast, most European national systems tested boys and girls at an early age and promoted only the best, a system they viewed as meritocratic.[4] But targeting talent at a young age, such as at age 11, privileged those with social standing and more educated parents.[5] Americans chafed at selection and deemed it elitist.[6] Their system was not improvident; it was egalitarian.

By the early twentieth century America educated its youth to a far greater extent than did most, if not every, European country. Secondary schools in America were free and generally accessible, whereas they were costly and often inaccessible in most of Europe. Even by the 1930s America was virtually alone in providing universally free and accessible secondary schools.

America's approach to schooling was critically important to its technological dynamism, rapid economic growth, more equal income distribution, assimilation of great waves of immigrants, and transition to mass college education. In this chapter we present the trends in formal educational attainment for the United States and make comparisons with European nations at various moments in the twentieth century. We also set forth a framework to understand the economic significance of human capital for individuals and for the nation. But first we must demonstrate how the twentieth century became the Human Capital Century, and why the Human Capital Century turned out to be the American Century.

Human Capital and Income across Nations

Worldwide Schooling Rates at the Beginning of the Twenty-First Century

By the end of the twentieth century no country could afford *not* to educate its citizens beyond the elementary grades. The technologies of richer nations have spread throughout the globe. Workers now have to

read complicated documents, master blueprints, work computers, solve formulas, and use the Internet, among other tasks. Simple literacy and numeracy are no longer sufficient. To be a full-fledged member of the global economy requires higher levels of education for most workers.

An educated citizenry does not guarantee rapid growth and inclusion in the "convergence club" of nations,[7] but the inverse of that statement is generally true.[8] Low levels of education nowadays prevent a nation from reaching the technological frontier and taking full advantage of the global economy. Most low-income nations today have schooling levels that are relatively high in contrast with historical standards. To demonstrate the enormous change in the role of education across the twentieth century, it is helpful to look at a cross section of nations at the start of the twenty-first century. We compare the secondary schooling rates and real income levels of more than one hundred nations at the start of the twenty-first century with the United States at various moments in its twentieth-century history.

Our demonstration reveals that at the beginning of the twenty-first century even nations with low income per capita have schooling rates that are high in comparison to the historical standards set by early twentieth-century America. When the United States had a per capita income level equal to that of many of today's low-income nations, its education rate was often less. The poorer nations of the current century appear to understand that they must endow their citizens with a secondary school education to operate in the world economy.

Consider Figure 1.1, in which the (net) secondary school enrollment rate is graphed against real gross domestic product (GDP) per capita, both in the year 2000, for 114 nations.[9] A positive correlation between income and schooling is revealed. Although the causality of the relationship (more education leads to greater income) has been the subject of much research, that is not the issue at the moment.[10] The main point of the figure is to demonstrate that secondary school enrollment rates in 2000 for low-income nations were substantial by the standards of the early to mid-twentieth century, even compared with higher-income nations of that period.

Because the United States was the leader in education in the twentieth century it will be the historical gold standard for this example. The solid vertical line that goes through the dot in Figure 1.1 gives U.S. real per capita GDP in 1900, immediately prior to the great

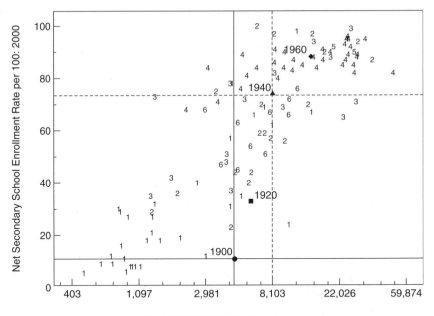

Figure 1.1. Net Secondary School Enrollment Rate and Real Per Capita GDP in 114 Countries: 2000. The numerical markers for each country refer to geographic groupings in the UNESCO data: 1 = Africa; 2 = Central and North America and the Caribbean; 3 = Asia, including the non-African Middle East; 4 = Europe; 5 = Oceania; and 6 = South America. The net secondary school enrollment rate nets out individuals who are not between the usual ages for students in these schools. The four markers with the circle (1900), square (1920), triangle (1940), and diamond (1960) refer to U.S. GDP/capita and net secondary school enrollment rates for the years given. Sources: Net secondary enrollment rates from United Nations Organization for Education, Science and Culture (UNESCO), http://unescostat.unesco.org/en/stats/stats0.htm. Real GDP/capita from the Penn World Table, http://pwt.econ.upenn.edu/ (Heston, Summers, and Aten 2002). "Real" in the Penn World Table means PPP (purchasing power parity).

increase of high school education in America. We will use the 1900 level of per capita GDP in year 2000 dollars, $4,596, as a loose definition of low per capita income in 2000.[11] The solid horizontal line, also going through the dot, represents the public and private secondary school rate that existed in the United States in 1900.[12] These two solid lines divide the graph into four quadrants. Our interest is in both the South-East and the North-West quadrants. Similar lines (dashed) have

been drawn for the point given by the triangle, which gives the data for 1940. A box corresponds to the data for 1920 and a diamond signifies 1960. Lines for those years have been omitted for clarity.

Our interest in the South-East quadrant is as follows. Any country situated in that quadrant has an *income per capita higher* than that existing in the United States in the year being considered but an *enrollment rate that is lower*. One might think of the area as the "bad education" outcome quadrant. There are no countries in the bad education quadrant when the year given is 1900 and just one when the comparison date is 1920.[13] For the 1940 comparison there are nine bad education countries out of the 114, but just five are clearly within the quadrant whereas the others are on the margin. And for 1960, when both the secondary school enrollment rate and income per capita are quite high, about a dozen nations are in the bad education quadrant, but only six are clearly within it.[14]

The point is that almost all countries with incomes greater than the United States in a particular year have a secondary school enrollment rate that exceeds that attained in the United States for that year. Something fundamental changed during the twentieth century in the relationship between income per capita and schooling.

The other quadrant of note is the North-West, which we term the "good education" quadrant.[15] Nations found here have *education levels that are higher* than that in the United States in the given year, but *income levels that are lower*. They are, in some sense, overachieving on the education front (or, alternatively, underachieving on the income front). The good education quadrant is generally more crowded than is the bad education quadrant, until the comparison year has both high income and high schooling levels.

Of the 42 nations in the year 2000 with per capita income below our 1900 low income standard, 15 (36 percent) had net secondary school enrollment rates that exceeded 0.4.[16] If the income standard is set higher, say U.S. per capita income in 1920, 53 nations are below that cutoff and 25 (47 percent) had enrollment rates in the year 2000 that exceeded 0.4. We have chosen a secondary school enrollment rate of 0.4 as the standard. Secondary school enrollment rates, including full-time pupils in either general or technical schools, attained in European nations in the mid-1950s were never more than 40 percent and they were generally far lower (see Figure 1.7, below). The low-income nations we just described using either the 1900 or the 1920 income standard, therefore, had secondary school enrollment rates that were considerable even by the standards of mid-twentieth-century Europe. From 36 to 47 percent of youth from those low-income nations were enrolled in secondary school although their per capita incomes were substantially lower in real terms than were those in the European comparison group. In real per capita terms the average European nation in 1955 had a per capita income that was between three to four times that of the median low-income nation in 2000, even using 1920 as the low-income benchmark.[17]

The assertion we made at the outset is borne out by these data. Today's low-income nations and their people invest in education to a far greater degree, in terms of secondary school enrollment rates, than did the richer countries of the past.[18] They do so to partake in a global economy.[19] Some may eventually succeed and attain high growth rates, while others have such serious structural problems that they may be swimming against a strong tide.

Gender Differences in Schooling across Nations

The less well off a country, not surprisingly, the lower is the secondary school enrollment rate for all youth. Of equal interest, but somewhat less obvious, is that low-income nations have higher relative enrollment rates for males. In fact, when income per capita is above that achieved by the United States in 1900 (the low-income standard we used previously), gender distinctions in enrollment evaporate. Yet almost all the nations with incomes below the low-income standard have relative enrollment rates that favor males and in about one third the advantages for males are substantial (see Figure 1.2). Many of the low-income nations with relatively low levels of female schooling are predominantly Muslim countries (as indicated by the size of the country marker in Figure 1.2) and most are the poorer nations of Africa. Gender neutrality seems to be a virtue purchased with higher incomes. Similarly, gender differences in enrollment favoring males are apparent only when the enrollment rate is below about 0.4 (Figure 1.3). Almost all the nations to the left of the 0.4 dotted line are low income nations, all but two of the 27 are in Africa, and most are predominantly Muslim.

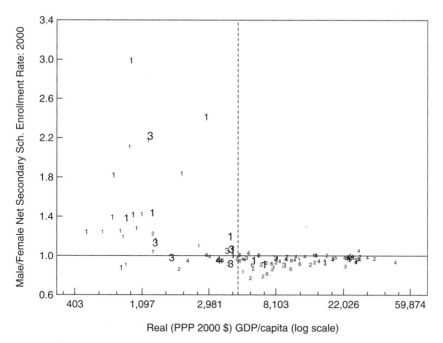

Figure 1.2. Ratio of Male to Female Secondary School Enrollment and Real GDP per Capita: 2000. The numerical markers for each country refer to the same regional groupings used in Figure 1.1. The size of the region marker indicates the fraction of the population that is Muslim. Sources: See Figure 1.1. Religion data by nation for the year 2000 are from Robert Barro (personal communication) who used Barrett, Kurian, and Johnson (2001).

We noted above that only the very poorest nations can afford not to educate their people. We now add to that concept by asserting that no nation today, save the very poorest, can afford not to educate its girls to the same degree as its boys.[20] The United States has led the world in both ideas. America began a major transformation to mass secondary schooling at the start of the Human Capital Century, and girls were educated in secondary schools to the same, and very often to a greater, degree than were boys.

The Human Capital Century became the American Century. Not only did the United States lead in education, it also began to lead in income per capita early in the twentieth century. It then expanded its lead in both education and income. Is the relationship causal or merely coincidental? Before we can judge the relationship between human capital and economic indicators, we first need estimates of the

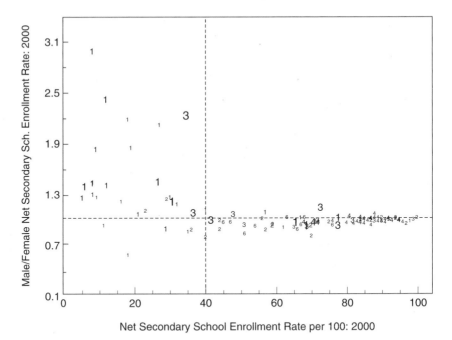

Figure 1.3. Ratio of Male to Female Secondary School Enrollment: 2000. Sources and Notes: See Figures 1.1 and 1.2.

educational stocks of the labor force built from educational attainment and other data.

America in the Human Capital Century

Educational Advance in the Twentieth Century

The United States expanded its lead in education in the twentieth century by instituting mass secondary schooling and then by establishing a flexible and multifaceted higher education system. These transformations to mass education are considered in detail in Part II. At this point, we summarize the broad outlines of schooling advances in the twentieth century by exploring the educational attainment of successive birth cohorts of native-born Americans.[21] In doing so we are observing changes in schooling across the twentieth century for those directly affected by the U.S. educational system. These estimates are then inputs to the computation of the educational stock of the U.S. labor force.[22] We focus on the educational attainment of each birth cohort measured at age 35, when almost all have completed their formal schooling.[23]

From the beginning of the twentieth century to the early 1970s, and for all groups considered here, the increase in years of schooling for the native-born population was substantial, as illustrated in Figure 1.4. For cohorts born between 1876 and 1951 (who were 24 years old from 1900 to 1975) the increase was 6.2 years, or 0.82 years per decade. The increase was sufficiently continuous and unbroken that a straight line would nicely fit the data, especially for the 1880 to 1940 birth cohorts.[24] After the 1951 birth cohort, however, a great slowdown ensued. Educational attainment barely changed for cohorts born between 1951 and 1965 (24 years old between 1975 and 1989), and for cohorts born from 1965 to 1975 (24 years old between 1989 and 1999), educational attainment started rising again but increased by just 6 months overall.

After increasing nonstop for the first three quarters of the twentieth century, educational attainment among the native-born population slowed considerably during the last quarter of the century. The educational attainment of a child born in 1975 was just 0.50 years more than that of his or her parents born in 1951, but the educational attainment of a child born in 1945 was 2.18 years more than that of his or her parents born in 1921. A well known dream of many American families is that their children will do better than they did. But that dream began to unravel in the latter part of the twentieth century, at least with regards to educational attainment.

Both men and women shared in the increase in educational attainment during the first three quarters of the twentieth century (Figure 1.5). Women gained more than did men both at the beginning and the end of the century. Men gained more during the middle decades. Women began the period with more education than did men, in large part because they attended and graduated from high school to a greater degree. In addition, women attended college at about the same rate as did men, although they graduated from four-year institutions to a lesser extent. Because college was but a small part of total schooling during the first few decades of the century and high school was far more important, native-born women accumulated more years of education than did their male counterparts.

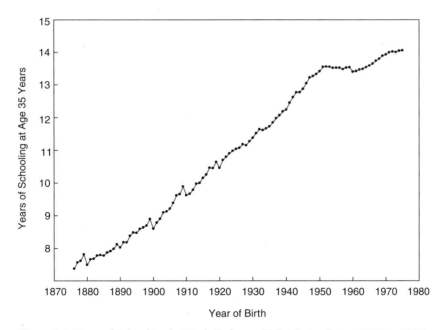

Figure 1.4. Years of Schooling by Birth Cohorts, U.S. Native-Born: 1876 to 1975. The figure plots the mean years of completed schooling by birth cohort adjusted to 35 years of age for the U.S.-born. For the 1940 to 1980 samples, years of schooling are given by the highest grade completed, top coded at 18 years for 1960 to 1980 and at 17.6 years for 1940 and 1950. The categorical education variable for the 1990 and 2000 samples was converted to years of completed schooling. Categories covering more than a single grade were translated as follows: 2.5 years for those in the first through fourth grade category; 6.5 years for those in the fifth through eighth grade category; 12 years for those with a general equivalency diploma or a high school diploma; 14 years for those with some college or with an associate's degree; 16 years for those with a bachelor's degree; 17.6 years for those with a master's degree; and 18 years for those with a professional or doctoral degree. The log of the mean years of schooling for a birth cohort-year cell is the dependent variable in the age-adjustment regression that includes a full set of birth-cohort dummies and a quartic in age as covariates. The age-adjustment regression is run on birth cohort-census year cells, pooling all the IPUMS for 1940 to 2000. The samples include all U.S.-born residents aged 25 to 64 years. For further details on the method, see DeLong, Goldin, and Katz (2003), notes to figure 2.1. Sources: 1940 to 2000 Census of Population Integrated Public Use Micro-data Samples (IPUMS).

The advantage women had in education disappeared rapidly with cohorts born in the 1910s and 1920s. Many men in these cohorts fought in World War II or in the Korean War and attended college on the G.I. Bill (other reasons for the vast relative increase in college among men are discussed in Chapter 7). Whatever the reason, the out-

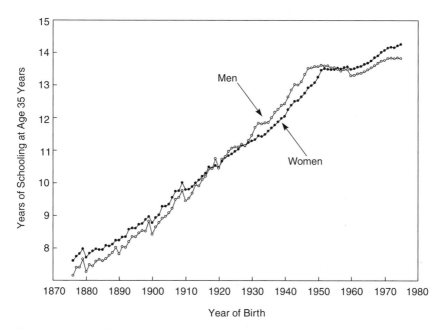

Figure 1.5. Years of Schooling by Birth Cohorts, U.S. Native-Born, by Sex: 1876 to 1975. This figure plots the mean years of completed schooling for U.S. native-born residents by birth cohort and sex, adjusted to age 35 using the approach described in the notes to Figure 1.4. Sources: 1940 to 2000 IPUMS.

come that is clear is that by the birth cohorts of the 1920s men caught up with women in educational attainment, and for the cohorts from the 1930s to the early 1950s they exceeded women in educational attainment. These gains, however, were reversed with cohorts born in the 1960s, as women rapidly increased their attendance at and graduation from college. At the end of the twentieth century women's educational attainment exceeded that of men just as it had during the early decades of the century.

Educational gains for African Americans were far greater than for the total population because their educational attainment began at so low a level. At the start of the period (for cohorts born in the late 1870s), the gap in educational attainment between whites and African Americans was 3.7 years. On average, white students spent nearly twice as long in school as did black students. Furthermore, differences in the actual level of schooling are understated by the attainment figures because there were wide discrepancies in the quality of schooling between the races. Beginning with the cohorts born around 1910 the absolute gap in years of schooling began to close (Figure 1.6). This convergence slowed with cohorts born around 1940 and slowed even more for those born after 1960. The black-white schooling gap for cohorts born in the 1970s was 0.8 years, about one-fifth its level a century earlier. The gap in educational attainment between non-Hispanic whites and U.S.-born Hispanics was 1.1 years for the 1970s birth cohorts, somewhat larger than the difference (0.7 years) between non-Hispanic whites and non-Hispanic blacks.[25]

Educational attainment during the twentieth century expanded by 5.27 years for those born in the United States from 1895 to 1975. About 50 percent of that gain was attributable to the rise of high school education (60 percent or more for cohorts born from 1895 to 1935), 30 percent was due to the increase in college and post-college education, and 20 percent to continued increases in elementary education. Thus the spread of mass secondary schooling, a movement that began in earnest around 1910, was responsible for much of the increase in the schooling of native-born Americans in the twentieth century. (For this reason and others, we devote several chapters in Part II to the push for universal secondary school education.)

In sum, rapid educational advance characterized cohorts born from the late nineteenth century to 1950, and that was especially true for black Americans, but those advances stagnated for cohorts born after 1950. Although an increase in years of schooling is discernible for cohorts born after 1965, for cohorts born from 1950 to 1975 the increase was only about one tenth of that for cohorts born from 1876 to 1950.

A full century of educational advance can therefore be divided into two parts. During the first three quarters of the century educational attainment rose rapidly, but during the last quarter of the century it stagnated. As we will see in Chapter 2, twentieth-century trends in economic inequality also occurred in two parts—first declining and then rising. The relationship between the two trends is made clear in Chapter 8.

European Comparisons

To understand the slowdown in U.S. educational attainment, one must seek comparisons with other nations. Is a slowdown a natural occurrence as educational levels increase to very high levels? Is the phenom-

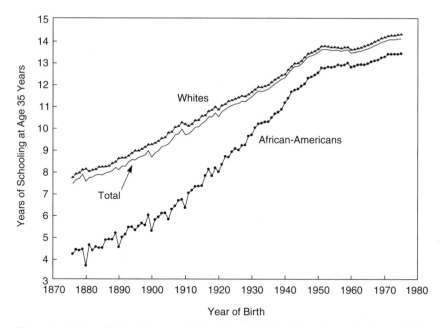

Figure 1.6. Years of Schooling by Birth Cohorts, U.S. Native-Born, by Race: 1876 to 1975. This figure plots the mean years of completed schooling for U.S. native-born residents by birth cohort and race, adjusted to age 35 using the approach described in the notes to Figure 1.4. Sources: 1940 to 2000 IPUMS.

enon unique to the United States? Many other nations now have education rates for their younger cohorts that exceed those in the United States. And, in the past few decades, they have caught up to and some have even surpassed the high education levels that had once been set by the United States.

At the start of the twenty-first century, young people in many European nations attended college to about the same extent as did young people in the United States. Among 25- to 34-year-olds, 39 percent in the United States had attained two- or four-year college degrees. Four European nations equaled or exceeded that figure in 2004, and eight others were close behind.[26] At the same time, a much smaller fraction of the 55- to 64-year-olds in these 12 countries had attained college or university training relative to the same group in the United States; their educational attainment represents the state of higher education four decades earlier. For these 12 European nations, the average college completion rate for the older group was 56 percent of the U.S. figure.

Part A: Europe and the U.S. 1955/56

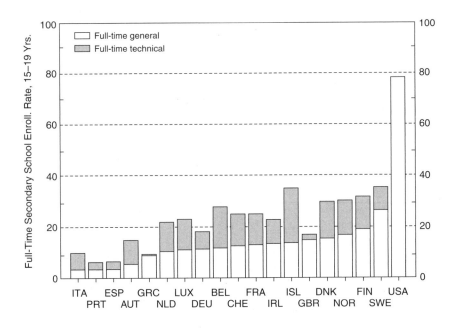

Part B: Europe and the U.S. 1955/56

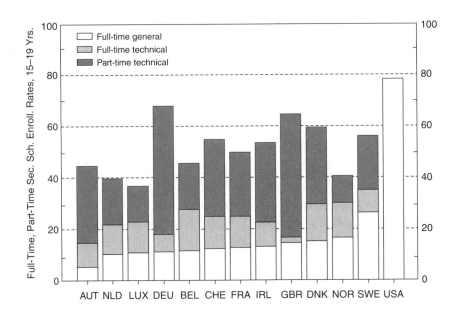

The educational attainment in these nations advanced rapidly in the last several decades of the twentieth century. In almost all cases the college rates achieved by the 25- to 34-year-olds in these European nations were more than 1.5 times the rates of the 55- to 64-year-olds, and in four of the twelve cases the increase was more than twofold. In contrast, the college degree rate of U.S. 25- to 34-year-olds (39 percent) was only 1.08 times that for U.S. 55- to 64-year-olds (36 percent) in 2004.

In essence, many of the nations with college degree rates that were considerably behind that of the United States in the last third of the twentieth century had rates placing them ahead or very close to the United States by the start of the twenty-first century. The slowdown in the educational attainment of young Americans at the end of the twentieth century is especially striking when compared with the acceleration of schooling among many nations in Europe and in parts of Asia, where educational change has been exceedingly rapid.

The fact that Europe was far behind the United States in higher education until the last several decades of the twentieth century

---

Figure 1.7. (opposite page) Secondary School Enrollment Rates for European Nations and the United States, c.1955. The data refer to the number of youths in public and private upper and lower secondary schools (of the types listed) ranging from those who turned 15 years old during the school year to those who turned 19 years old during that year. Thus, the age group under consideration is approximately all 15- to 18-year-olds, plus one-half of 14- and 19-year-olds. No youths in elementary schools or colleges and universities are included even if they were in the included ages. The procedure ensures consistency but implicitly favors countries, such as the Nordic nations, that have late starting ages and penalizes those, such as France and the United States, that have earlier starting ages. The computation for the United States assumes 100 percent enrollment for the 14-year-olds and then adds those enrolled in ninth through twelfth grades and divides by the age group given above. The six nations that did not give data on part-time technical schools are excluded from Part B. All data are for c.1955. Abbreviations are: Italy (ITA), Portugal (PRT), Spain (ESP), Austria (AUT), Greece (GRC), Netherlands (NLD), Luxemburg (LUX), Germany (DEU), Belgium (BEL), Switzerland (CHE), France (FRA), Ireland (IRL), Iceland (ISL), Great Britain (GBR), Denmark (DNK), Norway (NOR), Finland (FIN), and Sweden (SWE). Sources: European nations: Dewhurst et al. (1961, tables 10-2 and A). The data for England and Wales, France, Germany (including the Saar and West Berlin), and Sweden, have been checked against the original administrative records and small errors have been corrected. United States: U.S. Department of Education (1993, tables 1 and 9).

should not be surprising. European nations had lagged considerably in providing secondary schooling. According to a careful analysis of data from 18 European nations in the 1950s, none enrolled more than 30 percent of older teen youth in full-time academic (general) secondary schools and most were below the 20 percent mark (see Figure 1.7, Part A). In addition, enrollment did not exceed 40 percent in both full-time academic and full-time technical schools. Even the inclusion of enrollment in part-time technical schools does not materially alter the conclusion (Figure 1.7, Part B). At the same time, the United States enrolled more than 70 percent of its teen youth in secondary schools. The United States was a clear leader in mass secondary school education in the mid-twentieth century.[27]

A possible response to these facts is that Western Europe was set back by World War II and had not fully recovered by 1955. Yet levels of schooling for teenagers in Britain, for example, seriously lagged those in the United States during the first half of the twentieth century. As shown in Table 1.1, the fraction of 17-year-olds attending secondary schools in England and Wales was abysmally low for all years shown, including those in the 1950s. High school graduation rates in the United States, on the other hand, reached almost 30 percent nationwide as early as 1930 and were in the 60 percent range in the 1950s.[28] By 1960 Great Britain was about 35 years behind the United States in the educational attainment of its high-school aged youth even though it had had a fully supported public secondary school system since 1944.

Another possibility is that Europe was far behind America in secondary education because it was generally less wealthy than was the United States, but secondary school rates in Europe in the 1950s trailed those that had existed in the United States at the same level of real per capita income. Although real per capita income in the United States in 1940, as the nation was just coming out of a decade-long depression, was about equal to the incomes in much of Europe in 1955,[29] secondary school enrollment rates in the United States were more than twice those in the Europe of 1955, including students in full-time technical schools.[30] One would have to return to the 1910s to find levels of secondary school enrollment in the United States that match those in 1950s Western Europe.

A clear statement of the differing commitments to publicly provided and funded education can be seen in the response of the governments

Table 1.1. Schooling of 14- to 18-Year-Olds in Great Britain and the United States: 1870 to 1960

British School Attendance Rates (in percent)				U.S. High School Enrollment and Graduation Rates (in percent)		
Year	14-year-olds	17-year-olds	15- to 18-year-olds[a]	Year	Enrollment/14- to 17-year-olds	High school graduates/17-year-olds
1870	2	1	n.a.	1870	n.a.	2.0
1900/02	9	2	n.a.	1900	10.6	6.4
1911/12	12	1	n.a.	1910	14.5	8.8
1931	n.a.	n.a.	9.4–10.6	1930	51.1	29.0
1936/38	38	4	n.a.	1938	67.7	45.6
1950	100[b]	10.5	12.6 [14.4][c]	1950	74.5	59.0
1956	100	n.a.	14.9	1956	83.5	63.1
1957	100	9.0	16.0			
1960/62	100	15	17.5	1960	86.9	69.5

Sources and Notes: The term "attendance" for England and Wales is from Ringer (1979); the term "enrollment" for the United States is from the *Biennial Reports*. "Attendance" in England and Wales often meant "on the registers," which is similar to enrollment. Attendance figures are almost always lower than enrollment data within a country, but for these comparisons, the difference may be slight.

England and Wales: Ringer (1979) all years, except those that follow. The 15- to 18-year-old group in attendance is 15 years and over divided by the 15- to 18-year-old group in England and Wales only. The 14- and 17-year-old percentages are for full-time attendance at any school and refer to all of Great Britain.

1911/12: U.S. Bureau of Education, *Biennial Reports* (1916–18, p. 25), the Lewis Report 1917, is given as the source. See also Matthews (1932) whose data are derived from the Lewis Report.

1931: Board of Education for England and Wales (1932) provides information on "students on the registers" by age for schools receiving public funds. The number of youths 15 to 18 years old in entirely private institutions is estimated from university enrollments and makes assumptions regarding the proportion of secondary students who continued to university based on evidence from 1948. A set of reasonable assumptions results in the 9.4 percent figure and 10.6 percent is the upper bound estimate.

1956: Carr-Saunders et al. (1958, p. 60) give the number of pupils on school registers by age and the total in the age group. The figure will not include youths at university.

1957: Dewhurst et al. (1961). The 15- to 18-year-old group is a simple average of the four years.

United States, all years: U.S. Department of Education (1993).

a. The 15- to 18-year-old group includes those who are just 15 years old but excludes those who are just 19 years old.

b. The 100 percent figure is assumed for 14-year-olds because their attendance at some school was mandatory after 1944.

c. The 14.4 percent figure is estimated for males only, using data for 1948.

in Britain and the United States during World War II. When the United States entered World War II, the median American 18-year-old was a high school graduate and, outside the South, more than 60 percent had recently graduated from high school. When President Franklin Roosevelt signed the GI Bill of Rights into law in 1944, the average GI was able to attend college for he had already graduated from high school. Nothing more clearly demonstrates the difference in schooling between the United States and Britain than the simultaneous passage by the Labour Government of the long-awaited 1944 Education Act. The GI Bill paid the tuition and a stipend for military veterans attending college, while the Education Act did no more than guarantee to all British youth a publicly-funded grammar or secondary school education.

We referred earlier to statements by Western European commentators in the early twentieth century that Americans wasted resources in educating the masses in secondary schools. These nations, instead, promoted a cadre of extraordinary children and gave the residual a fair elementary education. Great Britain, France, and Germany all tested children, generally before their twelfth birthday, to see who would be tracked to enter secondary school. Many European nations of the early twentieth century followed this general scheme, although their educational systems were structured differently. These nations provided three different models or templates that the United States could have followed.

The British emphasized classical training for those who were allowed to go beyond the compulsory grades. The French system produced a small group of civil servants and well-trained professionals for technical and scientific fields. The German system contained a number of tracks—for industrial work, for commercial pursuits, and an elite course for students who would attend universities. The American system can be characterized (as we do in Chapter 4) as open, forgiving, lacking in universal standards, and having an academic yet practical curriculum. The European system, in contrast, was generally closed, unforgiving, with uniform standards, and an academic curriculum for some and an industrial one for others. One system was egalitarian; the other was elite.

Why Was America Different?

Why did the United States pioneer a novel and distinctively American form of secondary and higher education and break from the educational templates of Europe? Why did Americans engage in investments in human capital that Europeans viewed as wasteful of resources?

Part of the answer can be found in considering the choice between engaging in *general* training, such as formal schooling, and engaging in *specific* training, such as an apprenticeship or on-the-job training. Investment in general schooling may be more costly than an apprenticeship, but it produces skills that are flexible and thus transferable across place, occupations, and industries. Thus formal education is more highly valued when geographic mobility and technical change are greater. It is favored when the costs of specific training, such as for family businesses or apprenticeships, are higher, possibly because community ties are fewer. When education is publicly funded, formal schooling, moreover, has a lower direct cost for students and their parents.[31]

General schooling for the masses fit American circumstances more than it did those of early twentieth-century Europe. Many insightful historians and demographers have commented on the generally accepted notion that Americans were more geographically mobile than were Europeans, within their nations.[32] Extensive evidence exists for this greater geographic mobility for the post-1960s.[33] Although evidence for the nineteenth and early twentieth century is less plentiful, a careful analysis of inter-county migration for the mid-nineteenth century demonstrates that two-thirds of all American adult males shifted county at least once but that only one-quarter of all British men did.[34] Moves within the United States, furthermore, were considerably longer.

Formal, school-based education enabled American youths to change occupations over their lifetimes, to garner skills different from those of their parents, and to respond rapidly to technological change.[35] Apprenticeships and highly specific training were more cost effective for individuals who expected to spend their lives in the same place and in the same industry and occupation, but apprenticeships were not as valuable for others and clearly not for their employers.[36] As economic historian Stanley Lebergott noted: "incessant mobility [of Americans] made it thoroughly unwise for any employer to invest much in training his employees" (1984, p. 372).

Europeans may have been correct in their assessment that an American educational system was wasteful of resources, at least in their circumstances. But it was not wasteful in the technologically dynamic, socially open, and geographically mobile New World setting. And it assuredly enhanced the dynamism.

Human Capital and Economic Growth

Human Capital Stock of the Workforce

Measuring how much human capital enhances the dynamism of an economy would be extremely valuable. Measuring how much human capital increases income by making individual workers more productive is equally important and conceptually simpler. We will do the latter and compute how growth was increased by a more educated workforce.

To produce estimates of the impact of education on growth, we first need to measure the *human capital stock of the workforce* at various points in time. Human capital includes a broad class of inputs such as education, on-the-job training, and health. We use a definition that includes only formal education and does not adjust for potential school quality differences across cohorts and years.[37] A well-established growth accounting framework is employed to guide the analysis of the direct effect of a more educated workforce on labor productivity. The estimates of the educational attainment of the native-born population by birth cohort, discussed above, provide a starting point for constructing estimates for the human capital stock of the workforce and its evolution over the twentieth century.

The human capital stock of the workforce differs in several important ways from the educational attainment of the native-born population by cohort. For instance, the human capital stock depends also on the foreign-born population, which was a large fraction of the workforce in both the earlier and later parts of the twentieth century. The two measures also differ because the stock estimates aggregate the educational attainment data by the relative sizes of the cohorts. Yet another difference is that the stock estimates are for the workforce, whereas educational attainment by birth cohort was presented for the entire population. The educational attainment of the workforce can differ from that of the population to the extent employment (and labor force participation) rates vary across schooling groups.

We are able to construct estimates of the human capital stock of the workforce for the years after 1940 because the U.S. Population Census asked about educational attainment starting with 1940. For the previous period we employ a state census that asked detailed information (more comprehensive than in any federal population census) about educational attainment. The Iowa State Census of 1915 is a unique group of records, and is discussed at greater length in Chapter 2. Suffice it to say here that we have collected a large and representative sample from original documents, consisting of some 60,000 individuals living in both cities and rural areas. Although we make no claim that Iowa was a microcosm of America, the educational attainment of its workforce was not much different from that of the nation's in the last several decades of the twentieth century. It was, however, a more educated place than was the rest of the country earlier in the twentieth century. We primarily use the Iowa data to obtain changes in the educational attainment of the workforce.

From 1915 to 2005, the period shown in Table 1.2, the increase in the educational stock of the U.S. workforce, as measured by mean years of schooling, was just short of 6 years or 0.66 years per decade.[38] The educational stock progressed with great speed from 1940 to 1980, when the increase was 0.86 years per decade and as better-educated young people replaced lesser-educated older cohorts in the workforce. Progress slowed thereafter, from 1980 to 2005, when the increase barely exceeded one year, or 0.43 years per decade.

The rapid growth of the educational attainment from 1915 to 1980 was driven largely by the sharp increases in the schooling of successive cohorts of the U.S native-born population through the early 1950s birth cohorts (illustrated in Figure 1.4) Similarly, the slowdown of growth of the human capital stock of the workforce since 1980 mainly reflects the slower rate of increase of educational attainment for post-1950 birth cohorts.

What about the possibility that changes in the share of workers who are foreign-born drove the aggregate human capital stock changes? In 1910, amidst the great migration from Europe, the fraction of foreign-born among U.S. workers reached 22 percent. The vast majority of immigrants came from countries that had far lower levels of education

Table 1.2. Educational Attainment of the Workforce, 1915 to 2005

	United States					Iowa				
	1915	1940	1960	1980	2005	1915	1940	1960	1980	2005
Mean years	7.63	9.01	10.53	12.46	13.54	8.45	9.83	10.87	12.49	13.55
Fraction, by years										
0–8	0.756	0.522	0.303	0.087	0.034	0.726	0.476	0.289	0.077	0.020
9–11	0.129	0.174	0.218	0.154	0.070	0.129	0.165	0.184	0.126	0.060
12	0.064	0.185	0.262	0.346	0.309	0.083	0.229	0.316	0.424	0.322
13–15	0.028	0.061	0.121	0.228	0.290	0.037	0.076	0.128	0.210	0.338
16+	0.026	0.058	0.096	0.185	0.297	0.026	0.055	0.083	0.164	0.261

Sources: 1915 Iowa State Census; 1940, 1960, and 1980 Integrated Public Use Microsamples (IPUMS) of the U.S. federal population censuses. 2005 Current Population Survey (CPS), Merged Outgoing Rotation Groups (MORG). 1915 U.S. data are extrapolated from the 1915 Iowa data. U.S. mean years of education for 1915 is given by the U.S. mean for 1940 minus the difference between 1940 and 1915 means for Iowa. The fraction, by years, is an extrapolation from the Iowa data that is scaled to sum to one.

Notes: Samples are restricted to those aged 16 or older and exclude those who were in the military or institutionalized. The workforce in each year from 1940 to 2005 consists of those who are employed at the survey reference week. The workforce for Iowa in 1915 includes those reporting occupational earnings for 1914; each individual is weighted according to the number of months worked in 1914. Years of schooling for 1940 to 1980 are measured using the same approach as that used for Figures 1.4, 1.5, and 1.6. Years of schooling in 2005 are measured using the approach to post-1991 CPS MORG of Autor, Katz, and Krueger (1998). Measures of years of schooling and months worked for Iowa in 1915 were constructed using the methods of Goldin and Katz (2000). For 1960 to 1980 we follow Autor, Katz, and Krueger (1998) in including all those who attended 13 years of schooling (whether or not they completed the final year) in the 13–15 schooling category. Sampling weights are used for all samples.

than did the United States. In the 1915 Iowa Census, the typical foreign-born worker had almost 1.5 fewer years of schooling than the typical native-born worker. Because Congress restricted immigration in the 1920s, the labor force began to age without a steady influx of immigrants, and the fraction of foreign-born in the workforce fell.

If we look at the overall and U.S.-born workforces, years of schooling increased by 0.64 and 0.59 years per decade, respectively, from 1915 to 1960. The net effect of the decline in the foreign-born share of the workforce, from 21 percent in 1915 to 7 percent in 1960, was to raise the educational attainment of the workforce by 0.05 years per decade. The immigrant workforce share barely changed from 1960 to 1980, and thus there was a negligible impact from immigration during that period. In contrast to the earliest period examined, the foreign-born share of the workforce increased from 7.6 percent in 1980 to 16.3 percent in 2005. From 1980 to 2005 the schooling of the overall and U.S.-born workforces increased by 0.43 and 0.48 years per decade, respectively. The impact from immigration during 1980 to 2005, therefore, was to reduce the growth in workforce schooling by 0.05 years per decade, or the same magnitude but opposite in sign for the 1915 to 1960 period.[39]

Thus, changes in the schooling levels of those born and educated in the United States not the relative shares of immigrants in the population and the labor force, have been the driving force behind the evolution of the educational stock of the U.S. workforce. Although the declining foreign-born share of the workforce from 1915 to 1960 contributed to an increase in the educational attainment of the workforce, and the large rise in immigration during the last 25 years placed a small drag on educational attainment both of these changes have been modest compared with the changes for the native-born population.

During the first half of the twentieth century the major change in the educational attainment of the labor force was a result of the replacement of workers who had less than a high school education with workers who had completed high school (see Figure 1.8).[40] In the late twentieth century, this change was furthered by the entrance of workers with at least some college education. The fraction of the U.S. workforce with no more than a common or elementary school education exceeded 75 percent in 1915, declined to 30 percent by 1960, and decreased to a trivial 3 percent by the early twenty-first century. In

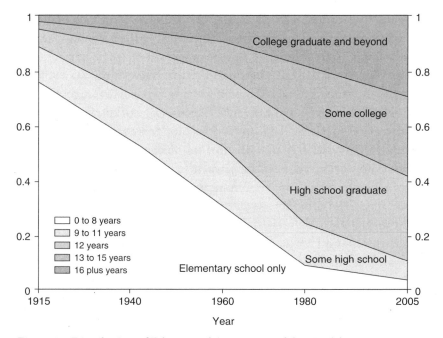

Figure 1.8. Distribution of Educational Attainment of the Workforce: 1915 to 2005. Sources: See Table 1.2.

1915 barely more than 5 percent had some college or university training, but 22 percent did in 1960 and more than 58 percent did by 2005. Finally, whereas less than 3 percent had a college degree in 1915, 30 percent did by 2005.

Measuring the Contribution of Education to Growth

HISTORY AND THEORY

The twentieth century, as we noted earlier, was both the Human Capital Century and the American Century. It was the century when education became the dominant factor determining the wealth of nations and it was the century when America was first to discover that notion. It was the century when America began to lead the world economically and it was the century during which America remained at the top. Armed with our estimates of the human capital stock of the workforce, we can now return to the relationship between human capital and economic growth at various moments in U.S. history. Later, in

Chapter 2, we address who shared in the benefits of economic growth and the relationship between changes in distribution and education.

Across the twentieth century the American economy grew at a remarkable pace. In per capita or per labor hour terms, economic growth was moderate at best, certainly relative to the impressive rates of various transition and emerging nations in recent decades. But the impressive growth of the American economy is to be found in its continuous nature over the century and thus in its compounding over the century. Productivity (or output) per labor hour grew at an average of 2.2 percentage points annually across the entire century. There were periods of a quickening in the pace of growth, such as the 1920s, the 1940s, and the 1960s, and there were moments of rather sluggish performance, as in the late 1970s and 1980s (and, of course, the 1930s). By and large, however, the growth was moderate but steady over long periods, and thus was impressive overall.

The impressive growth of the United States at mid-century amidst the flagging development of many parts of the world provided a strong impetus for economists to find the factors that caused economic growth. Some believed they could create a recipe for economic success or an inoculation against economic disaster, akin to the "magic bullet" for disease. Great minds worked in various ways on the problem. A few believed that economic growth began in certain capital-intensive sectors, such as transportation or manufacturing, and that a particular infrastructure would provide the key to success.[41] Others modeled the growth process in a more general fashion. Output, they posited, is a function of the various inputs in the economy and, therefore, economic growth occurs when these inputs increase.

Growth can also occur when the inputs become more productive or if the functional relationship between the inputs and the output changes. For example, capital could become more productive, and technological change could be embodied in capital. Laborers themselves could become smarter, more able, healthier, and simply better workers. Advances in this manner would be embodied in human capital. Furthermore, technological change could be neutral with regard to the inputs. Output could increase even if none of the inputs changed, as in Adam Smith's famed pin factory in which workers became more productive due to the division of labor rather than an increase in individual skill. By formalizing these notions of inputs, outputs, and the production function we are able to estimate the impact of educational advance on economic growth.

Following a long literature in economic growth, assume that output (Y) is a function of a set of inputs, which will be hours of labor (L) and capital (K). Other inputs, such as land, could be added with no loss in generality. The functional form chosen is additive in the logs of these factors, with a multiplicative component known as total factor productivity (A), since it reflects the productivity of all the factors, taken together. The functional form just described is known as the Cobb-Douglas production function. We consider the case when the exponents sum to one:

$$Y = A \, K^{(1-\alpha)} L^{\alpha} \qquad (1.1)$$

which implies the per unit labor input or intensive form of

$$y = A \, k^{(1-\alpha)} \qquad (1.2)$$

where $y = Y/L$ and $k = K/L$. In rate of change form, where an asterisk (*) denotes dlog {·}/dt:

$$y^* = A^* + (1 - \alpha) \, k^* \qquad (1.3)$$

Eq. 1.3 is the formulation described in Robert Solow's pioneering papers on economic growth (1956, 1957). One of the many virtues of the formulation is that it places few demands on the data since y^* is labor productivity change, k^* is the growth of the capital stock per worker (or capital to labor ratio), and the exponent on labor (α) in the production function, under reasonable assumptions, is the share of labor in national income.[42] Total factor productivity change, A^*, is derived as a residual.

The early growth empiricists fitted data to eq. 1.3 and found, to their surprise, that most of the labor productivity change was due to changes in total factor productivity, and not to an increase in the measured inputs, mainly the change in the ratio of capital to labor. The conclusion was that the residual, dubbed the "measure of our ignorance," was the driving factor. If growth theorists were searching for a recipe to provide to the world's poor nations, their advice would have been simple: "Do something economists are not very good at measuring."

But economists soon realized that the simple formulation used crude measures of the inputs.[43] Labor, for example, was measured simply as hours of work when labor actually had two main components: raw labor in hours (L) and labor in efficiency units (E). The full, augmented labor input was ($L \cdot E$). The augmented labor input could change if labor hours changed and it could also change if labor efficiency per hour changed. The change in efficiency units could be due to changes in formal education, on-the-job training, age, health, or a host of other factors that enhance the effectiveness of workers.

Augmenting the labor input implies that eq. 1.1 becomes:

$$Y = A'K^{(1-\alpha)}(L \cdot E)^{\alpha} \tag{1.1'}$$

and that eq. 1.3, in turn, becomes:

$$y^* = (A')^* + (1-\alpha)\,k^* + \alpha E^* \tag{1.3'}$$

Whereas growth empiricists focus on all components of eqs. 1.3 and 1.3′, our interest here is in the relationship between αE^* and y^*. We want to know the degree to which changes in the efficiency units of labor, in particular those due to changes in formal education, can explain changes in labor productivity over the course of U.S. history. Put simply, we want to know the direct effect of education on economic growth. We turn now to the estimation of that relationship.

DIRECT EFFECT OF EDUCATION ON GROWTH

There are three components of the estimation. The first is y^* or the growth in labor productivity (output per hour). For the period since 1947 we use the official Bureau of Labor Statistics (BLS) estimates and for the previous years we use data based on the work of John Kendrick (1961).

The second component is the change in the educational productivity index, E^*. We compute the index $E_t = \Sigma_i w_{it} S_{it}$, where w_{it} is the (adjusted) wage of education group i (relative to a reference education group) in base period t, and S_{it} is the share of education group i in total hours in year t.[44] When differences in earnings by education reflect the impact of schooling on productivity, the growth in the index measures the contribution of educational upgrading to aggregate labor-input growth through improvements in the average human capital or quality of the workforce. The estimation of S_{it} was discussed in the previous section and is given in Table 1.2. Educational wage differentials, the w_{it} part of the formula, will be only briefly summarized here because Chapter 2 will concern how the economic returns to education evolved across the twentieth century. In 1915, at the start of the period under examination here, the private economic return to a year of either high school or college was substantial, upwards of 11 percent per year. Educational wage differentials then narrowed considerably from 1915 to 1950. They expanded modestly in the 1950s and 1960s before narrowing again in the 1970s. Educational wage differentials increased significantly in the 1980s, with a modest advance in the 1990s. At the end of the twentieth century educational wage differentials remained quite high, although at the start of the period they were higher still.

The third component is α, which in a competitive pricing economy is the share of labor in national income. Compensation of labor (wages plus fringe benefits) accounts for approximately 70 percent of production. On the assumption that labor is paid its marginal contribution to output and that output is proportional to inputs, a 1 percent increase in effective labor through an increase in the average human capital of the workforce directly boosts output by 0.7 percent.[45] The results of the computation are shown in Table 1.3. We find that on average across the 90-year period from 1915 through 2005 increases in educational attainment boosted the effective size of the workforce by 0.48 percent per year.[46] Thus, education *directly* contributed an average of 0.34 percentage points a year to economic growth (0.7×0.48) over the 90-year span. Differences in the contribution of human capital exist across the four subperiods we employ. Educational advance contributed 0.49 percent per year to labor productivity growth from 1915 to 1960, but from 1960 to 1980 the contribution increased greatly to 0.59 percent per year and then sharply decreased in the remaining 25 years to 0.37 percent per year. Similar changes, given in last column of Table 1.3, can be seen in the average years of schooling of the workforce, especially the slowing of growth since 1980.

One might wonder how immigration affected educational productivity. We noted before that immigration slightly increased the growth in workforce education on average from 1915 to 1960, that it had almost no effect from 1960 to 1980, and that it reduced schooling

Table 1.3. Educational Growth Accounting, 1915–2005

	(1)	(2)	(3)	(4)
	Average annual percentage point change in:		Fraction "explained" by educational change	Change in mean years of workforce education
Period	y^*	E^*	$\alpha \cdot E^*/y^*$	
1915–40	2.45	0.50	0.143	1.38
1940–60	2.92	0.49	0.118	1.52
1960–80	2.41	0.59	0.171	1.93
1980–2005	2.18	0.37	0.119	1.08
1915–2005	2.47	0.48	0.136	5.91

Sources: Col. 1: *Historical Statistics, Millennial Edition* (2005), table Cg265–272, series Cg265 for 1915–40 and table Cg273–280, series Cg273 for 1940–60, and U.S. Bureau of Labor Statistics (BLS), "Major Sector Productivity and Costs Index," series PRS84006093, for 1960–80 and 1980–2005 from the BLS Internet site (www.bls.gov). Col. 2: 1915 Iowa State Census; 1940, 1960, and 1980 IPUMS; 2005 CPS MORG. Col. 3 multiplies col. 2 by 0.7 and divides by col. 1. Col. 4: see Table 1.2.

Notes: The labor productivity measure (y) in col. 1 is real gross private domestic product per labor hour for 1915 to 1960 and output per hour of the business sector for 1960 to 2005. The construction of the educational productivity index used in col. 2 follows the procedures used in DeLong, Goldin, and Katz (2003). The index covers the civilian workforce (ages 16 or older) in each year. The reported educational productivity changes (E^*) are based on chain-weighted prices. (Fixed-weighted prices give similar results.) Changes from 1915 to 1940 are for Iowa; changes for the other time periods cover the entire United States. The education groups used are 0–4, 5–6, 7–8, 9–11, 12, 13–15, and 16 or more years of schooling. The chain-weighted index covering years t to t' uses the average educational hourly wage differentials for t and t'. The index is hours-based and weights workers by the product of their sampling weight and hours worked in the survey reference week. Because hours data are not available in the 1915 Iowa state census we use employment weights for the 1915–40 educational productivity change and educational wage differentials based on monthly earnings for 1915. Mean years of workforce education used in col. 4 are employment weighted. See DeLong, Goldin, and Katz (2003, appendix 2B) for further details on the methodology.

growth within the workforce by a small amount from 1980 to 2005. Similarly, immigration had only a modest impact on educational productivity growth. The falling immigrant share of the workforce increased educational productivity by 0.03 percent per year from 1915 to 1960. The immigration effect was almost nil from 1960 to 1980. From 1980 to 2005 the rising immigrant share reduced educational productivity by 0.03 percent per year, from 0.40 percent per year for the U.S.-born workforce to an overall rate of 0.37 percent per year with immigration.[47] The immigration effects are considerably smaller than one might have thought given past and current debates over immigration.

Across the entire period from 1915 to 2005, the direct contribution of educational advance within the workface of 0.34 percent per year explains about 14 percent of the average annual increase in labor productivity of 2.47 percent (see Table 1.3).[48] The differences by subperiod are slight. Generally, the explanatory power of the direct effect of education is greater when labor productivity is lower except for the most recent period when labor productivity was relatively sluggish and the increase in the educational attainment of the workforce was low. If we used, instead, output per capita as our benchmark, educational advance within the workforce would account for about 15 percent of the 2.23 percent per year gain in real GDP per capita from 1915 to 2005, or about the same as in the case of labor productivity.[49]

Worker characteristics other than education also affect labor productivity and can be incorporated into the analysis. Such characteristics include work experience, sex, nativity, and race. If wage differentials by these characteristics largely reflect differences in worker productivity, the broader group of worker characteristics and associated wage differences can be used to construct an augmented measure of labor force quality. We find that labor force quality, using the augmented set of characteristics, grew on average by 0.42 percent per year from 1915 to 2000, whereas that for education alone grew by 0.48. In other words, educational upgrading accounted for nearly all of the secular improvement in measured labor force quality since 1915.[50]

INDIRECT EFFECT OF EDUCATION ON GROWTH

There are various ways in which education increases productivity and thus economic growth. We have estimated the direct effect, which is the increase in productivity for a given technology and capital stock through an increase in the *quality* (efficiency units) of the workforce. But there are also various indirect effects. The higher income generated by the direct effect of education indirectly contributes to labor productivity by raising physical capital investment and consequently increasing the capital-to-labor ratio. A better educated workforce facilitates the adoption and diffusion of new technologies.[51] Finally, education contributes to innovation and technological advance because scientists, engineers, and other highly educated workers are instrumental to the research and development (R&D) sector as well as to the creation and application of new ideas.[52] Although it is difficult to quantify these indirect contributions of education to economic growth, they are bound to have been quite large.

Suggestive evidence exists that the magnitude of the indirect effects of education on labor productivity is substantial. For example, firms and establishments with more educated workers have long been found to be earlier adopters of new technologies and have been shown recently to garner greater productivity benefits from information–technology investments.[53] Furthermore, highly educated labor is the primary input to R&D, and some estimates suggest that rising R&D intensity in the United States and other advanced economies has been a significant (and possibly the largest measurable) contributor to growth in U.S. labor productivity over the last 50 years.[54]

America at the End of the Human Capital Century: A Summary

At the end of the twentieth century almost all nations have discovered what America knew at the beginning of the century. Human capital, embodied in one's people, is the most fundamental part of the wealth of nations. Other inputs, such as natural resources and financial capital, can be acquired at world prices in global markets, but the efficiency of one's labor force rarely can be. Not only does more education make the labor force more efficient, it makes people better able to embrace all kinds of change including the introduction of new technologies. And for some extraordinary individuals, more education enables them to create new technologies.

The Human Capital Century rapidly became the American Century. The United States became and remained the most economically advanced nation in the world. Was the relationship causal or was it merely coincidental? We have shown in this chapter that advances in education across the twentieth century account for almost 15 percent of the labor productivity change. That is, labor productivity increased by 2.47 percent, on average, each year for 90 years from 1915 to 2005. Education directly increased worker efficiency by 0.48 percent per year, thus directly increasing labor productivity by 0.34 percent per year (0.7×0.48). The actual role of more education must have been considerably greater because of omitted indirect effects, primarily faster technological diffusion and more innovation.

Thus the relationship between the Human Capital Century and the American Century is not at all coincidental. The fact that virtually all other nations have followed suit and invested heavily in the education of their people is testimony to how important human capital became in the twentieth century and how important it is in the twenty-first.

The educational attainment of the American people even near the start of the twentieth century was substantial, particularly in comparison with other nations whose income levels were fairly comparable. The increase during much of the twentieth century was large. We estimated educational attainment by birth cohort for Americans born in the United States from 1876 to 1975 to gain a measure of what the U.S. educational system produced. We then added foreign-born workers and weighted by cohort size, age, and labor force status to obtain the human capital stock of the nation at various moments in U.S. history.

For both the human capital stock of the nation and that of each cohort, we found that the twentieth century was a tale in two parts with regard to educational change. For the first three-quarters of the century, educational attainment advanced by 6.2 years or 0.82 years per decade. But change in the subsequent decades was not all that rosy. During the 15-year period from 1975 to around 1990, there was almost no increase at all. A gain was recorded during the next decade, but it was just half a year.

The gain in educational attainment for the entire century, from the cohort born in 1876 to that born in 1975, was 6.7 years. About 50 percent of this increase was entirely due to increased attendance at and graduation from high school; for cohorts born from 1876 to 1935, fully 60 percent of the increase was due to high school. The high school, not the college, was primarily responsible for some of the largest gains in educational attainment in U.S. history. In Part II we investigate the factors that drove the increase in high school attendance and graduation.

Although recent and past debate over foreign-born workers would suggest that the aggregate impact on education was large, it was, in fact, relatively small during the twentieth century. For the 1915 to 1960 period, when the foreign-born were becoming a smaller share of the workforce, the gain from less immigration was just 0.05 years per decade. Similarly, the loss from 1980 to 2005, when the foreign-born increased their share of the workforce, was also just 0.05 years per decade.

Educational attainment in the United States was high for most of the twentieth century by the standards of other nations, and the increase in years of education was substantial for most of the century. However, gains in educational attainment in Europe and parts of Asia in the past three decades have been simply staggering. Younger cohorts in these nations have considerably more education than do older cohorts, and many of the younger cohorts have higher education levels than exist in the United States. The U.S. educational system would appear to be flagging not only in terms of quantity but also with respect to quality. These are subjects to which we return in Chapter 9.

Great educational change occurred in the United States during the first three-quarters of the Human Capital Century, but these achievements gave way to a rather flaccid performance in the subsequent decades. Likewise, the history of inequality in the twentieth century contains several distinct turning points. In Chapter 2 we explore who gained from economic growth and when.

~ 2

Inequality across the Twentieth Century

Late Twentieth-Century Angst

A once buoyant and ebullient America plunged into the doldrums in the late 1970s. Even after the oil price shocks and high inflation of the decade abated, many analysts through the early 1990s continued to believe that something was seriously awry with the U.S. economy.[1] Three economic facts were put forward as signs of U.S. economic malaise. The first was lagging productivity growth. The second was economic convergence among nations and America's possible loss of economic supremacy. And the third was rising and high levels of economic inequality.

The decrease in productivity growth appeared to be a serious economic problem. Labor productivity (meaning output per hour worked), which had been rising at a whopping 2.77 percent average annually from 1947 to 1973, began to slow and rose at a mere 1.39 percent average annually from 1973 to 1995.[2] Had productivity change from 1973 to 1995 remained at its previous level, output per hour would have been 35 percent higher in 1995 than was actually the case—we had discarded 35 percent of potential output. Although the productivity evidence indicated that we were doing something terribly wrong, there were inconsistencies with that assessment. Virtually all other high-income nations also experienced a post-1973 productivity slowdown and many had more severe declines than the United States.[3] More importantly, the U.S. productivity change made an abrupt (and welcomed) about-turn in the mid-1990s and rose at a very healthy 2.92 percent average annually from 1995 to 2005. America, it appeared, had made a comeback. Maybe nothing was wrong after all.

The second fact—economic convergence—was not new and probably not a real issue of concern. Greater income per capita in other nations has generally helped Americans by providing markets for our goods and services and being a bulwark for a more stable international environment. A more meaningful question is whether other economies were doing something right that we were doing wrong. That, too, does not appear to be the case. During the 1990s the U.S. economy had stronger employment performance and, since the mid-1990s, more rapid productivity growth than most European nations.

But the third reason for America's anxiety has not gone away. Economic inequality in terms of income, wages, consumption, and wealth rose rapidly from the late 1970s to the mid-1990s. Inequality has continued to rise for wages and incomes over the last decade, although it appears to have abated somewhat for wealth and consumption (but not reversed itself).[4] The degree of inequality resulting from these trends is one that the United States has not seen since before the 1940s and has left the country with the most unequal income and wage distributions of any high-income nation.[5] Although some degree of economic inequality may be desirable to spur incentives for hard work and innovation, too much can contribute to social and political discord. Many commentators believe that U.S. inequality has reached that level.[6]

During much of the first three-quarters of the twentieth century rapidly rising productivity translated into widely shared prosperity and enormous increases in the standard of living, straight across the income distribution. But lagging productivity growth from 1973 to 1995 and rising inequality meant that the incomes of a substantial fraction of American families stagnated or even declined. Incomes for most U.S. families did not increase as much as one would have expected in response to the productivity surge of the last decade. The strong historic connection between standard of living growth for typical families and overall productivity change has weakened substantially during the past three decades. Only the economic elite—the top 10 percent of the income distribution—has had real earnings that increased at least as fast as productivity growth on average in recent decades.[7]

The Recent Rise in U.S. Inequality

The measurement of economic inequality is the study of the distribution of economic resources (e.g., income, consumption, and wages) across economic units (e.g., households, families, and individuals). The analysis of inequality typically involves examining the distribution of these resources across households or individuals.[8] Our summary of recent trends in U.S. economic inequality begins with a discussion of the evolution of family income inequality and related trends in the distribution of consumption. We then explore changes in wage inequality, a driving force behind the recent rise in U.S. income inequality.

Income and Consumption Inequality

Relatively consistent data on the distribution of U.S. family incomes are available starting in 1947 from the U.S. Census Bureau's Current Population Survey (CPS). We illustrate trends in family income inequality and real family income levels using the standard measure of pretax, post-transfer money income.[9] Two well-known summary measures of family income inequality are plotted in Figure 2.1: the Gini coefficient and the log of the ratio of family income at the 95th and 20th percentiles.[10] Both measures indicate a modest decline in income inequality from 1947 to the early 1970s, followed by a sharp rise during the last three decades (particularly so in the 1980s).[11]

From 1947 to 1973 family incomes grew rapidly; they also grew closer together. In contrast, since 1973 incomes have grown slowly and have grown apart. These patterns are clear in Figure 2.2, in which real income growth is compared across the family income distribution for the postwar period before and after 1973. For the pre-1973 period, real income growth was fastest near the bottom of the income distribution and slowest near the top, making the change modestly equalizing. For the post-1973 period, family incomes virtually stagnated for the lowest quintile but grew over three times more rapidly for the top 5 percent than for the middle group. In fact, only that top group experienced average real income growth that was nearly as rapid as in the pre-1973 period.[12]

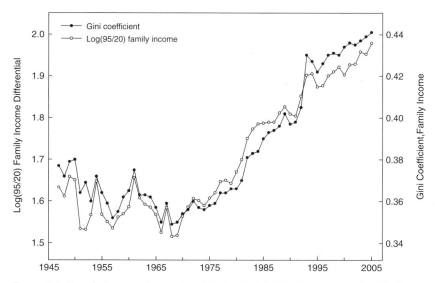

Figure 2.1. Family Income Inequality: 1947 to 2005. The figure plots the Gini coefficient for family income and the log of the ratio of the incomes of the 95th to the 20th percentile families for 1947 to 2005. The income concept used is the official U.S. Census Bureau measure of pre-tax, post-transfer money income. The Gini coefficient varies from 0 to 1, where 0 is complete equality and 1 is complete inequality; for a definition of the Gini coefficient, see text. Sources: Gini coefficient series from U.S. Census Bureau, Historical Income Tables, table F4, http://www.census.gov/hhes/www/income/histinc/f04.html, updated September 15, 2006. Incomes of the 20th and 95th percentile families from the series for the upper limit income of the lowest quintile and lower limit income of the top 5 percent from U.S. Census Bureau, Historical Income Tables, table F1, http://www.census.gov/hhes/www/income/histinc/f01.html, updated September 15, 2006.

The timing of change in real income growth across the distribution is further illustrated in Figure 2.3. Real family income at the 20th, 50th, and 95th percentiles, normalized for each group to 100 in 1973, is plotted from 1947 to 2005. All three groups rose together from the 1950s to the late 1970s. Real income growth was equally shared. The three lines then spread apart, showing the enormous growth of inequality since the late 1970s.

Although CPS data are adequate for measuring incomes across much of the distribution, they are imperfect for incomes at the very upper end (top 1 percent). Better data to measure the upper end of the distribution are available from IRS tax return information. These data reveal a large rise in the share of income accruing to the top part of the

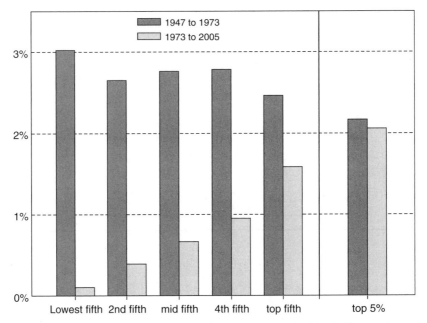

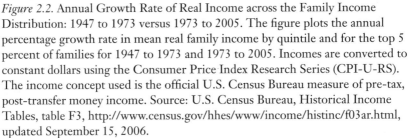

Figure 2.2. Annual Growth Rate of Real Income across the Family Income Distribution: 1947 to 1973 versus 1973 to 2005. The figure plots the annual percentage growth rate in mean real family income by quintile and for the top 5 percent of families for 1947 to 1973 and 1973 to 2005. Incomes are converted to constant dollars using the Consumer Price Index Research Series (CPI-U-RS). The income concept used is the official U.S. Census Bureau measure of pre-tax, post-transfer money income. Source: U.S. Census Bureau, Historical Income Tables, table F3, http://www.census.gov/hhes/www/income/histinc/f03ar.html, updated September 15, 2006.

income distribution in the post-1970s period and a decrease of top end shares in the pre-1970s period.[13]

When rising inequality became noticed by researchers in the 1980s, some initially doubted its significance. Some questioned whether the facts would stand up to closer scrutiny and to a wide range of measures. But the large increase in U.S. economic inequality since the late 1970s is robust to a host of alternative measures and is revealed by many data sources. Other researchers were concerned that income inequality changes reflected no more than a rise in the transitory variation in household income that was offset through saving and borrowing, but that does not appear to have been the case. The sharp rise of income inequality of the 1980s is echoed in the large increase in the inequality

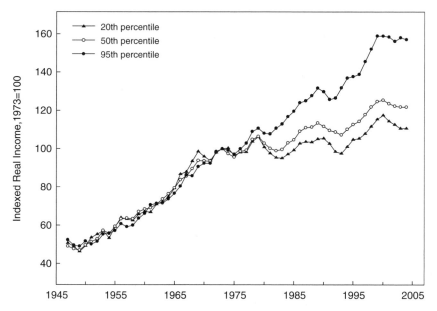

Figure 2.3. Evolution of Low, Middle, and High Family Incomes, 1947 to 2005. The figure plots the indexed real family incomes of the 20th, 50th, and 95th percentile families with real family income indexed to 100 in 1973 for each group. Incomes are converted to constant dollars using the CPI-U-RS. The income concept used is the official U.S. Census Bureau measure of pre-tax, post-transfer money income. Source: U.S. Census Bureau, Historical Income Tables, table F1, http://www.census.gov/hhes/www/income/histinc/f01.html, updated September 15, 2006.

of consumption per adult equivalent among U.S. households and in long-run measures of family incomes and labor market earnings.[14] Rising economic inequality since the end of the 1970s is a very real phenomenon.

U.S. economic growth has recovered over the last decade, but the benefits of economic growth are now far less equally shared than in the past. Only the top part of the U.S. income distribution has seen income gains in recent decades as strong as in the pre-1973 period. Because labor income makes up the vast majority of national income, and since most American families make their living from work, the story behind rising inequality is one about the labor market and changes in the inequality of labor market earnings.[15] We now turn to documenting recent trends in U.S. wage inequality.

Wage Structure Changes

Wage inequality and educational wage differentials have increased sharply in the United States since the late 1970s.[16] Although there is debate concerning the causes of changes in the wage structure and earnings inequality, substantial agreement exists on the facts. These facts are, in a nutshell, that virtually *every* aspect of earnings by education, occupation, experience, age, and so forth has widened in the post-1970s era.[17] For example, the wage premium earned by young college graduates (with exactly 16 years of schooling) relative to young high school graduates (with exactly 12 years) more than doubled from 1979 to 2005.[18] Within-group wage inequality, known as "residual inequality," has also expanded. That is, the wages of individuals of the same age, sex, education, and job experience are far more unequal today than 25 years ago.

The distribution of wages did not widen in the same manner at all times. For example, upper tail inequality (the 90–50 wage gap, meaning the log difference between the wages of individuals at the 90th and 50th percentiles of the wage distribution) has increased steadily and rapidly since the late 1970s, whereas lower tail inequality (the 50–10 wage gap) grew sharply in the 1980s but has changed little since around 1990. The only part of wage inequality that has not increased concerns gender differentials: women gained on men, particularly in the 1980s.

The substantial overall widening of the U.S. wage distribution during the four decades from 1963 to 2005 is well summarized in Figure 2.4. The figure plots the change in log real weekly wages by percentile from 1963 to 2005, for men and women separately.[19] A sizable expansion in both the male and female wage distributions is evident with the 90th percentile earners rising by approximately 48 log points (about 62 percent) more than the 10th percentile earners for both men and women. The entire wage distribution spreads out monotonically (and almost linearly) for women and for men a similar monotonic expansion is seen above the 30th percentile. The female line is above the male line at every point in the distribution, showing that women have gained substantially on men during the four decades shown and that their gain occurred in every part of the wage distribution.

The timing and the key components of the recent rise in U.S. wage inequality are highlighted in Figure 2.5. Three aspects of wage inequality are shown: the 90–10 overall log wage differential (for males), the

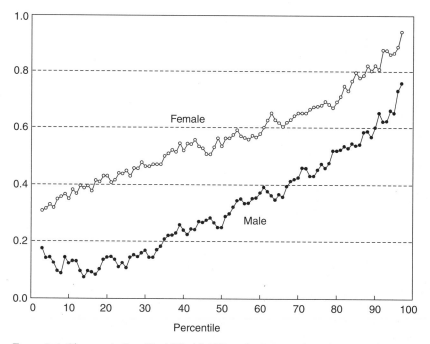

Figure 2.4. Changes in Log Real Weekly Wages by Percentile and Sex: 1963 to 2005. Source: Autor, Katz, and Kearney (2007, figure 1) based on data from the March CPS for 1964 to 2006 on the weekly earnings in the previous calendar year of full-time, full-year workers (those working at least 35 hours a week and at least 40 weeks in the year). The sample includes those aged 16 to 64 years in the earnings year.

90–10 residual log wage differential (for males), and the college–high school log wage differential (for males and females combined). Large increases in all three measures of inequality occurred since 1980, with the sharpest increases occurring in the 1980s. Although all three inequality measures (overall, residual, and educational) expanded rapidly and in tandem during the 1980s then flattened modestly in the 1990s, the series diverged in both the 1970s and the 1960s.

A key and often neglected fact about the evolution of U.S. wage inequality over the past four decades is that the rise of inequality was *not* a single phenomenon. Rather, it encompasses several elements that have not always moved together. Specifically, while overall and residual inequality rose modestly during the 1970s, the college wage premium declined sharply and then rebounded rapidly during the 1980s. Similarly, the college wage premium expanded considerably during the 1960s,

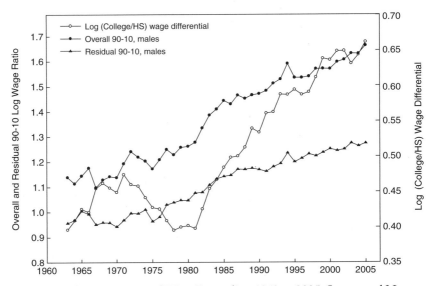

Figure 2.5. Three Measures of Wage Inequality: 1963 to 2005. Source and Notes: Autor, Katz, and Kearney (2007, figure 2a), based on data from the March CPS. This figure plots the (college plus/high school) log wage differential for males and females combined; the male log 90–10 wage differential; and the male log 90–10 residual wage differential for full-time, full-year workers from 1963 to 2005. Details on the construction of these series are contained in the data appendix to Autor, Katz, and Kearney (2007).

even while aggregate inequality was relatively quiescent. A more detailed probing of the data shows that that the 90–50 wage gap continued to rise rapidly over the last 15 years but the 50–10 flattened out.[20]

In sum, the sharp rise in U.S. wage dispersion since 1980 has involved a large increase in between-group wage differentials driven by a rise in the returns to education and a large increase in within-group (residual) wage inequality. Four explanations, complementary in part, may account for these striking trends.

The first reason, which we examine in detail in Chapter 3, attributes the sharp rise in wage differentials to increases in the rate of growth of the relative demand for highly educated and more-skilled workers arising from skill-biased technological changes driven largely by the diffusion of computer-based technologies. The second reason concerns the role of globalization forces, particularly increased trade with less-developed countries and greater international outsourcing, in reducing production employment and shrinking the relative demand for the less educated. The third factor emphasizes a slowdown in the rate of growth of the relative supply of skills (that we documented in Chapter 1) arising from a slower rate of increase in the educational attainment of successive cohorts born after 1950, changes in entering labor force cohort size, and an increased rate of unskilled immigration. The fourth explanation involves changes in labor market institutions, such as the decline in unionization and the reduction in the real value of the minimum wage.

A full assessment of these explanations requires putting the recent rise of U.S. wage inequality into a longer-term historical perspective. We address that task here and return in Chapter 8 to evaluate the alternative explanations in a supply-demand-institutions framework of the labor market.

Wage Inequality since 1939

Individual-level data on labor market earnings are available from the decadal population censuses extending back to 1939.[21] The information can be used to produce estimates of wage inequality and educational wage differentials. The 1940 Census was the first to inquire about labor income and educational attainment and although the data have various deficiencies, including the absence of self-employment income, the census provides the first view of incomes covering virtually all U.S. workers. We know a considerable amount about inequality trends since 1939 from the decadal censuses combined with the annual micro data available since the early 1960s from the Current Population Surveys.

These data clearly demonstrate that the period from 1939 to the present contained two opposing trends. From 1939 to the early 1970s, the wage distribution either narrowed or was relatively stable. From the late 1970s to the present, as we saw in the previous section, the distribution widened considerably, more than wiping out all the narrowing that had taken place since 1939. These two opposing trends are displayed in Figure 2.6; the trends in overall wage inequality from 1939 to 2005 are summarized by the log ratio of wages at the 90th percentile to those at the 10th, and the economic returns to education given by the college wage premium. Inequality and the returns to education were not always rising, as they have been from the 1970s to the

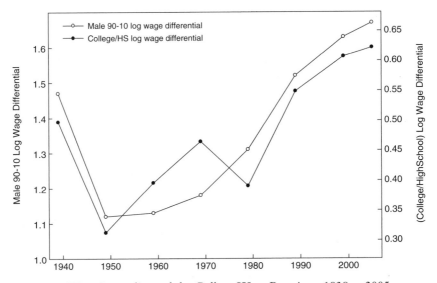

Figure 2.6. Wage Inequality and the College Wage Premium, 1939 to 2005.
Sources and Notes: Male 90–10 Log Wage Differential: Census IPUMS for 1940
to 2000 (covering earnings years 1939 to 1999) and the CPS MORG samples for
1999 and 2005. From 1939 to 1999 the male 90–10 log wage differential uses
weekly wages for full-time, full-year male wage and salary workers, 18 to 64 years
old, in the earnings year. Full-time, full-year workers are those working 35 or
more hours per week and at least 40 weeks in the earnings year. We include only
those earning at least one-half the federal minimum wage in the earnings year.
The change in the male 90–10 log wage differential from 1999 to 2005 is based
on hourly wages from the CPS MORG samples. (College/High School) Log
Wage Differential: Autor, Katz, and Krueger (1998, table I), updated to 2005.
The log college/high school wage differential is a weighted average of the college
(16 years of schooling) and post-college (17 plus years of schooling) hourly wage
premium relative to high school workers (those with 12 years of schooling) for the
year given. The weights are the employment shares of college and post-college
workers in 1980. Educational wage differentials in each year are estimated using
standard cross-section log hourly earnings regressions for wage and salary
workers in each sample. The college wage premia for 1939 to 1999 are estimated
from the 1940 to 2000 IPUMS and the change from 1999 to 2005 uses the
estimates from the 1999 and 2005 CPS MORG samples. For further details see
Autor, Katz, and Krueger (1998).

present. As reflected in both series, inequality decreased in the 1940s
and the reductions were substantial. The narrowing of the wage structure during the 1940s has been termed the "Great Compression."[22] It
involved a world war, inflation, tight labor markets, rising union
strength, and substantial government intervention in the labor market.

The fact that the wage structure today is at least as wide as it was in 1939, almost a decade into the Great Depression, may be disconcerting. But the comparison between 1939 and the more recent period has been made largely out of necessity and owes little to its potentially interesting chronology. As we just noted, the 1940 Census was the first at the federal level to collect information on annual earnings, weeks worked, and education. Thus, that census became the starting point for most discussions of the wage structure, the income distribution by skill, and the returns to education over the long-run.[23]

But was 1939 a typical year for the wage structure and the returns to education and skill? After ten years of record-high unemployment, those at the bottom of the skill distribution might have acceded to extremely low real wages. If so, the wage distribution below the median would have been abnormally and temporarily stretched by 1939. Using this logic, the narrowing of the 1940s would have returned the wage structure back to its pre-Depression level. The more recent period would not then be accurately construed as a return to the wide income distribution of a distant past. Rather, the more recent wage structure could be interpreted as one that was *never* before experienced in the United States in a time of national prosperity and moderate unemployment. That conclusion would be more disconcerting.

However, many research findings that we present in this chapter establish a good case for the opposite conclusion. The high level of inequality in 1940 was not due solely to the Great Depression. Rather, the wage structure observed in 1939 was similar to that of the 1920s. The evidence marshaled to support this conclusion comes from a set of wages for occupations that were relatively skilled, such as ordinary white-collar jobs (e.g., clerks) and blue-collar craft positions (e.g., machinists), as well as those that were unskilled (mainly laborers). In each case, the ratio of the hourly, weekly, or monthly wage of the skilled position to that of a laborer in the late 1930s was virtually identical to its level in the late 1920s. The evidence on the wages or earnings of skilled relative to unskilled workers suggests that the levels in the late 1930s were not anomalous by the standards of the 1920s.[24]

A related phenomenon is apparent in the often cited income distribution series assembled from income tax returns, first used by Simon Kuznets (1953) and later revised and extended by Thomas Piketty and Emmanuel Saez (2003, 2006). These data give the proportion of national income earned by the top echelon of income earners (the top decile and higher) from 1913 to 2004.[25] Only the fraction earned by the richest can be estimated prior to the 1940s because only the very top part of the income distribution paid income tax and had to file a tax return.

The income tax series show a precipitous decline in inequality in the 1940s, reinforcing the findings from other series on wages, earnings, and income.[26] The income share (excluding capital gains) of the top 10 percent, which was at 43.8 percent in 1929 and 44.6 percent in 1939, declined to 33.8 percent in 1949.[27] The important point is that the income tax series for the top decile hardly rose at all during the 1930s, certainly not relative to its large decline in the 1940s, and for the super rich (the top 1 percent) the share actually fell. The income tax data, therefore, reveal nothing particularly unusual about the late 1930s in comparison with the 1920s.

The evidence to date, therefore, is in agreement that 1939 was not an oddity or an anomaly with respect to various measures of inequality. It is even possible that the income distribution was more compressed in 1939 than in the 1920s. Because there is considerable evidence that the wage structure and income inequality among the rich in 1939 were not unusual in comparison with the late 1920s, we ask whether the wage structure was even wider earlier in the century than it was in 1939 and also whether the returns to education were yet higher.

Our answer is that the wage structure *was* even wider earlier in the century than it was in 1939 and that the monetary returns to a year of schooling were also greater than in 1939. We are able to provide an answer because of the recent retrieval of data from a remarkable and unique document—the 1915 Iowa State Census. We also used several less obscure materials but ones, nonetheless, that have long remained dormant. Using all these sources, we found that the wage structure narrowed several times in the first half of the twentieth century and that the returns to education were in fact *higher* in 1914 than in 1939, when they were, by historical standards, quite high.

We assembled data from a wide variety of sources and found that there was, during the first half of the century, a substantial decrease in various measures of inequality. The wage structure in manufacturing narrowed, the premium to various white-collar occupations decreased as did that for many craft trades, and the return to years of post-elementary education fell. These declines, moreover, both came during wartime periods and were subsequently sustained. Not only was there a wage and income compression in the 1940s, about which much has been written, but there was also a narrowing in the late 1910s. Both periods of reductions in the premium for skilled labor and decreases in the pecuniary return to education coincided with expansions in education, first for secondary schooling and later at the college level.

Inequality Trends before the 1940s

We have shown using relatively rich and complete data that the wage structure, and also the distribution of income and the returns to a year of schooling, declined substantially in the 1940s. After a several decade lull of semi-stability, these measures then rapidly expanded from the 1970s to the 1990s. We also demonstrated that the compression in the wage structure in the 1940s did not return it to one that existed just before the Great Depression. That is, there is no indication that the wage structure in 1939 was anomalous and was simply a product of the 1930s unemployment. Rather, there is evidence that the wage structure in the 1920s was just as wide, possibly even wider, than it was in 1939. We now turn to a closer examination of inequality in the first half of the twentieth century.

Because no comprehensive, national sample of the wage structure and income distribution is available for the pre-1939 period, we must piece one together using a variety of sources. To do so, we extend the more recent data on the wage structure and the returns to education back to the early part of the twentieth century (and for some series, to the late nineteenth century). We use large, representative samples, including one from the manuscripts of the 1915 Iowa state census and others from less obscure documents that have previously eluded investigators.

The data series we unearthed and compiled revealed that the wage structure and the returns to education and skill all moved in the direction of greater equality decades before the better known Great Compression of the 1940s. The wage structure narrowed, skill differentials were reduced, and the return to education decreased sometime between 1890 and 1940, most likely in the late 1910s. The entire compression of the wage structure across the twentieth century, therefore, was larger in magnitude, lengthier in duration, and more complicated in its reasons than has been previously recognized.

We used data for specific sectors, such as manufacturing, and for particular occupations, such as professors, engineers, laborers, operatives, and mechanics. We chose the occupations for uniformity across time and the sectors because of their large relative size and the availability of consistent data over the period. We often used data that reveal one aspect of the distribution rather than the entire distribution. The ratio of the earnings of the more highly skilled and educated to those who are less skilled or educated gives but one part of the distribution, although we often could not pinpoint exactly where in the distribution these occupations were located. The salient point is that the evidence shows that wages and earnings compressed in several stages even *before* the Great Compression of the 1940s.

Wage Structure in Manufacturing and for Manual Workers

An extensive literature exists on the pre-1940 wage premium to skill for manual workers.[28] The research, mainly done in the immediate post–World War II period, was largely motivated by the wage compression of the 1940s. Several of the studies measured skill premiums by constructing the ratio of the earnings of skilled production workers in manufacturing to lower-skilled workers, such as laborers, helpers, janitors, and teamsters.[29] Others examined changes in wages by narrowly defined occupations.[30]

Almost all the researchers found a narrowing of the wage structure in the pre-1950 period. Since we already know that there was a narrowing in the 1940s, the question we must address is, what occurred before? The existing literature provides some clues, and here we have built on them. Our answer with regard to the manufacturing sector and manual work is that there is unambiguous evidence that the wage structure narrowed and compressed in several stages before 1940.

Several labor economists, including some who worked at the Bureau of Labor Statistics in the mid-twentieth century, added considerably to the literature on the pre-1950 wage structure. Harry Ober (1948) analyzed annual information on skilled and unskilled building tradesmen (union wage scales) from 1907 to 1947 and a set of skilled and unskilled occupations for five years between 1907 and 1947. In both cases, and also in his related work on the printing trades (Ober, 1953), Ober uncovered two periods of persistent narrowing, one in the late 1910s and the other from the late 1930s to 1947, the endpoint of his study.[31] The narrowing of skilled manual worker wages to all manufacturing worker wages during the 1910s and its persistence through the 1920s is shown in Figure 2.7. In explaining the changes Ober emphasized the role of inflation, changes in fairness norms in setting wages at the lower end, and automation in rendering many unskilled jobs superfluous. Stanley Lebergott (1947) also examined wages by occupation for various industries from 1900 to 1940 and found strong evidence of compression in the wage structure for manual workers prior to the 1940s. His evidence also revealed that the timing of the change was sometime between 1913 and 1931.

Although the literature on the wage structure for the manual trades is in agreement with our finding that compression occurred before the 1940s, the conclusions rest on the construction of ratios for the mean wages of skilled and unskilled workers. We now present evidence on the entire distribution of wages in the manufacturing sector, which leads to a similar conclusion with regard to the compression of the wage structure.

Wage Structure in Manufacturing for 1890 and 1940

We discovered new data supporting the notion that the wage structure for manual workers compressed sometime between 1890 and 1940. The data provide the wage structure for production workers in 1890 and around 1940 in various manufacturing industries matched between the two years. Rather than estimating the ratio of wages for craft workers to those for laborers or for a range of occupations in particular industries, as was done in the literature just summarized, we produced summary statistics for the *full distribution* of wages for manual workers in manufacturing.

The data for 1890 come from special tabulations of the 1890 Census of Manufactures. The 1890 schedules included a question on the number of employees by weekly wage brackets, but the data were not published in the volumes containing the national data by industry. Instead, the wage distribution data were published in a volume on urban manufacturing (covering the 165 largest cities in 1890) and in special industry reports.

The data for 1940, or around 1940, were derived from studies comprising the "wage and hours" series that have been executed by the

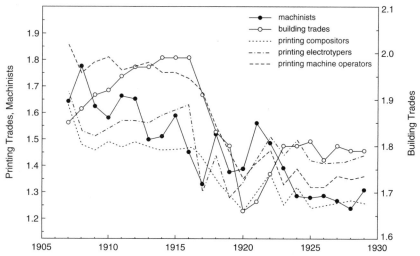

Figure 2.7. Skill Ratios in the Manual Trades, 1907 to 1929. Sources and Notes:
Building and printing trades: Ober (1948). Machinists: U.S. Department of Labor
(1934); Goldin and Margo (1991, 1992), who use the Interstate Commerce
Commission reports. Manufacturing earnings: Historical Statistics (1975), series
D 740. For details on all sources and for notes, see Goldin and Katz (2001a,
figure 2.2). The ratios given are for the annual earnings of skilled manual workers
in various trades to an average of all manufacturing workers.

Bureau of Labor Statistics ever since the 1890s.[32] The series changed
form over the years, shifting in 1907 to union wage scales and then in
the 1930s to all workers. At some point in the 1930s the surveys began
to report the full distribution of weekly or hourly wages by industry. In
the years just after passage of the Fair Labor Standards Act (1938), the
reports often noted the impact of the minimum wage on the bunching
of employment by wages. During and after World War II the surveys
occasionally provided information on the impact of war industries, collective bargaining, and extensions of the minimum wage.

In most respects, the comparability between the data for 1890 and
1940 is fairly good. In both years we can compare the distribution of
wages for male workers (older than 16 years in 1890) in relatively
narrow brackets. The one potentially important difference is that data
for 1890 refer to weekly wages, whereas those for 1940 are for hourly
earnings. Because workers with lower hourly earnings often worked a
greater number of hours than those with higher hourly earnings, the

bias should make the 1890 distribution more, rather than less, compressed in comparison with 1940.[33] We located 12 reasonably similar industries with data for both years.[34]

For most industries in 1890, the wage distributions include both production and nonproduction workers (e.g., officers, managers, clerical workers), whereas in 1940 the data include only production workers. We cannot add nonproduction workers to the 1940 data, but we can subtract them from the 1890 data. To construct a wage distribution for production workers in 1890, we assumed that each nonproduction worker was paid more than the highest paid production worker. Thus we subtracted *all* nonproduction workers from the *top* of the wage distribution, an assumption that biased the results toward a narrower distribution of wages in 1890, particularly at the upper end. The extent of the bias depends on the fraction of nonproduction workers in the industry, as well as the degree of overlap in the wage distributions of production and nonproduction workers. The nonproduction employment shares for 1890 ranged from 2.6 percent in cotton goods to 40 percent in cigars.[35]

Our matched-industry data for 1890 and 1940 not only has unique evidence on the wage structure, it also represents a large fraction of all male production workers in manufacturing. The 12 industries in the sample included 28 percent of all male (time-rate) production workers in manufacturing in 1890 and 25 percent of all wage earners in manufacturing in 1940.[36]

For almost all industries and inequality measures the wage structure in our sample was wider in 1890 than in 1940 (see Table 2.1). The finding is most apparent using the 50–10 measure, but also holds for most of the other summary statistics, such as the 90–50, 90–10, and 75–25.[37] The 75–25 and 90–50 statistics change the least with time, and in several industries these measures remain virtually unchanged. One or two reveal some widening; but on average, using 1940 employment weights, the distribution narrows for all measures considered.[38] It should be recalled, in addition, that we made various assumptions to exclude white-collar workers from the sample in 1890. These assumptions, by necessity, bias the upper end of the distribution to be more compressed in 1890 than in actuality. Thus it is not surprising that the measures that place more weight on the upper end show the least compression.

Table 2.1. Wage Distribution for Male Production Workers in Manufacturing, 1890 and 1940

	Log wage differentials							
	50–10		90–50		90–10		75–25	
Industry	1890	1940	1890	1940	1890	1940	1890	1940
Cotton goods	1.64	1.33	1.67	1.48	2.75	1.97	1.63	1.46
Dyeing and finishing textiles	1.71	1.47	1.61	1.39	2.76	2.04	1.59	1.51
Flouring and grist (grain, 1940) mill products	1.47	1.69	1.51	1.60	2.22	2.69	1.43	1.90
Foundry and machine shop products	1.72	1.51	1.58	1.52	2.72	2.30	1.93	1.55
Furniture, factory product	1.75	1.43	1.63	1.68	2.85	2.40	1.70	1.67
Iron and steel	1.41	1.25	2.04	1.48	2.88	1.85	1.72	1.40
Lumber and planing mill products (no logging)	1.80	1.30	1.52	1.93	2.73	2.51	1.91	1.97
Shipbuilding (private shipyards, 1940)	1.72	1.45	1.47	1.32	2.52	1.92	1.74	1.46
Silk and silk goods (and rayon, 1940)	2.06	1.38	1.61	1.62	3.32	2.23	1.80	1.62
Soap (and candles, 1890)	1.97	1.51	1.48	1.33	2.90	2.01	1.55	1.35
Tobacco: chewing, smoking, and snuff (and cigarettes, 1940)	1.55	1.37	1.81	1.46	2.81	1.99	1.79	1.54
Tobacco: cigars (and cigarettes, 1890)	2.01	1.49	1.54	1.66	3.11	2.48	1.70	1.68
Weighted average (using 1940 employment weights)	1.66	1.35	1.71	1.60	2.81	2.15	1.74	1.60

Sources and Notes: 1890: U.S. Census Office (1895b, c); c.1940: U.S. Department of Labor, Bureau of Labor Statistics, *Monthly Labor Review* (various issues, 1938 to 1942). See Goldin and Katz (2001a, appendix table 2.1). The abbreviations 50–10, 90–50, 90–10, and 75–25 refer to the log wage differentials at these various points of the distribution. The 1940 figures are for approximately the year 1940. Spring 1937 figures are used for "cotton goods." The 1941 figure is used for "furniture." An average of the two 1890 figures (with and without white-collar workers) is used for "tobacco: cigars." Cigarettes were unimportant in 1890 and were included with cigars for that year. By 1940, cigarettes had the majority of tobacco industry employment.

Our review of previous studies and the new evidence on the manufacturing wage structure bolsters the evidence that the wage distribution significantly compressed for production workers sometime between 1890 and 1940.[39]

It is useful to consider how the narrowing between 1890 and 1940 compares with that of the Great Compression of the 1940s. We can do that by comparing changes in the wage distribution from 1890 to 1940, with the information available for 9 of the 12 industries for the late 1940s to the early 1950s. On average, the narrowing in the 90–10 log wage difference from 1890 to c. 1940 was more than twice as large as it was from around 1940 to the early 1950s. The weighted average for the change in the 90–10 log wage difference is 27.9 log points for the 1890 to 1940 period, whereas the weighted average for the 1940 to early 1950s period is just 11.6 log points.[40]

The pre-1940 compression in the wages of manufacturing workers was substantial. Although we discuss the reasons for change in the distribution of economic returns in Chapters 3 and 8, several factors may be of particular importance in the manufacturing and manual job sectors. The most important were the reduction of immigration flows beginning in the mid- to late 1910s, the increase in secondary schooling starting in the first decade of the twentieth century, and rising union influence in the 1910s and 1930s. Another factor concerns compositional changes. Factory electrification during the late 1910s and the 1920s and the installation of hoisting and moving equipment eliminated many low-wage workers, such as common laborers, who hauled goods around the factory floor.[41]

Occupational Wage Ratios: Non-Manual Occupations

The compression in wages within the manufacturing sector and for manual workers was also uncovered for the white-collar sector by the University of Chicago labor economist, and later U.S. Senator, Paul Douglas. Douglas (1926) used wage data for clerical employees and lower-level managers, known often as "ordinary white-collar workers," to explore the premium to white-collar work. He found a substantial decrease in the earnings of these white-collar workers relative to manual workers from around 1900 to the late 1920s.[42]

According to Douglas, persons eligible for white-collar jobs had, before the expansion of public secondary schools, comprised what he

termed a "non-competing" group. But with the "high school movement" of the early 1900s and the vast increase in proprietary commercial schools, the market became flooded with literate and numerate
young people who had skills applicable to the commercial workplace.
Thus, the increase in both formal education and technical training led
in the late 1910s to a plummeting in the wage premium of various
white-collar positions.

Douglas was keenly aware, as well, of another set of forces. Technical change was fast moving in the factory, office, and home from
1900 to 1920. The new technologies raised the demand for skilled
workers at the same time that the high school movement increased the
supply. The demand for more able and educated workers increased in
the blue-collar sector at the same time that it did in the white-collar
sector. Thus, in both sectors the demand for and the supply of educated and skilled workers was increasing. But white-collar work employed a far greater fraction of workers who were more educated than
did blue-collar work. Thus, Douglas surmised, the overall impact of
the increase in the supply of educated workers would have been greater
in the white-collar sector, thus producing the reduction in their relative earnings.

Although Douglas broke new ground in his work, various factors
complicate his story of the wage structure. These factors concern large
changes in the composition of ordinary white-collar workers during
the period when Douglas found a decrease in their relative earnings.
The magnitude of the changes necessitates that we make adjustments
to his estimates to ensure that the findings do not rest on compositional changes.

Most important was the substantial increase in the female share of
ordinary white-collar workers. In 1890, women were just 20 percent of
all office workers, but in 1930, that number had increased to 50 percent.[43] There were, as well, large changes in the composition of office
jobs. High-ranking secretaries—the keepers of the officers' "secrets"—
became less numerous. Lower-level clerks, typists, and stenographers
proliferated. Hand-bookkeepers gave way to machine-operators armed
with "comptometers." The Douglas series does not separate the clerical group by sex and does not present data on separate office occupations. Thus the Douglas series could well overstate the decrease in relative wages due to compositional changes by sex and occupation.

Furthermore, the findings Douglas reported could have been transitory. The wage structure among the manual trades in the late 1910s experienced large changes due to the increased relative demand for unskilled workers during World War I and wartime inflation; these effects were generally not maintained through the 1920s. Since Douglas' data do not extend beyond 1926, they do not reveal whether the prewar premium to ordinary white-collar workers was later reinstated, as appears to have been the case for some of the skilled blue-collar trades.

To resolve these issues, we constructed several white-collar earnings series and earnings data by sex for detailed occupations using all the sources Douglas consulted, adding others, and extending the data forward in time to 1940. The results are reported in Table 2.2 and the series is expanded to 1959 using the public-use micro-data samples from the federal population censuses for 1940, 1950, and 1960.

Rather than overturning Douglas' conclusions on the trend in relative earnings, our additions serve mainly to confirm and extend them. The earnings of ordinary white-collar employees declined relative to those of production workers in manufacturing and the decreases are evident by sex and by occupation. That is, Douglas' results were not due solely to compositional changes. While clerical occupations became feminized and occupations shifted away from the more highly skilled and better remunerated, the earnings of each white-collar occupation declined relative to those of production workers, by sex, in manufacturing.

Although we are in agreement with Douglas on the broad outlines of the pre-1930 period, our extensions serve to amend the timing of the decline. The decline in the earnings of ordinary white-collar workers relative to manual workers in the Douglas series occurs just after 1900 and the ratio may even have increased from 1890 to 1900. In our series, for males and females separately, the decline in the relative earnings of ordinary clerical workers occurs in the late 1910s and early 1920s. The resulting lower level persists to 1939, after which it declines once again. The results are robust to distinctions by sex and by separate occupations. A very important conclusion is that the decline exists even for tasks that did not experience much technical advance during the period.

The conclusions based on our extensions to 1959 serve to place the earlier results in a longer-term perspective. The decrease in the premium to ordinary white-collar work that occurred in the early part of the

Table 2.2. Ratios of Clerical Worker Earnings to those of Production Workers, by Sex and Occupation: 1890 to 1959

Year	All Clericals		Clerks		Typists and Stenographers		Bookkeepers and Cashiers	
	Females	Males	Females	Males	Females	Males	Female	Male
	(1)	(2)	(3)	(4)	(5)	(6)	(7)	(8)
1890	1.848	—	—	—	—	—	—	—
1895	1.936	1.691	1.798	1.388	2.099	1.638	2.001	2.278
1909	1.956	1.652	—	—	—	—	—	—
1914	2.073	1.696	—	—	—	—	—	—
1919	1.525	1.202	—	—	—	—	—	—
1923	1.413	1.099	—	—	—	—	—	—
1924	1.399	1.097	—	—	—	—	—	—
1925	1.466	1.101	—	—	—	—	—	—
1926	1.480	1.113	1.177	1.084	1.641	1.319	2.205	1.604
1927	1.501	1.131	—	—	—	—	—	—
1928	1.546	1.117	—	—	—	—	—	—
1929	1.527	1.128	—	—	—	—	—	—
1939	1.557	1.150	1.499	1.088	1.652	1.100	1.613	1.268

Year	All Clericals		Typists, Stenographers, and Secretaries		Bookkeepers, Cashiers, and Accountants	
1939	1.369	1.187	1.430	1.288	1.309	1.341
1949	1.137	1.076	1.166	1.333	1.131	1.236
1959	1.133	1.019	1.171	1.168	1.097	1.188

Sources: 1890 to 1939: Goldin and Katz (1995, tables 5 and 6). 1939 to 1959: Integrated Public Use Micro-data Samples (IPUMS) of the U.S. federal population censuses.

Notes: "All clericals" excludes supervisors. "Clerks" includes all clerks, except chief and senior clerks, file clerks, and mail clerks for 1895 and 1926. "Typists and stenographers" in 1895 includes secretaries, but excludes male secretaries with very high earnings. "Bookkeepers and cashiers" includes chief and senior clerks, accountants, and assistant bookkeepers for 1895 and 1926, and includes tellers in 1939. Occupational categories for the 1939 to 1959 series use the census definitions in each of the years given. "All clericals" excludes "clerks working in stores" in 1939. The production worker wages for the 1939 to 1959 series use only those working in the manufacturing sector. The wage ratios for 1939 to 1959 are for annual earnings of full-time, full-year workers (those working 35 or more hours per week and 50 or more weeks in the previous calendar year). Weekly wage ratios for all full-time workers produce similar estimates in all cases for 1939 to 1959.

twentieth century (to 1939) was of a greater magnitude than that which took place later (1939 to 1959). Over the full period, from the start of the twentieth century to 1959, the premium to ordinary white-collar workers declined by about 42 percent for female clerical workers and 53 percent for male clerical workers.[44] About 55 percent of the decrease in the premium for females occurred up to 1939, and 45 percent took

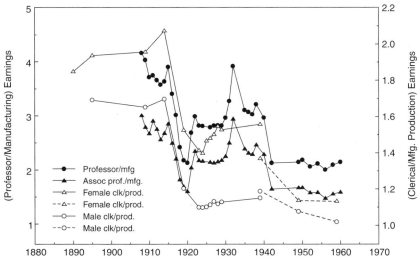

Figure 2.8. Ratios of White-Collar to Blue-Collar Earnings. Sources: See Tables 2.2 and 2.3.

place between 1939 and 1959. For males, about 72 percent of the decrease occurred up to 1939 and just 28 percent took place subsequently.

PROFESSIONAL OCCUPATIONS

A few additional white-collar occupations exist for which earnings data can be compiled in a consistent fashion from the early 1900s to the immediate post-1940s period. The occupations include college professors and engineers.[45] These data reveal trends that are almost identical to those found in the data for ordinary white-collar workers.

The data on college professors were originally compiled by Viva Boothe (1932) and later extended by George Stigler (1956). We made further extensions, revisions, and robustness checks to the Boothe-Stigler series using the original data collected and published by the U.S. Office of Education.[46] The findings, given in Table 2.3, part A, reinforce those in Table 2.2 concerning ordinary white-collar workers. Relative to production employees in manufacturing, the earnings of college professors decreased in the late 1910s to early 1920s. The reduced level for the relative earnings of professors to production workers persisted through the 1920s and into the 1930s (see Figure 2.8).[47] The premium to professors decreased again in the 1940s. These conclusions hold for all professorial ranks.

Table 2.3. Earnings of Professionals Relative to Manufacturing or Low-Skilled Workers, 1900 to 1960

Part A: College Professors

Year	Annual Earnings (Professors/Average Worker in Manufacturing)			Annual Earnings (Professors/ Average Low-Skilled Worker)		
	Full Professor	Associate Professor	Assistant Professor	Full Professor	Assistant Professor	Highest Full Professor Salary
1908	4.159	3.004	2.648	4.460	2.840	5.765
1909	4.032	2.788	2.386	4.658	2.756	5.939
1910	3.713	2.668	2.209	4.539	2.700	—
1911	3.747	2.902	2.362	4.675	2.948	—
1912	3.659	2.751	2.309	4.616	2.913	5.961
1913	3.575	2.559	2.193	4.544	2.788	—
1914	3.635	2.675	2.251	4.694	2.907	6.460
1915	3.903	2.849	2.396	4.845	2.975	6.659
1916	3.406	2.491	2.156	3.713	2.350	5.035
1917	3.014	2.202	1.866	3.098	1.919	3.972
1918	2.418	1.818	1.548	2.468	1.580	3.364
1919	2.175	1.688	1.366	2.360	1.482	3.319
1920	2.129	1.597	1.320	2.511	1.557	3.310
1921	2.686	2.039	1.734	3.566	2.302	4.868
1922	2.989	2.344	1.936	3.778	2.447	5.408
1923	2.817	2.173	1.816	3.548	2.287	4.994
1924	2.809	2.161	1.786	3.578	2.274	5.060
1925	—	—	—	—	—	5.256
1926	2.786	2.141	1.782	3.543	2.266	5.017
1927	2.816	2.128	1.781	3.594	2.273	4.914
1928	2.821	2.150	1.786	3.622	2.293	5.102
1929	2.818	2.177	1.744	3.564	2.206	5.349
1930	2.962	2.248	1.865	4.025	2.534	6.094
1931	3.272	2.497	2.056	4.672	2.935	7.055
1932	3.917	2.938	2.435	6.222	3.867	8.559
1935	3.104	2.387	2.014	4.121	2.674	—
1936	3.070	2.310	1.932	3.951	2.486	—
1937	3.028	2.285	1.858	3.718	2.281	—
1938	3.212	2.461	2.000	4.028	2.508	—
1940	2.964	2.285	1.819	3.551	2.179	—
1942	2.127	1.643	1.307	2.569	1.580	—
1949	2.145	1.667	1.363	—	—	—
1950	2.183	1.670	1.361	—	—	—
1952	2.061	1.577	1.283	—	—	—
1954	2.115	1.579	1.281	—	—	—
1956	2.003	1.472	1.197	—	—	—
1958	2.093	1.551	1.258	—	—	—
1960	2.145	1.585	1.277	—	—	—

Table 2.3. (continued)

Part B: Engineers

| | Annual Earnings (Engineers / Manufacturing Workers) | | | | |
| | (1) | (2) | (3) | (4) | (5) |
Year	Starting Engineers	Starting Engineers (Index)	First Year Engineers	Second Year Engineers	All Engineers, Median Monthly × 12
1900	—	—	1.643	—	—
1901	—	—	—	2.104	—
1904	1.338	—	—	—	—
1905	—	—	1.604	—	—
1906	—	—	—	2.080	—
1909	1.202	—	—	—	—
1910	—	—	1.382	—	—
1911	—	—	—	1.899	—
1914	1.149	—	—	—	—
1915	—	—	1.513	—	—
1916	—	—	—	1.598	—
1919	1.005	—	—	—	—
1920	—	—	1.175	—	—
1921	—	—	—	1.486	—
1922	1.029	—	—	—	—
1923	1.026	—	1.283	—	—
1924	1.034	—	1.261	1.472	—
1929	—	1.037	—	—	2.248
1932	—	1.037	—	—	2.452
1934	—	1.024	—	—	2.186
1939	—	1.008	—	—	2.439
1943	—	0.997	—	—	1.706
1946	—	0.985	—	—	1.950
1947	1.048	—	—	—	—
1948	0.987	—	—	—	—
1949	1.012	—	—	—	—
1950	0.945	—	—	—	—
1951	0.898	—	—	—	—
1952	0.955	—	—	—	—
1953	0.962	—	—	—	1.534
1954	1.004	—	—	—	—
1955	0.994	—	—	—	—
1956	1.030	—	—	—	—

Sources: The data for the professor series are from American Association of University Professors (various years), Boothe (1932), Stigler (1950), and U.S. Bureau of Education (various years). The data on manufacturing employees and low-skilled workers are from *Historical Statistics* (1975), series D 740, 778, 841. Data for engineers are from Blank and Stigler (1957). For details see Goldin and Katz (2001a, table 2.3).

Note: The year given is the end of the academic year. Col. (2) index data are spliced to col. (1) numbers around 1929.

Full professors had annual earnings around 1910 that were almost four times those of an average manufacturing employee. In the 1920s, the annual earnings of full professors were less than three times those of manufacturing employees, and in the 1950s they earned about double. Across the half-century from 1910 to 1960, professors at all ranks saw their relative earnings cut nearly in half.[48]

The Bureau of Labor Statistics, in conjunction with several professional societies, surveyed engineers in 1935 on their earnings starting with 1929. The BLS also conducted a survey in 1946. Prior to 1929, however, the data are retrospective and come from a survey of an engineering society. Several series, therefore, can be produced and are given in Table 2.3, part B. The series on engineers is less complete than that for professors. Even though there is a break in the series between 1924 and 1929, and the data from 1904 to 1924 are retrospective, the results follow those for the other white-collar series. The premium to engineering relative to production work decreased from 1904 to the 1920s and again from the late 1940s to the mid-1950s.[49]

Summary on Trends in the Wage Structure: 1890 to 1940

We have marshaled considerable evidence that the wage structure compressed in several stages from the late nineteenth century to the mid-twentieth century. Among manual or blue-collar workers, the evidence on the entire wage structure suggests that there was compression sometime between 1890 and 1940. Because various historical wage series on blue-collar workers do not reveal a decrease in the premium to skilled workers between the early 1920s and 1940, it seems realistic to presume that the narrowing predates the early 1920s.[50] Another compression of the wage structure occurred during the 1940s. Of the two periods, the first one contains a greater narrowing than that in the 1940s, at least in terms of the 90–10 log wage differentials for various manufacturing sectors.

Because data on inequality among manual workers have generally begun with 1940, and because the compression of the 1940s appeared to have been substantial, the findings that we report are novel and important. Not only was there a narrowing of the wage differential that predated the 1940s but it was, as well, sizeable in magnitude.

There were two periods of compression for the three white-collar series presented. One occurred sometime before the 1920s and the other was situated in the 1940s. A summary of the various findings for the white-collar occupations relative to production workers in the manufacturing sector (or all wage and salary earners in manufacturing) is presented in Figure 2.8. With the exception of the increase in the relative earnings of professors during the height of the Great Depression, all series decrease in two giant steps and are level in between. The two periods of compression, moreover, occurred during war, inflation, and blue-collar union activity; however, the narrowing remained long after the wars, inflations, and burst of union activism.

Returns to Education

The return to years of schooling is another aspect of inequality differences by skill. It has generally, though not always, tracked changes in the wage structure (Figure 2.6). The college wage premium decreased in the 1940s, rose in the 1950s and 1960s, fell in the 1970s, and since that time has increased substantially.[51] Because the 1940 Census was the first to ask highest grade attained and also the first to inquire of wage and salary income, there have been few estimates of education returns for the period prior to 1940 and none using a large representative sample.

The previous discussion of the premium to ordinary white-collar work around the turn of the twentieth century suggested that returns to secondary school were considerable. The ratio of the earnings of a clerical worker, who generally would have gone to and possibly graduated from a high school, to those of a factory operative, who generally would have stopped school at age 14, is a reasonable proxy for the returns to four (or fewer) years of high school.[52]

The returns were probably considerable because, as Douglas noted, clerical and other office workers, before advances in secondary school, formed a "non-competing group." Prior to 1900, or about that date, youths who went to and graduated high school most likely came from families with sufficient means to forego their earnings, who had enough income to afford a private institution or lived in geographic proximity to a public high school, and who had some foresight. The increase in high school enrollment and graduation served to flood the labor market with literate and numerate workers whose skills enabled them to move into white-collar office jobs. It also increased the supply of those capable of filling blue-collar positions that required the reading of manuals, deciphering of blueprints, computing of formulae, and use of elementary science.

Estimating the returns to formal schooling prior to 1940 is important because of the vast increase in schooling in the first several decades of the twentieth century. From 1910 to 1940 secondary schools mushroomed all over the nation and youths began to go to high school in ever-increasing numbers to learn skills for life, not necessarily just for college. Certain parts of the nation experienced the high school movement earlier than others and were educational leaders. The states in the West North Central portion of the United States were among the leaders and one of them—Iowa—figures prominently in our analysis.

The rapid increase in secondary school enrollment and graduation in the 1910 to 1940 period raises the question of why this grand educational transformation began around 1910. Although we address this question in more detail in Chapter 5, we evaluate related questions here. Was the rate of return to high school (and college) substantial around 1910? If the return was high in 1910, did it decline over the course of the twentieth century as cohorts of educated Americans entered the labor force?

The Iowa State Census of 1915

The federal census originated from the need to seat Congress. Similarly, states took censuses prior to 1940, generally at the mid-point between the federal decennial censuses, to aid the seating of their legislatures.[53] State censuses were generally sparse documents and rarely included information other than that needed to count residents by age, sex, race, citizenship, and possibly ethnicity. A few contained other information, but just two of them (Iowa and South Dakota) asked a question on educational attainment prior to 1940. It is not coincidental that the states that pioneered surveying their citizens about educational attainment were in the West North Central region, a region that led the nation in the high school movement (discussed further in Chapter 6). Among the state censuses that asked revealing questions, the Iowa State Census of 1915 stands out for its detail and comprehensiveness.

Of most importance to our use of the document in this chapter are the questions on educational attainment, income, and occupation. Like most other state censuses, the primary use of the Iowa State Census of 1915 was to seat the representatives of the state legislature and for that purpose it surveyed *all* residents of the state.

There are many fortunate aspects concerning this particular census. One is that it was taken in 1915, on the cusp of the high school movement. Another is that it requested information on occupational earnings in 1914, just before the large increase in industrial demand due to World War I. Also fortunate is that the manuscripts from the census survived—on more than 2 million separate index cards—and that they were microfilmed (in 1986) by the Genealogical Society of Salt Lake City. A facsimile of the document appears in Appendix A, together with the details of the sample we used.

The 1915 Iowa State Census requested detailed information from its residents concerning educational attainment, current schooling, occupation, income from occupation, wealth, unemployment, and church affiliation, to mention just a few of the topics. No federal population census, not even the more recent Current Population Survey, has asked for such a range of information.

The questions in the Iowa census regarding educational attainment are exceptionally detailed and reflect the wide array of educational institutions in which Iowa's residents in 1915 had received instruction. Individuals from Iowa's cities could have attended a graded elementary school and a regular high school, at least since the 1870s. Those who had grown up in rural Iowa or who migrated to Iowa from Europe probably attended only common school. Iowa also contained numerous small colleges and several larger universities. In recognition of this diversity of educational experiences, individuals were asked to give their years of education by type of school: common, grammar, high school, business college, and college. The 1940 Census did not give respondents the ability to differentiate among the types of school they had attended, although Americans nationally would have attended an equally wide range of schools at that time.

We collected a cluster sample of almost 60,000 individuals, about equally divided among Iowa's cities (Davenport, Des Moines, and Dubuque) and ten rural counties, where "rural" indicates that the counties did not have a city of more than 25,000 people (see Appendix

A). The data set is large (approximately a 1 in 40 sample of Iowa's 1915 population) and representative.

In 1915, Iowa's labor force was more agricultural than was that in the rest of the United States (41 percent versus 31 percent) and was employed less in manufacturing (20 percent versus 29 percent). But in the sectors of trade, public, professional, and clerical employments, Iowa's white-collar labor force was proportionally equal to that in the rest of the United States (23 percent versus 22 percent). Iowa was an agricultural state, yet its population was just as urban as the rest of the United States if all incorporated towns are included. Iowa had few cities of modest size and none that would have been considered large, but it contained an enormous number of tiny incorporated towns, the quintessential "central towns" of location theory.[54] Situated at the crossroads of a prosperous agricultural economy, these towns housed the grain elevator, railroad, retail stores of various kinds, and, often, retired farmers and their immediate relatives. The Iowa rural sample, therefore, contains both farm and town people.

One important difference between Iowa in 1915 and the rest of the nation is that Iowa was a leading state in the high school movement. The high school movement was only beginning to gather momentum in 1915 and, at that time, by high school graduation rate Iowa ranked tenth in the nation and by high school enrollment rate it ranked fourteenth.[55] It would soon rank much higher, especially in terms of its graduation rate. Why Iowa became one of the leading states in the high school movement is a question we consider in Chapter 6. To presage that discussion and give a further idea of Iowa's economy, it should be realized that in 1912 Iowa had the second highest value of per capita taxable property in the nation. Prairie land was very fertile and thus highly valued.

Iowans were an exceptionally well educated group compared with others in the United States and in 1915 had an educational attainment about equal to that of the U.S. population in 1940 (see Table 2.4). Thus Iowa was 25 years ahead of its time in asking a question on educational attainment in its census and was, as well, 25 years ahead of its time in the education of its people. For example, Iowa women 25 to 59 years old in 1915 had a mean highest grade completed of 8.86 years (using our version II estimates in Table 2.4), which is exactly the same number as that attained in 1940 by a similar age group of women in the entire

Table 2.4. Formal Schooling Indicators in 1915 Iowa and 1940 United States, by Sex for 25- to 59-Year-Olds

Schooling indicators (version)	Males, 25 to 59 Years Old		Females, 25 to 59 Years Old	
	1915 Iowa	1940 United States	1915 Iowa	1940 United States
Mean highest grade completed (I)	8.40	8.60	8.68	8.86
Mean highest grade completed (II)	8.56	8.60	8.86	8.86
Average years of education	8.61		8.98	
Fraction with less than 8 years	0.235	0.311	0.185	0.278
Fraction with some high school (I)	0.233	0.410	0.290	0.462
Fraction with some high school (II)	0.379	0.410	0.446	0.462
Fraction graduating high school (I)	0.152	0.248	0.179	0.287
Fraction graduating high school (II)	0.156	0.248	0.184	0.287

Notes and Sources:

1915: 1915 Iowa State Census Sample. See Appendix A.

1940: IPUMS of the U.S. federal population census of 1940. The enumerators were asked to inquire "what is the highest grade of school completed?" (ICPSR 1984, 6.40–6.41). The highest grade of school is restricted to be less than 18.

Mean highest grade completed [1915]: reconstructs the 1915 data to approximate the 1940 instructions to enumerators. In line with the 1940 Census, the highest grade is restricted to be less than 18. Values were constructed to be consistent with the instructions to enumerators. For example, if an individual attended eight years of grammar school and four years of college but no high school, the individual received 16 years of schooling, rather than 12. For version (I) no individual in 1915 is given years of education exceeding eight for the sum of common and grammar school years; for version (II) the maximum for this case is nine years.

Average years of education: the sum of years of schooling in the various school categories with the highest grade restricted to be less than 18.

Fraction with some high school [1915]: fraction with years of education equal to at least nine. Version (I) and (II) differ in the same manner as for *mean highest grade completed* [1915].

Fraction with some high school [1940]: fraction with highest grade completed equal to at least nine.

Fraction graduating high school [1915]: fraction with years of education equal to at least 12. Version (I) and (II) differ in the same manner as for *mean highest grade completed* [1915].

Fraction graduating high school [1940]: fraction with years of education equal to at least 12.

United States. Among that same age and sex group, 44.6 percent in Iowa had attended some secondary school; in the entire United States in 1940, 46.2 percent had. Not only was there a relatively high educational attainment among the adult population of Iowa in 1915, the youth of Iowa attained exceptionally high secondary school attendance rates. Although in 1915 most rural areas in Iowa were not served by a local secondary school, 26 percent of all 15- to 18-year-olds were attending some post-grammar school and 54 percent of the age group attended some type of school.[56]

Value of Schooling by Age and Sector: Iowa 1915

We estimate the return to years of formal education using a standard (Mincerian) log annual earnings equation augmented to allow the returns to vary by type of schooling.[57] For males 18 to 65 years old, the return to a year of high school was about 10 percent; for the younger group, 18 to 34 years old, the return was larger, about 12 percent (Table 2.5, cols. 1 and 7). Returns to college years were also large and were similarly higher for the younger group, 15 percent as opposed to 10 percent. The returns to high school and college for women were also substantial: 10 percent for a year of high school and 15 percent for a year of college for unmarried women, 18 to 34 years old (Table 2.5, col. 12).

Interestingly, the monetary returns to a year of high school (or college) are not much different for those across dissimilar occupations. Perhaps the most surprising result is that the return to a year of post-elementary schooling was substantial for those engaged in farm occupations (Table 2.5, cols. 4 and 9). We explored this result further and found that Iowa counties having a greater fraction of adults with post-elementary schooling had higher agricultural productivity in both 1915 and 1925.[58] Furthermore, counties with larger increases in the fraction of adults with post-elementary school education had larger increases in agricultural productivity. That is, the change from 1915 to 1925 in education by county was associated with an increase in the value of farm output given capital and land inputs.

The return to years of high school and college was garnered, in part, because individuals with more education could enter more lucrative occupations, such as those in the white-collar sector. The most remunerative nonprofessional jobs for men in Iowa in 1915 were various sales positions. Traveling salesmen, for example, were among those with the highest incomes in the sample. But more education enhanced earnings not simply by enabling individuals to shift from manual to nonmanual jobs. The return to a year of high school was high even *within* the white-collar group and it was also high *within* the blue-collar group. We estimate returns exceeding 8 percent per year within either blue-collar or white-collar occupations for males 18 to 34 years old (Table 2.5, cols. 10 and 11).

The role of within- and between-occupation returns to education can be demonstrated more effectively by adding a full set of occupation dummies to the earnings regressions.[59] For all males (18 to 65 years old), the addition of one-digit occupation dummies reduces the return to a year of high school from 0.103 to 0.062. The inclusion of a full set of 3-digit occupation dummies lowers the coefficient to 0.054. Comparable analyses for the blue-collar and white-collar groups separately produce similar results. Thus, for males, the monetary return to years of high school was about equally divided between that due to higher earnings within narrowly defined occupations and that due to a shift to higher paying occupations. The notable result from this analysis is that education enhanced earnings to a considerable extent within occupations and within even the manual job category.

Most important is the finding that the return to a year of secondary school (or college) was extremely high in Iowa in 1915. The return was considerable within sectors and, of note in this agricultural state, the return was substantial within farming. It is no wonder that the high school movement took off at this juncture in U.S. history and that many of the educationally progressive states were agricultural ones such as Iowa. (The precise reasons why these states were leaders in the high school movement are reserved for Chapter 6.)

Given the high relative wages of office workers, it should not be surprising that the return to a year of high school was substantial in 1914. But why was the return also considerable within the blue-collar sector and within farming occupations? Secondary school education helped provide the cognitive tools demanded for entry into the elite craft occupations, such as electrician and machinist. Many of the more educated blue-collar workers in 1915 Iowa owned shops and garages. The highly educated farmers read the progressive farming journals, were aware of animal inoculation, could fix various types of machinery, had

Table 2.5. Returns to a Year of Education by Type of Schooling, Occupational Grouping, Age, and Sex: Iowa 1915

	18- to 65-Year-Old Males					
	All Occupations		Non-farm	Farm	Blue-Collar	White-Collar
Years in school	(1)	(2)	(3)	(4)	(5)	(6)
Common school	0.0427		0.040	0.0375	0.0239	0.0275
	(0.00269)		(0.00300)	(0.00555)	(0.00314)	(0.00573)
Grammar school	0.0533		0.0647	0.0232	0.0585	0.0470
	(0.00292)		(0.00304)	(0.00800)	(0.00320)	(0.00591)
High school	0.103		0.102	0.114	0.0740	0.0609
	(0.00448)		(0.00401)	(0.0146)	(0.00584)	(0.00566)
College	0.103		0.106	0.132	0.0533	0.0783
	(0.00604)		(0.00520)	(0.0254)	(0.0151)	(0.00569)
Linear spline functions						
Common school ≤9 years		0.0452				
		(0.00336)				
Common school >9 years		0.0291				
		(0.00771)				
Grammar school ≤9 years		0.0547				
		(0.00340)				
Grammar school >9 years		0.0467				
		(0.0195)				
High school ≤4 years		0.111				
		(0.00491)				
Years of college		0.0958				
		(0.00729)				
Business school, dummy	0.379	0.371	0.393		0.441	0.202
	(0.0850)	(0.0849)	(0.0705)		(0.156)	(0.0776)
R^2	0.199	0.202	0.256	0.209	0.205	0.218
Number of observations	14,699	14,699	10,695	3,705	7,588	3,733

Table 2.5. (continued)

| Years in school | All Occupations (7) | Non-farm (8) | Farm (9) | 18- to 34-Year-Old Males | | Females[a] |
				Blue-Collar (10)	White-Collar (11)	All Occupations (12)
Common school	0.0483	0.0375	0.0637	0.0229	0.0438	0.00714
	(0.00395)	(0.00442)	(0.00837)	(0.00450)	(0.00889)	(0.00877)
Grammar school	0.0693	0.0671	0.0568	0.0634	0.0679	0.0454
	(0.00421)	(0.00443)	(0.0110)	(0.00458)	(0.00909)	(0.00913)
High school	0.120	0.114	0.132	0.0908	0.0826	0.101
	(0.00564)	(0.00516)	(0.0176)	(0.00738)	(0.00747)	(0.00760)
College	0.146	0.143	0.166	0.0575	0.131	0.151
	(0.00915)	(0.00799)	(0.0381)	(0.0195)	(0.00849)	(0.0122)
Business school, dummy	0.284	0.273		0.452	0.0825	0.508
	(0.0988)	(0.0831)		(0.180)	(0.0886)	(0.0969)
R^2	0.251	0.296	0.241	0.256	0.313	0.273
Number of observations	7,145	5,249	1,784	4,021	1,744	2,001

Source: 1915 Iowa State Census Sample. See Appendix A.

Notes: Sample excludes bottom 0.2 percent of the earnings distribution (less than $60) and is restricted to those out of school. Regressions also contain a quartic in potential experience, a race dummy, and a dummy variable for those missing "years in U.S." Potential experience is defined as min (age − 15, age − years of schooling − 7). All regressions are weighted by urban and rural sampling weights (Appendix A, for weighting information). Blue collar includes craft, operative, service, and laborer occupations. White collar includes professional, semiprofessional, managerial (but not farming), clerical, and sales occupations. Standard errors are given in parentheses below the coefficients.
   a. Includes only unmarried women.

knowledge of various crop varieties, and knew modern accounting techniques. Iowa's parents in the 1910s wrote of wanting secondary schools in their districts so that their children would not be left behind in the "new world" of business. Even though many of Iowa's educated children would leave the state as adults, and most would leave their home district, secondary school education was highly valued by the community as a public good.

One potential limitation to the results we have just presented is that individuals with greater innate ability garner more years of education but also have higher earnings because of their innate ability. Thus, in the absence of information on family background, the estimates we present could be too high, a product of what is known as "ability bias." Because these data are from a relatively rural population during a period of school diffusion, the luck of geography—more than factors concerning familial wealth and individual ability—would have greatly determined whether a youth could attend a secondary school.

Educational Returns over the Long Run: Iowa 1914 to 1959

To make further sense of the value of education in Iowa in 1914, we explore the change in the return to a year of high school and college in 1939, 1949, and 1959 using the IPUMS for 1940, 1950, and 1960.[60] For comparability across the years we restrict the 1940, 1950, and 1960 samples to full-year, non-farm male workers residing in Iowa.[61] The return to a year of high school or college, or simply to a year of school, declined between 1915 and 1950 (rows 1 and 2 in Table 2.6).[62] There is already substantial evidence that inequality measures, including the return to education, declined between 1940 and 1950; the more important issue is whether the return to a year of secondary school or college decreased before 1940, in particular between 1915 and 1940. The answer is that it did, but there are several complicating factors that we must first address.

The most important complication is that the 1940 Census inquired only of wage and salary income, not income from self-employment, whereas the 1915 Iowa State Census asked for income from the individual's occupation, which would have included that from self-employment.[63] The 1950 Census, on the other hand, asked for both wage and salary income and that from self-employment, given separately. We can, therefore, compare the returns to a year of schooling

Table 2.6. Returns to Education for Full-Year, Non-farm, Male Workers in Iowa: 1915, 1940, 1950, and 1960

Census year	Years of High School		Years of College		Linear in All Years of Schooling	
	18–65 years old	18–34 years old	18–65 years old	18–34 years old	18–65 years old	18–34 years old
(1) 1915	0.091	0.105	0.091	0.128	0.084	0.100
(2) 1950	0.051	0.067	0.073	0.086	0.054	0.069
(3) 1960	0.047	0.050	0.085	0.071	0.059	0.058
(4) 1940, wage and salary earnings only	0.064	0.097	0.081	0.086	0.064	0.075
(5) 1950, wage and salary earnings only	0.049	0.043	0.064	0.101	0.048	0.060
(6) 1960, wage and salary earnings only	0.040	0.049	0.064	0.057	0.046	0.050
(7) 1940, adjusted	0.064	0.097	0.094	0.095	0.068	0.079

Sources: 1915 Iowa State Census Sample (see Appendix A); 1940, 1950, and 1960 IPUMS.

Notes: See Goldin and Katz (2000, appendix table A3) for standard errors and for the other schooling coefficients. Coefficients listed for "years of high school" and "years of college" are those from a spline in years of education (1 to 8 years, 9 to 12 years, and 13 plus years) in a regression of (log) annual earnings. Coefficients listed for "linear in years of all schooling" are those from the sum of all years in school in a regression of (log) annual earnings. "Full year" is defined for 1940, 1950, and 1960 as more than 49 weeks of work; in 1915 it is defined as listing no unemployment. Controls in all regressions are: quartic in potential experience, whether native-born, and whether white. Each of the samples deletes the lowest 1 percent of earners. The samples for 1940, 1950, and 1960 include only those living in the state of Iowa.

The 1940 adjusted estimates, row 7, take the 1940 wage and salary earnings estimates (row 4) and add an adjustment factor to account for the absence of self-employment income for comparability with the estimates in row 1 for 1915. The adjustment factor, for each column, is constructed from national estimates because the sample used here (for Iowa) is small, particularly for 1950. We take the difference in returns to each type of schooling for the entire sample and for the wage and salary sample, averaging the estimates for 1950 and 1960, and then add this adjustment factor to the 1940 estimates in row 4.

for wage and salary earners in 1940 and 1950 (also 1960), and then make an adjustment for those with self-employment income in 1940. We do precisely that in producing the adjusted 1940 results given in Table 2.6, row 7.

In only one of the columns in Table 2.6 is the adjusted estimate of the returns for 1940 larger than the estimated figure for 1915. In all other cases the returns to a year of schooling were considerably greater in 1915 than in 1940. The anomalous case (given in row 7) is that of college years for the full population. The result is probably an oddity since many older, college-educated Iowans likely attended small denominational liberal arts colleges or bible schools rather than one of the two state institutions or a more modern liberal arts college.[64] The return to a year of schooling in one of the older-style and smaller colleges may not have been as high as in the more up-to-date and larger institutions.[65] The most important result in Table 2.6 is that the return to a year of post-elementary education was higher in 1915 than in 1940.

One may wonder whether the decrease in the estimated returns to a year of high school and of college that we have found reflects nothing more than greater selectivity on the basis of ability into the post-elementary grades in 1915 than in 1940. The existing literature on sorting by ability into secondary and higher education suggests just the opposite. There is evidence that, from 1917 to 1942, IQ test scores of high school students *rose* with the large increase in secondary school enrollment, and Iowa tests of achievement show increased high school student performance from 1940 to the early 1960s.[66]

The innate ability of high school students does not appear to have declined during the inter-war years, and the quality of secondary schooling does not appear to have been reduced during and for more than a decade after World War II. Furthermore, among high school graduates who continued to college, it has been shown that there was substantially *less* selectivity measured by cognitive test scores in the 1920s than in the 1930s and 1940s.[67] None of these findings should be surprising since during these periods both secondary and higher education were becoming more accessible to rural youths and the children of immigrants in large cities. These groups would not previously have had the ability to take full advantage of secondary and higher education.

Educational Returns over the Long Run: United States 1914 to 2005

By combining national estimates of returns to schooling from 1939 to 2005 with our data from Iowa for 1914, we can produce reasonably comparable estimates of the monetary gains to a year of high school or college from 1914 to 2005 for the entire nation. We do so by creating two 1914 variants of the national estimates.[68] Variants I uses the change in the Iowa estimates of the returns to schooling from 1914 to 1939 to construct the national estimate for 1914. Variant II uses the change in the returns from 1914 to 1959. The results of these calculations for young men and all men are given in Table 2.7 and graphed in Figure 2.9 for young men.

The justification for our assumption that the change in educational wage differentials in Iowa from 1914 to 1939 is a reasonable proxy for the change in the nation is as follows. Estimates of the return to a year of high school and college for Iowa in later years, such as 1939, 1949, or 1959, move closely with national estimates. In addition, occupational wage differentials for Iowa from 1914 to 1939 show a pattern of declining white-collar wage differentials similar to, albeit slightly more muted than, national estimates such as those depicted in Figure 2.8.[69] The higher educational attainment in Iowa than in the nation in 1914 suggests that the estimated decline in educational wage differentials in Iowa from 1914 to 1939 may, if anything, *understate* the national decline in the educational wage premium.

The full twentieth-century story of the returns to a year of schooling is that they were rather high at the start of the century. Schooling returns, for the most part, fell from 1914 to 1939 (but in the case of returns to college for all men, they were fairly constant). Returns to a year of high school and college plummeted in the 1940s. With increased educational access, returns were markedly reduced by the 1950s when, despite enhanced access to college, returns increased, though not to the levels achieved before or more recently. The return to a year of secondary schooling early in the twentieth century was higher than it is today and that to college about the same at least for younger workers. The high level of returns to skill achieved around 1939 was not anomalous; in fact, returns to education were lower in 1939 than they were 25 years earlier.

Table 2.7. Returns to Education for Male Workers in the United States: 1914 to 2005

| | Returns to a Year of: | | | |
| | High School | | College | |
Year	Young Men	All Men	Young Men	All Men
1914, variant I	0.110	0.112	0.148	0.097
1914, variant II	0.125	0.098	0.148	0.097
1939	0.102	0.085	0.115	0.100
1949	0.054	0.051	0.078	0.077
1959	0.070	0.054	0.090	0.091
1969	0.074	0.059	0.096	0.099
1979	0.081	0.066	0.084	0.089
1989	0.093	0.078	0.124	0.124
1995	0.096	0.081	0.133	0.129
2005	0.087	0.077	0.148	0.144

Notes and Sources: "Young" means 0 to 19 years of potential work experience. "All" means 0 to 39 years of potential work experience. The estimates for 1914 to 1995 are from Goldin and Katz (2001a, table 2.4). The estimates of returns to high school and college for 1939 to 1995 refer to composition-adjusted log weekly wage differentials by years of schooling for full-time, full-year male wage and salary workers. The changes in returns from 1995 to 2005 are based on composition-adjusted log hourly educational wage differentials for male wage and salary workers using the 1995 and 2005 CPS MORG samples.

Returns to High School: 1915 Iowa State Census Sample (Appendix A); 1940 to 1970 IPUMS; 1970 to 1996 March CPS; and 1995 and 2005 CPS MORG. The estimates from 1939 to 1969 equal the composition-adjusted log weekly wage differential between workers with exactly 12 and exactly 9 years of schooling divided by 3. The changes in returns from 1969 to 2005 equal one-half times the change in the composition-adjusted log weekly wage differential of workers with exactly 12 years of schooling and 10 years of schooling. The 1914, variant I estimate for each group equals the sum of our 1939 national estimate for that group and the corresponding estimated change in returns to a year of high school in Iowa from 1914 to 1939 (the difference between rows 1 and 7 for the relevant columns in Table 2.6). The 1914, variant II estimate equals the sum of our 1959 national estimate for that group and the corresponding estimated change in returns to a year of high school in Iowa from 1914 to 1959 (the difference between rows 1 and 3 for the relevant columns in Table 2.6).

Returns to College: 1915 Iowa State Census Sample; 1940 to 1990 IPUMS; 1990 and 1996 March CPS; and 1995 and 2005 CPS MORG. The estimates from 1939 to 2005 equal the composition-adjusted wage differential of those with exactly a college degree (16 years of schooling) relative to those with 12 years of schooling divided by 4. The 1914, variant I and variant II estimates of returns to college use the same methodology as used for the analogous 1914 estimates for returns to high school. The 1914, variant I estimates add the corresponding 1914 to 1939 changes in college returns for Iowa from Table 2.6 to our 1939 national estimates. The 1914, variant II estimates similarly add the corresponding 1914 to 1959 changes in college returns for Iowa from Table 2.6 to our 1959 national estimates.

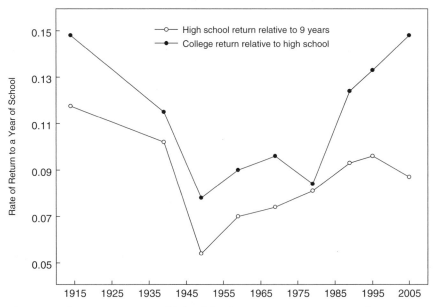

Figure 2.9. Returns to a Year of School for Young Men: 1914 to 2005. Source: Table 2.7. The average of variant I and variant II is used for 1914 for both the high school and college returns for young men.

Inequality Anxieties of the Past and Present

We began this chapter with our observation that economic anxiety increased in late twentieth-century America and that growing inequality was one of the reasons. Some have contended that rising inequality has far-reaching social and political effects if economic distance produces social distance, which makes political coalitions more difficult to assemble.

But even the most exaggerated allegations today concerning the negative effects of rising inequality do not approach those made in America more than a hundred years ago. Few today allege that rising inequality endangers our democracy or that a war between the rich and poor is imminent. These were, however, just some of the assertions made a century or more ago by a wide variety of individuals.

Important commentary on the distribution of income and wealth began with the economic downturns of the 1870s and 1880s. Edward Bellamy's overnight best-seller *Looking Backward* (1888) was an indictment of inequality and presented a vision of a futuristic egalitarian society with benign and efficient socialism. Bellamy's gruesome portrait of the lower classes physically oppressed by capitalists was given real meaning by the events of the following decade. The Homestead (1892) and Pullman strikes (1894), and the Haymarket riot (1886) that preceded them, were not simply tragic instances of labor unrest and expressions of labor's demands for shorter hours and higher wages. They were also important examples of U.S. military intervention to protect private property rather than to safeguard the right of labor to organize.

Mounting divisions between labor and capital were codified in the important U.S. Supreme Court case *Pollock v. Farmers Loan* (158 US 601, 1895), invalidating the income tax law of 1894. Speaking for the majority, Associate Justice Stephen Field asserted that the income tax would be the beginning of "a war of the poor against the rich."[70] The 1890s was a decade of growing fear of anarchists, later of syndicalists. These were moments of genuine concern regarding the creation of "classes" in America and the growing distance between them. America was beginning to look more like Europe, not just in terms of the distribution of income and wealth, but also in terms of the potential for political upheaval.

The era was also one of third-party movements, often a sign of upheaval and discontent. The Populist or People's Party was the most successful third-party movement in U.S. history, with the possible exception of the parties that joined together to create the Republican Party. At their peak, around 1896, Populists controlled one state legislature (Kansas) and ran William Bryan for president in 1896 and 1900; they also had championed James B. Weaver, less successfully, for president in 1892. Although neither the Populists nor the Progressives put forth proposals directly concerned with inequality, their platforms were motivated by growing differences in economic and political power among economic classes. Populists championed monetary liberalism (free silver) and regulation of the railroads and other business with agricultural ties. These policies were intended to redistribute resources from creditors to debtors and from the owners of various forms of capital to those purchasing their services. Progressives were concerned with government corruption, the trusts, and unfair labor practices, and they championed maximum hours laws, worker safety, and the minimum wage.

The timing of social and political expressions of fear of inequality coincides with our finding that the period around the late nineteenth and early twentieth centuries was probably a high point for the wage structure and also for the pecuniary return to education. But we have less evidence regarding income and wealth for this early period. Concern with the social consequences of inequality was most often directed at class divisions and the accumulation of great wealth by the few, rather than the distribution of labor earnings. Yet, as Paul Douglas recognized, the increased demand for educated workers during the rise of big business gave those fortunate enough to have received post-elementary schooling a large competitive edge and that such individuals formed a "non-competing group." An ameliorative policy, in the form of the high school movement, was embraced by thousands of individual school districts in one of the grandest grassroots movements in U.S. history. Perhaps it was mass secondary school education that checked the more extreme forms of socialism later embraced by Europe.

Two Tales of the Twentieth Century: A Summary

The history of inequality during the twentieth century is a tale in two parts. The first was punctuated by episodes of declining inequality, some quite sudden and rapid. Stable or slowly rising inequality marked other parts of the period. On the whole, the first three-quarters of the century were years of greatly diminished inequality and lowered returns to education. Americans grew together as economic growth was shared throughout the income distribution during much of the period.

Everything came to a halt in the 1970s. America started to grow more slowly and Americans began to grow apart. The last quarter of the twentieth century and the early twenty-first century have been distinguished by exploding inequality, chiefly at the upper end of the income distribution. Returns to education, particularly college, markedly increased. Economic growth slowed or was stagnant until the mid-1990s. Whatever growth occurred was unequally shared. With low or no growth and soaring inequality, the lower end often lost out altogether while the economic elites prospered.

We saw in Chapter 1 that the history of educational attainment in the United States is also a tale in two parts. For a long time cohorts of the American population and workforce increased their educational attainment rapidly relative to previous cohorts. But that trend halted and has remained on hold with the entering labor force cohorts of the late 1970s.

We are left with several questions. What accounts for narrowing inequality trends during the first part of the twentieth century and what could explain the possible failure of these trends during the second part? Did technological change accelerate between the first and the second parts of the twentieth century? Was the culprit the computer revolution? Alternatively, or in conjunction, did the supply of educated and skilled workers change? We turn to these issues in Chapter 3.

3

Skill-Biased Technological Change

Inequality and Technological Change

"Computers Did It"

Economic inequality since 1980 increased greatly, as we have just seen. The earnings of college graduates rose at a far greater clip than did the earnings of those who stopped at high school graduation. The incomes of top managers and professionals increased at a much faster rate than did those of ordinary workers.

The increase in inequality was more all-encompassing than a widening *between* different education levels or occupational groups. The expanding gap also occurred *within* groups, even within educational levels. Among college graduates, for example, those with degrees from institutions with higher standards for admissions earned relatively more over time. Those who went to more prestigious law schools did better relative to other law school graduates. The widening occurred within virtually all groups in a manner that is not easily explained by the usual observable factors such as years of schooling. At almost all educational and experience levels, for example, the earnings for those near the top of the distribution increased considerably relative to those near the middle or close to the bottom.[1]

The point that we made in the previous chapter was that widening inequality during the past 25 years has affected practically all Americans.

Few groups, by education, occupation, geography, and so forth, have been untouched. Some have gained, relatively, but far more have lost, at times in absolute terms. Widening inequality has been pervasive as well as rapid.

The pervasive and rapid increase in economic inequality has led many to search for explanatory factors that are themselves pervasive and rapid. A key suspect is *skill-biased* technological change, particularly that involved in the use of computers.[2] Chief among other factors that have been mentioned are increased international trade and outsourcing, the greater immigration of low-wage workers, the decline in private-sector unionization, the erosion of the real value of the federal minimum wage, and changes in social norms concerning the pay of executives and other top-end earners. Here we mainly discuss the role of technological change.

The central idea concerning the role of technology in affecting inequality is that certain technologies are difficult for workers and consumers to master, at least initially. Individuals with more education and higher innate abilities will be more able to grasp new and complicated tools. Younger individuals are often better able to master new-fangled equipment than are older individuals. Employers, in turn, will be more willing to hire those with the education and other observable characteristics that endow them with the capacity to learn and use the new technologies. Existing employees who are slow to grasp new tools will not be promoted and might see their earnings reduced. Those who are quicker will be rewarded.

The type of technological change that is necessary to explain the pervasive and rapid increase in economic inequality in the latter part of the twentieth century and the early twenty-first century must meet various criteria. First, it must have affected a large segment of the workforce, both production line workers and those in the office, both highly educated professionals and ordinary staff. As such, the innovation would probably have to be of the "general purpose technology" form.[3] A general purpose technology is one that is not specific to a particular firm, industry, product, or service. Instead, it is pervasive and omnipresent, cutting across various production methods and services. In addition, the technological innovation must have diffused during a fairly brief period. Finally, it must have required workers to think, adjust, and reconfigure the workplace. Computerization would seem to be the perfect culprit.

As we demonstrate in this chapter, it is clear that technological change—computerization in particular—is *part* of the explanation for rising inequality in the past 25 years. But, although computerization and other technological changes were culprits in fostering inequality, these "criminals" were not acting alone. The reasoning is simple.

New technologies alter the relative demand for different types of labor; however, the overall impact of a new technology on the wage structure reflects not only these demand shifts but also the supply responses by individuals attending various types of schools or obtaining skills on the job or in other ways. Just because a technology places increasing demands on the skill, education, and know-how of the workforce does not necessarily mean that economic inequality will rise and, if it does, that the increase will be sustained over a long period. If the supply of skills rises to accommodate the increase in demand for skill, then wage inequality need not change.

In other words, the evolution of the wage structure reflects, at least in part, a race between the growth in the demand for skills driven by technological advances and the growth in the supply of skills driven by demographic change, educational investment choices, and immigration.[4] This framework suggests that the rise in educational wage differentials and wage inequality since 1980 resulted from an acceleration in demand shifts from technological change, or a deceleration in the growth of the supply of skills, or some combination of the two.

The "computers did it" account also lacks historical perspective regarding technological change and inequality. Other critical moments existed in U.S. history when general purpose technologies swept the factory, office, and home. Consider, for example, the advent of motive power in the form of water wheels and later steam engines or, better yet, the electrification of the factory, home, and urban transportation. The notion that computerization provided the first or the most momentous instance in U.S. economic history of a complex technology that placed greater demands on the knowledge, ability, and flexibility of virtually all workers and consumers is gravely mistaken.

Lessons from History

In the early twentieth century, a wide range of industries, particularly the newer and more technologically dynamic ones, demanded more-educated workers. The workers to whom we refer were not necessarily of the professional class and they were not all working in an office, a boardroom, or on the sales floor. That is, they were not always white-collar workers. Rather, they included ordinary production line workers who were using more complicated and valuable machinery. For these workers, having more education meant having some high school and possibly a high school diploma. For the professional positions, the more-educated individuals would have gone to college. Our point is that new and more complex technologies have had a long history of transforming the workplace, as well as every day life, in ways that reward quick-thinking, flexible, often young, and educated individuals.

The most important historical point we want to make in this chapter concerns a unified explanation for long-term trends in inequality in America. It will be recalled from the previous chapter that the inequality story of the twentieth century contains two parts: an era of initially declining inequality and a more recent one of rising inequality. But can the two parts of the inequality experience have a unified explanation in the context of a demand-side framework? If technological change was skill-biased in the latter part of the twentieth century so that the more skilled and educated did relatively better, was the opposite true of the earlier part of the century so that the less skilled and educated fared relatively better?

The answers lie in the fact that skill-biased technological change was far more rapid and continuous during most of the twentieth century than has been previously suspected. Similar amounts of "skill bias" can be measured during much of the twentieth century. Thus, the demand-side argument, *by itself*, cannot explain both parts of the twentieth century inequality experience. One cannot fully resolve the divergent inequality experiences of the two halves of the twentieth century by appealing to a recent acceleration in the degree of skill-biased technological change brought on by the computer revolution. Computers, to be sure, have given us much that is novel, time-saving, informative, entertaining, and convenient. Yet, in terms of the skill bias to technological change and the increase in the relative demand for skill, the era of computerization has brought little that is new.

Nevertheless, something important did change in the market for skilled personnel. Since the demand-side changes were similar across the twentieth century, the change from the first part of the century to the last part concerns the other half of the inequality equation: supply. Supply fluctuations in educated and skilled labor are key factors in explaining changes in inequality. The growth in the relative supply of educated and skilled workers was rapid from the early to mid-twentieth century, but became lethargic in the late twentieth century. A unified explanation for the two halves of the twentieth-century inequality experience (at least with respect to educational wage differentials) does not require much (if any) acceleration in the skill-bias to technological change. Rather, the difference between the two halves came about primarily because of a change in the supply of educated and skilled labor.

The notion that technological change always and inexorably increases the relative demand for skill is mistaken. Even in the case of computers, it is widely conceded that the use of computers and their software have obviated the need for certain abilities and traits. One need only consider the role that computers have played in opening up new worlds for the blind, the physically challenged, and the deaf, as well as among a large group of more usual employees. Cashiers in stores, for example, are required to scan a bar code, not to do even simple arithmetic. Fast food workers need to know what a product looks like, not its name. However, we concede that the weight of the evidence is that computerization has, on net, increased the relative demand for various skills associated with higher levels of education and innate ability.

The possibility that technological change increased the relative demand for skill in the more recent past has led many to presume that the same was true for the more distant past, such as during the industrial revolution of the nineteenth century. But technological change across a longer-run historical period probably did not increase the relative demand for skill. The fact that more advanced technologies have, on net, increased the relative demand for education and various types of skill over the past century does not mean that they always did.

We assess when, during the past two centuries, increases in the relative demand for skill became quantitatively associated with technological advances raising productivity. Because advances in labor productivity may be associated with an increase in the amount of physical capital used by workers, we also address a related hypothesis—that an increase in the capital intensity of production is skill-biased. The former thesis is called "skill-biased technological change" and the latter thesis is termed "capital skill complementarity." The two are intricately related since more advanced technologies are generally more capital intensive.

The turning-point in the relationship between technological change and the relative demand for skill, as well as that concerning capital intensity and skill, was toward the end of the nineteenth century with the introduction of electricity and the adoption of various heavily capital-intensive technologies associated with batch and continuous-process machinery (terms that we later define more concretely). The reasons for these changes concern the substitution of motive power and machinery for brawn and also the increased demand for skilled mechanics, technicians, and various professionals who assembled and maintained the capital equipment.

But before we reach back into history and explore the origins of skill-biased technological change, we must first examine a period closer to the present. Why did inequality and educational wage differentials widen rapidly from 1980 to 2005 but not from 1950 to 1980? The "computers did it" hypothesis is correct for 1980 to 2005, but only in part, since technological change was similarly skill-biased during other moments in the past century. The difference between the recent episode and previous ones is primarily to be found in the slowdown in the supply of skills rather than in the speedup in the demand for skills.

Technological Change and the Relative Demand for Skill: 1950 to the Present

Skills and Skill Premiums: Some Facts and an Implication

To construct a convincing case that computers, or another technological change, played an important role in the growth in the college premium, and in widening wage inequality more generally, we need to introduce several facts. The set of facts concerns the relative supply of workers by skill level and their earnings. Both the relative supply of highly educated workers and the premium received by the more highly educated rose from 1980 to 2005. An important and powerful implication of these two facts is the following. If *both* the relative price and the relative quantity of a good (in this case, skill) increase, then the demand for the good (skill) must have been increasing at a rate greater than that of the supply of the good (skill).

If demand was increasing at a rapid clip, what could have caused it? One possibility is that skill-biased technological change was operating to increase the relative demand for more highly skilled workers. But there are other possibilities. For example, the United States has a comparative advantage in the production of high-skilled goods. If the growth of international trade increases the demand for such products, it can lead to substitution away from the U.S. production by using lower-skilled workers that can be purchased more cheaply abroad.[5] Although this is a possible explanation, our evidence provides more support for the skill-biased technological change theory than for the international trade story.

The only fly in the ointment is that skill-biased technological change has been operating at least since 1950, and probably for considerably longer. Therefore, "computers may have done it" from 1980 to 2005 but they had a sturdy accomplice in the slowdown in the growth of supply of skills.

The evolution of the college wage premium and the educational composition of the U.S. workforce from 1950 to 2005 are given in Table 3.1. The relative supply of college workers rose throughout the more than half-century era considered. The fraction of all full-time workers who were high school dropouts was almost 59 percent in 1950 and that for college graduates was barely 8 percent. In 2005, high school dropouts were but 8 percent and college graduates were almost 32 percent. Those with some college rose from 9 percent to 29 percent. At the same time, however, the wage premium for college graduates relative to high school workers more than doubled, from 36.7 percent in 1950 to 86.6 percent in 2005.[6] But most of the increase in the college premium occurred since 1980, with the decline in the college wage premium in the 1970s offsetting much of the earlier increases in the 1950s and 1960s. We have already gone over many of these details in Chapter 2, but the presentation in Table 3.1 has considerable meaning here, as we shall see.

The key implication of the two central facts of a rising college wage premium and rising relative supply of college workers since 1980 can be best understood with reference to Figure 3.1, which is a schematic representation of the market for skilled and unskilled labor.[7] In this simplified depiction, the workforce consists of two groups—the skilled

Table 3.1. U.S. Educational Composition of Employment and the College/High School Wage Premium: 1950 to 2005

	Full-Time Equivalent Employment Shares (%) by Education				College/High School Wage Premium
	High School Dropouts	High School Graduates	Some College	College Graduates	
1950 Census	58.6	24.4	9.2	7.8	0.313
1960 Census	49.5	27.7	12.2	10.6	0.396
1970 Census	35.9	34.7	15.6	13.8	0.465
1980 Census	20.7	36.1	22.8	20.4	0.391
1980 CPS	19.1	38.0	22.0	20.9	0.356
1990 CPS	12.7	36.2	25.1	26.1	0.508
1990 Census	11.4	33.0	30.2	25.4	0.549
2000 CPS	9.2	32.4	28.7	29.7	0.579
2000 Census	8.7	29.6	32.0	29.7	0.607
2005 CPS	8.4	30.9	28.9	31.8	0.596

Sources: Data for 1950 to 1990 are from Autor, Katz, and Krueger (1998, table I). Data for 2000 and 2005 are from the 2000 and 2005 Merged Outgoing Rotation Groups (MORG) of the CPS and 2000 Census IPUMS using the same approach as in Autor, Katz, and Krueger (1998).

Notes: The college/high school wage premium is expressed in logs. Full-time equivalent (FTE) employment shares are calculated for samples that include all individuals 18 to 65 years old in paid employment during the survey reference week for each census and CPS sample. FTE shares are defined as the share of total weekly hours supplied by each education group. The tabulations are based on the 1940 to 2000 Census IPUMS; the 1980, 1990, 2000, and 2005 CPS MORG samples. The log (college/high school) wage premium for each year is a weighted average of the estimated college (exactly 16 years of schooling or bachelor's degree) and post-college (17+ years of schooling or a post-baccalaureate degree) wage premium relative to high school workers (those with exactly 12 years of schooling or a high school diploma) for the year given. The weights are the employment shares of college and post-college workers in 1980. Educational wage differentials in each year are estimated using standard cross-section log hourly earnings regressions for wage and salary workers in each sample with dummies for single year of schooling (or degree attainment) categories, a quartic in experience, three region dummies, a part-time dummy, a female dummy, a nonwhite dummy, and interaction terms between the female dummy and quartic in experience and the nonwhite dummy. The levels of the log college wage premium can be compared across census samples and across CPS samples respectively. But the levels of the CPS and census wage differentials cannot be directly compared with each other due to differences in the construction of hourly wages in the two surveys. For further details see Autor, Katz, and Krueger (1998).

or highly educated (L_s) and the unskilled or less educated (L_u). Relative wages for the two groups (w_s/w_u) are determined by the intersection of a downward-sloping relative demand curve and an upward-sloping relative supply curve. The short-run relative supply of more-skilled workers is assumed to be inelastic, since it is predetermined by factors such as past educational investments, immigration, and fertility.

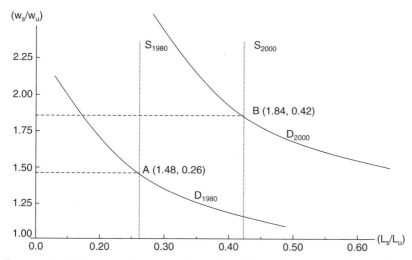

Figure 3.1. A Schematic Representation of the Relative Supply and Demand for Skill: 1980 and 2000. Point A is approximately the values for (w_s/w_u) and (L_s/L_u) from Table 3.1 (using the 1980 Census data and exponentiating the college/high school wage premium) and point B is the same for 2000 (using the 2000 Census data), where L_s/L_u is the ratio of college graduates to those with less than a college degree as per the selection criteria given in Table 3.1. Thus, point A is $e^{0.391} = 1.48$ for (w_s/w_u) and 0.26 for (L_s/L_u).

Although one cannot directly observe the entire demand and supply functions, one can observe the equilibrium relative wages and relative skills employed, as given in Table 3.1. Figure 3.1 illustrates the recent changes using data from 1980 and 2000. The actual labor market outcomes from Table 3.1 shifted from point A (year 1980) to point B (year 2000). Thus, the relative supply function of more-skilled workers shifted outward from 1980 to 2000, from S_{1980} to S_{2000}. It is also the case—and this is a major point we would like to make—that the relative demand function must also have *shifted outward*. We have drawn such a shift in Figure 3.1 as from D_{1980} to D_{2000}.[8] But why did the relative demand curve shift outward?

Evidence on Skill-Biased Technological Change

Just because the relative demand for more highly skilled workers increased does not necessarily mean that there was skill-biased technological change. Another possibility, and one that has received considerable attention, is that the manufacturing jobs taken by the less well educated in the United States have gone overseas. The relative demand for skill would then rise even if there had been no technological change. But there is very strong evidence for the technology explanation.

In the first place, relative employment of more-educated workers and of nonproduction workers increased rapidly *within* industries and *within* establishments during the 1980s and 1990s in the United States. The increased employment occurred despite the fact that the relative cost of hiring such workers greatly increased. Even though international outsourcing has been blamed for the decreased utilization of the less educated, the facts in this case argue against that explanation as being the primary factor. Large within-industry shifts toward more skilled workers occurred in sectors with little or no foreign outsourcing activity, at least in the 1980s and up to the late 1990s. Between-industry product demand shifts cannot be the main culprit. The magnitude of employment shifts to skill-intensive industries, as measured by between-industry demand shift indices, is simply too small.[9]

New technologies and greater capital intensity, as shown in many studies, are strongly and positively associated with higher relative utilization of more-skilled workers in firms and industries.[10] A clear positive relationship has been found between the relative employment of more-skilled workers and measures of technology and capital, such as computer investments, the growth of employee computer use, R&D expenditures, the utilization of scientists and engineers, and increased capital intensity.[11]

Case studies of the banking, auto repair, and valve industries show that the introduction of new computer-based technologies is strongly associated with shifts in demand toward more-educated workers.[12] Surveys of human resource managers reveal that large investments in information technology, particularly those that decentralize decisionmaking and increase worker autonomy, increase the demand for more highly educated workers.[13] The evidence is consistent with the "computers did it" view of widening inequality. But we have more direct confirmation.

How have computers increased the relative demand for educated and skilled workers? A multitude of reasons exist and often differ by workplace. Computerized offices have routinized many white-collar tasks, and the simpler and more repetitive tasks are more amenable to computerization than are the more complex and idiosyncratic. On the shop floor, microprocessor-based technologies have facilitated the automation of many production processes. Where hundreds of production workers once stood, often a handful remains together with a small team of workers operating the computer. Computers, the Internet, and electronic commerce have raised the returns to marketing and problem-solving skills.[14] The share of U.S. workers directly using computers on the job increased from 25 percent in 1984 to 57 percent in 2003.[15] Even though computers have been easy for younger and highly educated workers to use, they have been daunting for many others, at least initially.

The empirical and logical case for skill-biased technological change, as a substantial source of demand shifts favoring more-educated workers since 1980, would appear very strong. But that does not necessarily mean that the driving force behind rising wage inequality since 1980 was an alteration in the type or a quickening in the rate of technological change. The reason, we will soon demonstrate, is that capital-skill complementarity was present throughout the twentieth century as was rapid skill-biased technological change. These effects of technological change occurred even during periods of declining or stable educational wage differentials and narrowing economic inequality.

The evidence that skill-biased technological change has been ongoing for at least the last half century begins with a pioneering article by Zvi Griliches (1969), who found a substantial degree of capital-skill complementarity in U.S. manufacturing during the 1950s. Other researchers following Griliches' path documented strong within-sector skill upgrading, even in the face of rising educational wage differentials during the 1950s and 1960s, and a strong positive correlation of industry skill demand with capital intensity and technology investments throughout the 1950s to the 1970s.[16]

Technology Was Not Acting Alone: The Role of the Supply of Skills

Rapid skill-biased technological change has been in operation almost continuously for the past half century or more; however, for the period since 1980, an additional factor must be considered. The relative wages of more-educated workers increased sharply *and* the relative quantities of skill increased. There are two chief ways to explain this, but only one is consistent with the facts. The labor market impact of the computer revolution could have been quantitatively different from past technological changes if the rate of change in the relative demand for more-educated workers *accelerated* in recent decades. Alternatively, the labor market impact could have been different if there was a *slowdown* in the growth in the relative supply of skill, against a backdrop of skill-biased technological change.

We can assess these two possibilities by contrasting the evolution of the relative demand and supply for college workers during two different periods: the post-1980 era of large increases in the college premium and the 1950 to 1980 period when the college wage premium increased only modestly. It turns out that the second possibility is more consistent with the facts than is the first: there was a *slowdown* in the growth in the relative supply of skill at the same time that there was skill-biased technological change.

To accomplish the analysis, we return to the data in Table 3.1, which summarizes how the educational attainment of the U.S. labor force and the wage gap between the college and high school educated evolved from 1950 to 2005. The data are now used to estimate the growth in the relative supply of and demand for college workers (measured as college "equivalents") over selected periods since 1950.[17]

From 1950 to 2005 both relative demand and relative supply generally increased at a rapid clip, as can be seen in the upper part of Table 3.2. The large increase in the rate of growth of the relative supply of the college educated was the driving force behind the decline in the college wage premium of the 1970s. The 1970s surge in relative skill supply reflected the expanded access to higher education of previous decades (to be discussed in Chapter 7) and the large size of the entering labor market cohorts from the baby boom. The estimates in Table 3.2 show that the sharp increase in the college wage premium of the 1980s was the result of both a large slowdown of relative skill supply growth and a substantial acceleration in demand growth. Relative supply growth continued to decelerate in the 1990s and 2000s and, somewhat surprisingly given the continuing computer revolution, relative demand growth for college equivalents also slowed.[18]

A comparison of 1950 to 1980 with 1980 to 2000 in the lower part of Table 3.2 reveals almost no change in the average pace of relative demand growth during the second half of the twentieth century. On the

Table 3.2. Changes in Relative Wages by Education and the Supply and Demand for Educated Workers: 1950 to 2005 (100 × Annual Log Changes)

	Relative Wage	Relative Supply	Relative Demand
1950–60	0.83	2.91	4.28
1960–70	0.69	2.55	3.69
1970–80	−0.74	4.99	3.77
1980–90	1.51	2.53	5.01
1990–2000	0.58	2.03	2.98
1990–2005	0.50	1.65	2.46
1950–80	0.26	3.49	3.91
1960–80	−0.02	3.77	3.73
1980–2000	1.04	2.28	3.99
1980–2005	0.90	2.00	3.48

Sources: Autor, Katz, and Krueger (1998, table II), updated to 2005. The underlying data are presented in Appendix Table D.1 and are derived from the 1940 to 2000 Census IPUMS and the 1980 to 2005 CPS MORG samples.

Notes: The "relative wage" is the log (college wage/high school wage) or the college/high school wage premium from Table 3.1. The relative supply and demand measures are for college "equivalents" (college graduates plus half of those with some college) and high school "equivalents" (those with 12 or fewer years of schooling and half of those with some college). The implied relative demand changes assume an aggregate elasticity of substitution between college equivalents and high school equivalents of 1.64, our preferred estimate from Chapter 8, Table 8.2. The log relative supply of college equivalents is given by the log relative wage bill share of college equivalents minus the log relative wage series. The approach adjusts the relative supply measure for changes in the age-sex composition of the pools of college and high school equivalents. See the notes to Table 8.1 of Chapter 8 as well as Autor, Katz, and Krueger (1998) for details. To ensure data consistency across samples, changes from 1980 to 1990 use the CPS, changes from 1990 to 2000 use the census, and changes from 2000 to 2005 use the CPS.

other hand, the growth of the relative supply of skills substantially slowed after 1980. The slower relative supply growth since 1980 more than fully explains the enormous increase in the college wage premium from 1980 to 2005. In fact, the implied growth in the relative demand for college workers was slower from 1980 to 2005 than from 1950 to 1980. The most important conclusion to be drawn from this analysis is that the driving force behind the explosion of the college wage premium of the last 25 years was a sharp *decrease* in the growth rate of the relative supply of educated workers and not an increase in the growth rate of the relative demand for skill.[19]

Our conclusion does *not* imply that technological changes were unimportant to the wage structure. In fact, educational wage differentials would have declined substantially from 1950 to 2005 in the absence of rapid skill-biased technological change (as can be seen by the intersection of D_{1980} and S_{2000} in Figure 3.1). But our conclusion does mean that changes in the supply of skills are likely to be central to knowing why the wage structure behaved differently in the late twentieth century than it did earlier.

Technological Change and the Relative Demand for Skill: 1900 to 1950

New Technologies of a Bygone Era

It is often asserted that we are in the midst of extraordinary technological change today. That cannot be denied. Yet the technological advances that appeared and diffused in the two decades around 1915 were also extraordinary and may have had even greater economic consequence.

Consider the following. Manufacturing horsepower in the form of purchased electricity rose from 9 percent in 1909 to 53 percent in 1929; similar changes swept the residential use of electricity. New goods proliferated, including many that are still central parts of our lives (e.g., automobiles, airplanes, commercial radio, household electric appliances, and office machinery). Other goods that were invented and rapidly diffused during the period include aluminum, synthetic dyes, "artificial" (meaning manufactured) ice, motion pictures, and rayon. New techniques improved the production of rubber, plate glass, gasoline, canned condensed milk, and factory-made butter. Of the goods just mentioned all but the automobile disproportionately employed higher-educated production workers in 1940. Many of these industries also appear to have been the higher-skill industries in the 1910s and 1920s. Several of these industries used batch processes and continuous process machinery.

Because the terms "batch" and "continuous process" will reappear in our discussion, some definitions are in order. *Batch operations* are used for processing liquid, semi-solid, or gaseous matters (e.g., chemicals, liquors, dairy products, molten metals, wood pulp). *Continuous-process methods*, pioneered in the late nineteenth century, are used for products requiring little assembly and having few or no moving parts, such as oats, flour, canned foods (e.g., condensed milk, soup), soap, film, paper, matches, and cigarettes.

Continuous-process methods are well represented by Bonsack's famed cigarette-making machine, which eventually became a fully integrated and automated process. Tobacco leaves, rolling paper, foil, and cigarette boxes entered the machine at some point. Out came the finished product—a box of cigarettes that even had the tax stamp affixed. The Fourdrinier papermaking machine typifies many batch methods. Wood pulp was first created in batches and then sent to the next stage for the paper making. Both continuous-process and batch methods are "black-box" technologies. Raw materials are fed into the machine and finished products emerge, with few human hands intervening in production. A corps of machinists and mechanics attend the machinery, when needed. The non-robotized assembly line, in contrast, had vast quantities of human operatives taking part in a production process characterized by an extreme division of labor.

Labor Demand and Technology

As discussed in Chapter 1, educational advances in the twentieth century were, until a few decades ago, quite rapid. In the early part of the century the fastest growing educational category was secondary school. Thus a comparison between the impacts of technological change on the demand for more-educated workers today and in the past will involve a different level of education in each of the two periods. For the early twentieth century we concentrate on the increased demand for high school educated workers, whereas the discussion for the more recent period, as in the previous section, is most often in terms of college-educated workers. Furthermore, over the century, the nonagricultural part of the economy shifted from being dominated by manufacturing, transportation, and construction to being more service-oriented. Thus the occupations that we emphasize in the two periods also differ.

The point that we plan to make clear in the text that follows is that more-educated workers were in high demand in the early part of the twentieth century, particularly in the newer and more technologically dynamic industries and sectors. Furthermore, much of the increased demand came from the manufacturing sector. We present evidence regarding production workers that is not widely known, because the role of formal education among industrial workers in the past has escaped much notice. In fact, production workers for much of the twentieth century have been depicted as a rather undifferentiated group with regard to their formal education.[20] That was not the case.

In the early part of the twentieth century the majority of industrial workers were foreign-born or the children of the foreign-born. Many production workers were, for this reason, less educated than the average American. Although the average level of schooling was low, workers varied greatly in their formal education. Our focus is on the changing demand for educated *blue-collar* workers—craft workers, operatives, and laborers—mainly in the production part of the operation.

The white-collar workforce has always been more educated as a group than the blue-collar workforce. Thus it may seem odd to emphasize a group not known for being among the more highly educated in the population, but there are several reasons for our concentration on them. The blue-collar group was quantitatively large in the early twentieth century and was growing. It was 41.4 percent of the male work force in 1900, 45.4 percent in 1920, and 46.4 percent in 1940.[21] A further reason for their study is that we know a considerable amount about production methods, the capital stock, and technological change in the manufacturing sector during the early twentieth century.

Several types of evidence are used to demonstrate the role of technology in altering the demand for educated and skilled labor in the early part of the twentieth century. Although our emphasis in this chapter is on manual jobs, we also present evidence on the role of technological change in the demand for nonproduction workers.

Evidence from the 1940 Federal Population Census

We begin with evidence from the 1940 Census since it is the earliest federal census to contain information on educational attainment. We found that wide differences existed among blue-collar manufacturing workers in their educational attainment and that these differences were directly related to industry characteristics and thus to the technologies employed and the skills demanded of workers.

In our analysis of the 1940 IPUMS (Integrated Public Use Microdata Sample) Census data we limited the sample to men 18 to 34 years old employed in blue-collar occupations, such as craftsmen, operatives, laborers, and service employees, in the manufacturing sector.[22] Among this group 27.6 percent had completed 12 or more years of schooling, whereas 36 percent of all employed men in the 18- to 34-year-old group in 1940 had at least 12 years of schooling.[23] Not surprisingly, blue-collar workers were less educated on average, but there was considerable variation among them and many were well educated.

As a fraction of the total blue-collar workforce, the more-educated were disproportionately found in industries that produced high technology and recently innovated goods, such as aircraft, business machinery, and scientific and photographic equipment.[24] The industries employing the more-highly educated used continuous-process and batch technologies to a greater degree than did other industries, and these industries included petroleum refining, dairy products, paints and varnishes, and nonferrous metals.

As stated previously, to understand the role of education in the pre-1940 period, we generally use the completion of high school as the definition of more-educated, whereas for more recent times we use graduation from college (either four-year or a combination of two- and four-year). The reason for the different standard concerns changes in the average level of education across the century. In 1940, 34 percent of the U.S. male labor force 25 to 34 years old had 12 or more years of schooling, whereas in 2000 about the same fraction had a post-secondary degree.[25]

We list, in Table 3.3, industries by the percentage of their blue-collar male workers (18 to 34 years old) who were high school graduates, giving those in the top and bottom 20 percent by employment.[26] The industries clearly divide into two groups. At the low end of the education spectrum are the products of the first industrial revolution (cotton, woolen, and silk textiles; boots and shoes) and many that have been the mainstay of construction for centuries (lumber, stone, clay, and cement). At the high end are various products of the second industrial revolution (e.g., chemicals, petroleum), many in the machine-producing group, and some crafted in settings similar to that found in a traditional artisanal shop (i.e., clocks, watches, jewelry, and even aircraft.) Finally, there is a perennial among high-education industries: printing and publishing.[27]

Table 3.3. Percentage High School Graduates by Industry, 18- to 34-Year-Old Male Blue-Collar Workers: 1940

3-Digit SIC Manufacturing Industries	% High School Graduates	Number of Observations	3-Digit SIC Manufacturing Industries	% High School Graduates	Number of Observations
High-education industries (from high to low) Top 20% by employment			Low-education industries (from low to high) Lowest 20% by employment		
Aircraft and parts	52.7	541	Cotton manufactures	10.8	1512
Printing and publishing	44.7	1289	Tobacco	11.6	144
Office machinery	43.7	166	Logging	11.7	706
Petroleum refining	43.3	415	Sawmills and planing mills	14.1	1941
Dairy products	43.2	417	Not specified textile mills	15.6	128
Scientific and photographic equipment	40.8	227	Silk and rayon manufactures	16.6	350
Electrical machinery	40.5	977	Carpets and rugs	16.9	107
Nonmetallic mineral products	36.2	135	Misc. fabricated textiles	17.0	94
Paints and varnishes	35.9	107	Cut-stone and stone products	17.1	101
Clocks, watches, jewelry	34.7	197	Misc. textile goods	17.6	117
Ship building	34.4	528	Structural clay products	18.8	271
Miscellaneous machinery	33.5	1669	Cement, concrete, gypsum, and plaster products	19.2	263
Nonferrous metals	33.1	342	Hats, except cloth and millinery	20.5	60
			Dyeing and finishing textiles	20.6	191
			Misc. wooden goods	21.4	475
			Footwear industries except rubber	22.9	680
			Woolens and worsteds	23.1	368

Sources: 1940 IPUMS, 1/100; ICPSR (1984).

Notes: The sample is limited to 18- to 34-year-old, currently employed males in blue-collar occupations (craft, operative, laborer, service) in manufacturing. "High school graduates" is defined as 12 or more years of formal education. The mean of the high school graduation rate for the entire sample of 31,531 is 27.6 percent. Industry names are those given in ICPSR (1984). High-education (low-education) industries are obtained by ranking industries by their share of 18- to 34-year-old, male, blue-collar workers with 12 or more years of schooling and selecting industries from the top (bottom) until 20 percent of manufacturing employment (for all workers) is represented. The 1940 IPUMS sampling weights are used in all calculations.

Outside of the manufacturing sector in 1940, similar patterns of blue-collar hiring can be found. In the communications, transportation, and public utilities sectors more-educated blue-collar workers were found in the newer and more technologically advanced industries, such as telephone, wire and radio, air transportation, petroleum and gasoline pipe lines, electric light and power, and radio broadcasting and television. In retail trade the newer products and those that were time-sensitive or more valuable per unit were sold, delivered, and serviced by high school graduates. Drivers for jewelry stores and drug stores were more educated than were drivers who worked in other industries. Blue-collar workers in radio stores, and even gas station attendants, were far more educated than the average blue-collar worker. Our point is that in manufacturing, as well as in many other sectors, blue-collar workers using more advanced technologies and being entrusted with more expensive capital and goods were more educated than were others with similar occupational titles.

Because our findings regarding high- and low-education industries are somewhat novel, we must address potential concerns. The first is whether our results are primarily generated by geographic differences in *both* education and industry. Certain industries, because of the cost of non-labor inputs such as power or raw materials, could have been located in areas that had higher levels of education. In addition, age could matter for both employment and educational attainment. To assess these concerns we regress an indicator variable for high school graduation (12 or more years of schooling) on a full set of age, state, urban status, and industry dummies using our sample of young, male, blue-collar manufacturing workers from the 1940 IPUMS. The rankings of the adjusted industry coefficients and the mean industry residuals (neither is shown in Table 3.3) are almost identical to those given by the tabulations in Table 3.3. Furthermore, differences in educational attainment of blue-collar workers across industries are substantial even after adjusting for differences in urbanization, regional location of production, and age structure.[28]

RELATED EARLY TWENTIETH-CENTURY EVIDENCE

Even though the 1940 Census was the first at the national level to collect information on educational attainment, various earlier sources yield useful information on the utilization of more-educated workers in the manufacturing, service, and transportation sectors, particularly in blue-collar occupations. The 1915 Iowa State Census, to which we referred in Chapters 1 and 2, contains detailed information on educational attainment and occupation. Although Iowa did not have significant manufacturing employment, it did have a considerable group working in various blue-collar service jobs. Among these, the more highly educated men between the ages of 18 and 34 were disproportionately employed in the newer and higher technology industries, such as automobile repair, and they were often mechanics, machinists, electricians, and "engineers." Of course, the vast majority of the well-educated young men were in white-collar positions.

Another early piece of evidence concerning the relationship between education and the more technologically advanced industries is contained in a study of thousands of young men living in New York State who had recently left school during World War I.[29] Among 16- to 18-year-old boys who were working in a craft, operative, and laborer positions, the fraction employed in the metal trades rose with education, whereas the fraction employed in wood, leather, clothing, and textiles declined with education. The metal trades were considered among the more technologically advanced in manufacturing, whereas trades in the other industries mentioned were older and less dynamic. Of the young men with 12 years of schooling who were employed in blue-collar jobs, 54.4 percent were in the metal trades. But among those who left school after nine years, 44.4 percent were in the metal trades, and among those who left after six years just 30.3 percent were in the metal trades.[30]

We have established that firms in certain industries disproportionately hired educated blue-collar workers in 1940. Comparing the high- and low-education industries in Table 3.3 reveals that the industries that employed the highly educated workers produced newer products and used more advanced technologies, including continuous- and batch-processes.[31] It would appear that technology-skill complementarity was alive and well in 1940. The complementarity between technology and skill existed even earlier in the twentieth century and was associated with the introduction of electricity and the more extensive use of capital per worker. We now turn to that evidence.

Skill-Biased Technological Change in the Early Twentieth Century

Skill, Capital Intensity, and Electricity

To investigate technology-skill complementarity in the early twentieth century, we must use detailed data by industry that contain information on capital intensity. Because the existing early twentieth-century sources do not have data on educational attainment, we have merged education data by industry from the 1940 Census with that on industry attributes from the 1909, 1919, and 1929 censuses of manufactures.[32] To investigate whether more-educated workers used greater amounts of both capital and electric power, in Table 3.4 we present industry-level regressions of the educational attainment of 18- to 34-year-old male blue-collar workers in 1940 on the ratio of capital to labor (in 1909 and 1919), horsepower electrification (averaged over 1909, 1919, and 1929), and other controls for worker and industry characteristics.[33]

The ratio of capital to wage-earners in 1909 and 1919 is positively related to the education of workers by industry in 1940 (see Table 3.4, cols. 1, 2), and the effect is economically significant. Increasing the capital-to-labor ratio by the equivalent of the difference between, for example, the lumber and timber industry and the oleomargarine industry in 1909 increases the high school graduation rate by 7 percentage points, or by 25 percent, in 1940.[34] Thus, more capital-intensive industries in the early 1900s employed a more highly educated labor force some 20 years later. The implication, therefore, is that capital-intensive industries employed a more highly educated workforce earlier in the century, even though we do not know their educational attainment with any certainty.

Not only did educated blue-collar employees work with more capital, they also worked with greater levels of purchased electricity. Electricity use in manufacturing grew rapidly from 1909 to 1929. The percentage of all horsepower in manufacturing driven by electric motors was 23 percent in 1909 but soared to 77 percent by 1929.[35] Electric power was purchased from power plants and it was also generated by firms using their prime-movers (e.g., steam engines, water wheels). Motors powered by purchased electricity grew the faster of the two, rising from 9 percent of all horsepower in 1909 to 53 percent in 1929.

Table 3.4. Education, Capital Intensity, and Electricity Usage, 1909 and 1919

	Adjusted Fraction High School Graduates among 18- to 34-Year-Old Males in Blue-Collar Occupations, by Industry			
	Capital Intensity		Capital and Electrification	
	1909	1919	1919	1919
	(1)	(2)	(3)	(4)
log(K/L)	.0589	.0496	.0632	.0592
	(.0169)	(.0202)	(.0194)	(.0205)
% hp purchased electricity			.199	
			(.0531)	
log(hp purchased electricity/L)				.0359
				(.0151)
log(other horsepower/L)				−.0405
				(.0088)
log(total horsepower/L)			−.0043	
			(.0149)	
d log(employment)$_{1909, 1929}$			.0313	.0311
			(.0126)	(.0128)
% artisan	.187	.189	.118	.122
	(.0336)	(.0355)	(.0295)	(.0295)
% female	.142	.0932	−.0442	.0086
	(.0537)	(.0524)	(.0636)	(.0652)
% children	−1.56	−1.56	−.660	−.804
	(.490)	(.515)	(.463)	(.487)
Constant	.203	.193	.0921	.185
	(.0238)	(.0361)	(.0366)	(.0307)
Number of observations	57	57	57	57
R^2	.482	.428	.711	.703

Sources: U.S. Bureau of the Census (1913, 1923a, 1933) supplemented with data provided by Arthur Woolf. See Woolf (1980). 1940 IPUMS, 1/100; ICPSR (1984).

Notes: Standard errors are in parentheses. The unit of observation is a 1940–industry group. Industries from the 1909, 1919, and 1929 censuses of manufactures are aggregated up to the 1940 groupings, e.g., beverages in 1940 contains the 1909 categories of distilled, malt, and vinous liquors, and mineral and soda waters. See the data appendix to Goldin and Katz (1998) for how we matched industries across samples. Each observation is weighted by the industry share of total blue-collar employment in manufacturing averaged over 1909, 1919, and 1929. The dependent variable is the adjusted percentage of 18- to 34-year-old, male blue-collar workers in the industry in 1940 who graduated high school. The adjustment is as follows. A regression of an indicator variable for high school graduation (12 or more years of schooling) on a full set of state dummies, year-of-age dummies, indicator variables for central city and metropolitan area residence, and a full set of 3-digit 1940 Census industry dummies was run on all employed 18- to 34-year-old, male blue-collar workers in the manufacturing, communications, transportation, and utilities sectors in the

Table 3.4. (continued)

1940 IPUMS (sample size = 38,940, weighted by the 1940 IPUMS sampling weights). The coefficient on each of the 57 industry dummies is the adjusted high school graduate share of young, male blue-collar workers for each 1940 industry.

Variable definitions:

log (K/L): log of capital stock (000, in current dollars) per wage earner in each industry for 1909 and 1919, respectively, as indicated by the column headings.

% hp purchased electricity: fraction total horsepower run by purchased electricity, averaged over 1909, 1919, and 1929.

log (hp purchased electricity/L): log of the horsepower of motors run by purchased electricity per wage earner averaged over 1909, 1919, and 1929.

log (other horsepower/L): log of total horsepower of prime movers per wage earner averaged over 1909, 1919, and 1929; the total horsepower of prime movers includes the horsepower of steam engines, steam turbines, water wheels, internal combustion engines, and so on.

log (total horsepower/L): log of total horsepower of prime movers + the horsepower of motors run by purchased electricity per wage earner averaged over 1909, 1919, and 1929.

d log (employment)$_{1909, 1929}$: change in log of total employment from 1909 to 1929.

% artisan: fraction of wage earners in the 1940 industry categories who were in a disaggregated industry (in 1909, 1919, 1929) classified as an artisanal trade, averaged over 1909, 1919, and 1929. "Artisan" is defined as working in gold and silver, leaf and foil; jewelry; photo-engraving; stereotyping and electrotyping; cardcut design; wood and die engraving; glass; glass cutting, staining & ornamenting; instruments; optical goods; and statuary art. Printing and publishing is also included because of its special feature of demanding a literate labor force.

% female: average fraction of wage earners female in 1909, 1919, and 1929.

% child: average fraction of wage earners child in 1909 and 1919.

Even though the growth of purchased electricity was greater than that of all electricity, generated electricity was still a sizable fraction of all horsepower in 1929 at 24 percent. Using the best estimates provided for the two types, we find a far greater impact on education levels from the use of purchased electricity than from generated electricity.[36] The reason for the difference concerns how electricity was used and the newness of plants that used purchased electricity.

During the period of the greatest diffusion of electricity, 1909 to 1929, more purchased electricity meant that a firm was using newer equipment, separate (or unit-drive) motors, and more technologically advanced machinery. In addition, industries using purchased electricity hired a more highly educated blue-collar workforce, similar to our findings for the ratio of capital to labor. In our analysis, the electricity variable is entered either as the fraction of horsepower that was run by purchased electricity or as the log of the electricity variable per wage earner (Table 3.4, cols. 3, 4). The effect, moreover, is present even when the growth of industry employment from 1909 to 1929 is held constant.

Our results show that not only was production-worker skill positively correlated with industry growth and the resulting newness of plant and equipment, but also that skill was associated with the use of purchased electricity independent of industry growth. An 18 percentage point increase in the purchased electricity share of horsepower (a one standard deviation change) was associated with a 3.6 percentage point increase in the share of young blue-collar workers who were high school graduates. Increases in total horsepower did not have a comparable effect. The type of power, not simply its amount, affected the skill level. Electricity—the latest and most flexible form of motive power during the period—mattered, not just the use of horsepower.

We do not mean to imply that individual workers in 1909 (or 1919) were identical those in 1940. Education levels in Table 3.4 refer to men 18 to 34 years old in 1940. These individuals could not have been in the 1909 labor force since the oldest would have been born in 1906. What we are claiming is that the technologies used in certain industries increased the value of a secondary-school education during the 1909 to 1940 period.

There are many potential reasons for the findings relating technology and blue-collar education or skill. One possibility is that purchased electricity, rather than self-generated electricity, was associated with larger changes in factory machinery. Electricity and separate motors for each machine (termed "unit drive") enabled firms to automate conveying and hauling operations, thereby eliminating substantial numbers of common laborers who simply moved items around the job floor. Many industries were prompted to introduce labor-saving methods with the onset of World War I, including iron and steel, brick manufacturing, pottery, portland cement, pulp and paper, rubber tires and tubes, slaughtering and meat packing, lumber manufacture and woodworking, and mining.[37] Ample and cheap electricity rendered feasible the production of various materials, such as aluminum and other electrochemicals, which disproportionately used skilled labor. Cheap electricity also encouraged a more intensive use of machines, thereby increasing demand for the skilled personnel who maintained them.[38] However, purchased electricity may also have been simply associated with newer factories and technological advances built into a newer capital stock.[39]

Role of High School Curriculum

As we demonstrated in Chapter 1 and will elaborate further on in Part II, secondary schooling spread rapidly in the United States after 1910. Less than 10 percent of youths had high school diplomas in 1910, but by the mid-1920s to mid-1930s 30 percent to 50 percent did, depending on the region. The increase in formal education expanded the supply of skilled manufacturing workers and altered their training. Before the spread of high schools, most individuals in skilled machine-maintenance occupations (e.g., machinist, electrician, and technician) learned relevant cognitive skills (e.g., algebra, geometry, trigonometry, mechanical drawing) on-the-job. But these skills were precisely those that were taught in high schools. Formal education, therefore, substituted for a combination of raw ability and job training. The expansion of secondary schooling, therefore, greatly increased the supply of individuals who could become skilled manufacturing workers.

That high school graduate blue-collar workers were employed in particular industries (more capital-intensive, using a greater fraction of horsepower run by electricity, often producing newer goods and those made with continuous-process and batch methods), may come as a surprise. Rarely is the education of production workers mentioned in the labor history literature. Yet there is ample qualitative evidence, complementing our empirical findings, that certain cognitive skills were highly valued in various trades (see, for example, the discussion in Chapter 5 on corporate schools).

High school graduates were sought because they could read manuals and blueprints, knew about chemistry and electricity, could do algebra and solve formulas, and, we surmise, could more effectively converse with the professionals, such as chemists and engineers, in high-technology industries. Blue-collar positions requiring some years of high school or a diploma were described by the U.S. Department of Labor as needing cognitive skills such as "good judgment," "skilled in free-hand drawing," "special ability to interpret drawings," "[familiarity] with the chemical formulas," "general knowledge of chemicals used," "[ability] to mix the chemicals." More technical skills, such as "knowledge of electricity" and "of electric wire sizes and insulation," "technical knowledge of the properties of glass," "general knowledge of photography," were also valued. Printing establishments required that beginners be "well versed in grammar, spelling, punctuation," and noted "an elementary knowledge of Latin and Greek is helpful."[40] High school educated youths were hired into skilled occupations, but they were also sought for ordinary positions in many of the "high-education" industries.

Reinforcing the shift in manufacturing was an increased demand for educated labor to sell, install, and service technologically advanced products. It may be hard to believe that purchasing a radio once took skilled sales personnel, but it did. Early radios had to be installed in people's homes; customers had to be taught to tune in frequencies. Certain youths, often termed "radio nuts," built their own radios and were occasionally hired by radio shops to help customers or were employed less formally. Richard Feynman, who received the Nobel Prize in physics, recalled that he fixed radios as a boy. "The main reason people hired me was the Depression. They didn't have any money to fix their radios, and they'd hear about this kid who would do it for less. So I'd climb on roofs to fix antennas, and all kinds of stuff" (Feynman 1985, p. 19). Because they often shared their expertise with those who sold radios, sales personnel were soon warned to stop fraternizing with the wireless amateurs and increase sales instead.[41] Ordinary household electrical devices, such as toasters, irons, vacuums, and the like, also needed more educated and experienced sales personnel when they first appeared. Similar changes have occurred more recently in personal computing. When personal computers first appeared, "computer nerds" sold them. As the public became more familiar with the technology and as software became more user friendly, the nerd was replaced by the slick salesperson.

Skill and Earnings among Blue-Collar Workers

It is perhaps not surprising that the better-educated blue-collar employees were paid commensurately more than others. The real surprise is that the rate of return to a year of schooling for a blue-collar employee was not much below that for an ordinary white-collar worker. Using 1940 Census data, we estimated that the rate of return to a year of schooling for young, male blue-collar workers (white, 18 to 34 years old) was 8.3 percent (in a standard human capital log earnings equation), whereas the return for a similar group of ordinary white-collar workers was 9.1 percent.[42] To differentiate whether this return came from the type of industry or from the type of worker, we estimated the same equation with a full set of industry dummies. The coefficient on years of schooling in this estimation is 0.065, just a bit lower than our original estimate of 0.083.[43] What this tells us is that the positive earnings differential for more educated blue-collar workers in 1939 arose in part because they ended up in the higher-paying industries. But most of the return was because they earned a substantial educational wage premium *within* industries.

Because we also wanted to estimate the returns to schooling among blue-collar workers in the period before 1940, we used a technique similar to that employed in Table 3.4 and assigned to each industry the mean education level of its blue-collar workers in 1940. We found that the average blue-collar worker's wage in 1909, 1919, and 1929 is strongly and positively related to the constructed education levels. The coefficient on years of schooling was 12.5 percent in 1909, almost 10 percent in 1919, and 17.6 percent in 1929 in the Table 3.5 regressions of the (log) average wage in each year on average years of schooling among blue-collar workers in 1940. These estimates can be interpreted as a combination of the rate of return to an individual's education and the return to working in an industry having more highly educated workers.

Further and more direct evidence of the relationship between the earnings and education of blue-collar workers comes from our sample from the 1915 Iowa State Census (see Chapter 2 for more details). In the Iowa sample, we identified non-farm, blue-collar workers (laborers, operatives, and craft workers), although we do not know their industry. The coefficient on years of schooling in a log (annual) earnings regression containing potential experience and its square and an urban area dummy was 0.082 (s.e. = 0.0042) for 18- to 34-year old, non-farm, (white) male blue-collar workers in 1915. When we separated years of education by type of schooling (common school, grammar school, high school, and college), the coefficient increased to 0.105 (s.e. = 0.00831) for a year of high school.[44]

Thus the evidence from several sources indicates that employers valued the education of blue-collar workers in the 1909 to 1939 period and that the return to schooling, even for blue-collar workers, was substantial.

Table 3.5. Relationship between Earnings (in 1909, 1919, 1929) and Education (for 1940) among 18- to 34-Year-Old Blue-Collar Males, by Industry

	Log (Average Annual, Current $, Wage) in 1909, 1919, or 1929 Industry		
	1909	1919	1929
Average years schooling among	.125	.0995	.176
blue-collar, 18- to 34-year-old	(.0111)	(.0128)	(.0190)
males in 1940 industry grouping			
Percentage (women + children)	−.494	−.605	−.497
among wage earners in 1909 or	(.0681)	(.111)	(.0917)
1919 industry; percentage female			
in 1929 industry			
Constant	5.21	6.24	5.63
	(.102)	(.114)	(.184)
Number of observations	191	191	191
Weighted mean of dependent	6.24	7.00	7.13
variable			
R^2	.699	.644	.657

Sources: U.S. Bureau of the Census (1913, 1923a, 1933) supplemented with data provided by Arthur Woolf. See Woolf (1980). 1940 IPUMS, 1/100; ICPSR (1984).

Notes: The number of 1940 industries is less than the number of 1909, 1919, or 1929 industries. Regressions are weighted by the number of wage-earners in each 1909, 1919, 1929 industry. Numbers in parentheses are Huber (White) standard errors allowing for grouped errors within 1940 industries. The education variable (average number of years of school) is from the 1940 IPUMS. The average wage is computed as the (wage bill) / (average annual number of wage earners) for all years. See the data appendix to Goldin and Katz (1998) for further details.

Educational attainment provides a useful proxy for the skill levels of production workers, but it does not fully capture skill differences related to on-the-job training, apprenticeship training, and other sources. Earnings provide another (indirect) summary measure that may capture all aspects of worker skill rewarded by employers. The data on average earnings of blue-collar workers by industry from the census of manufactures for 1909, 1919, and 1929 allow us to look at contemporaneous correlations of an indicator of skills and industry characteristics. Table 3.6 details the relationship between average earnings per wage earner and the capital and electricity variables for 1909, 1919, and 1929 that we have previously discussed.

We found a positive relationship between the ratio of capital to labor and wages and, similarly, a positive correlation between wages and the

Table 3.6. Relationship between Production-Worker Earnings and Industry Characteristics, 1909, 1919, 1929

	Log (Average Annual, Current $, Wage) in 1909, 1919, or 1929 Industry			
	1909	1919	1929	Means, 1929
log(K/L)	.0910	.0417	.0480	1.44
	(.0151)	(.0169)	(.0262)	(.510)
% hp purchased electricity	.439	.266	.546	.637
	(.0556)	(.0374)	(.0548)	(.211)
log(total horsepower/L)	−.0213	−.00149	.0184	1.09
	(.0115)	(.0131)	(.0189)	(1.07)
log(employment/number of establishments)	.0622	.0780	.0638	4.51
	(.00633)	(.00577)	(.0103)	(1.08)
% female	−.427	−.308	−.307	.210
	(.0563)	(.0613)	(.0881)	(.225)
% child	−3.41	−6.91	−6.41	.014
	(.377)	(.697)	(.927)	(.016)
Artisan	.136	.144	.273	.0563
	(.0338)	(.0336)	(.0481)	(.231)
Constant	5.99	6.65	6.56	
	(.0306)	(.0386)	(.0666)	
Number of observations	228	225	228	
Mean of weighted dependent variable	6.23	7.03	7.15	
R^2	.791	.813	.667	

Source: U.S. Bureau of the Census (1913, 1923a, 1933) supplemented with data provided by Arthur Woolf. See Woolf (1980).

Notes: Standard errors (standard deviations for means, 1929) are listed under the coefficients. The average wage is computed as the (production-worker wage bill)/(average annual number of wage earners) for all years. Independent variables are defined in Table 3.4. Log(K/L) by industry for 1919 is used in the 1929 regression; % child for 1919 is used in the 1929 regression. Regressions are weighted by the number of wage earners in each year, as are the means for 1929. See the data appendix to Goldin and Katz (1998) for further details.

percentage of all horsepower from purchased electricity, consistent with the previous results.[45] Because World War I caused a transitory compression in production-worker wages, we prefer to concentrate on the 1909 and 1929 coefficients. The magnitudes implied by the coefficients are substantial, particularly with regard to purchased electricity use. The difference between the capital-to-labor ratios in the oleomargarine and lumber and timber industries, the two industries we previously used, implies a 5 percent premium in wages for oleomargarine; the difference in their purchased electricity use implies a 23 percent wage difference.[46] The wage premiums we measured are largely due, we suspect, to compositional effects. That is, industries with more capital per worker and with more horsepower coming from purchased electricity had relatively more educated blue-collar workers.[47]

Technology-Skill Complementarity and Nonproduction Workers

The relative size of the nonproduction (white-collar) group provides an alternative way to measure skill. A larger nonproduction worker share of employment is likely to be associated with greater skill required of all workers because white-collar jobs tend to have higher educational requirements and because technical nonproduction workers (engineers and chemists) tend to work with more-educated production workers.

The new continuous-process and batch methods of production of the early twentieth century required more managerial and professional employees relative to production workers.[48] Such processes also required relatively more skilled blue-collar workers (as shown in Table 3.3).

The data from the census of manufactures for 1909 and 1919 allow us to further assess the importance of technology-skill complementarity early in the twentieth century through an examination of whether the relative utilization of nonproduction workers increased with capital intensity and reliance on purchased electricity. Cross-industry comparisons for both 1909 and 1919 indicate robust and strong positive partial correlations between the nonproduction worker share of employment (or labor costs) and both the capital-to-labor ratio (in logs) and the fraction of horsepower from purchased electricity (in regressions analogous to those in Table 3.4, including controls for horsepower per worker [in logs], demographics, and other industry characteristics). We found, similarly, that within-industry changes in the relative utilization of nonproduction workers (as measured by the change in the nonproduction worker share of labor costs) from 1909 to 1919 were strongly positively correlated with increases in capital-output ratios and in electricity utilization, even after controlling for many other relevant factors.[49] Thus, the evidence from the U.S. manufacturing sector in the early twentieth century suggests that more advanced and capital-intensive technologies were associated with increased relative demand for occupations with higher educational requirements and for more-highly skilled workers within occupations.

Was Skill Bias Greater after 1980?

Various researchers have claimed that technology became *more* skill-biased across the last century. The "Taylor-Fordist" mode of production, according to this view, shifted to a more flexible organization form, which raised the demand for skill.[50] Others have argued that the skill-bias of technological change is endogenous; as the workforce becomes more skilled, technologies that use greater skill are adopted. The effect, moreover, can spiral, producing an ever-increasing change in the skill-bias of technological change.[51] But these claims do not appear to be consistent with much of the historical evidence.

As discussed above and documented in Table 3.2, there does not appear to be much (if any) acceleration in the rate of growth in the relative demand for more-educated workers since 1950. The relative demand for college workers from 1950 to 1980 increased at a pace rather similar to that from 1980 to 2000. Even though the growth of demand for educated workers across the last half of the twentieth century appears reasonably stable, there is still the possibility that it accelerated from the first half of the twentieth century to the second half. How can one assess that important issue since there are no nationally representative samples with information on education and earnings prior to 1940?

It is difficult to evaluate overall shifts in the relative demand for more-educated workers before 1940 in a manner comparable to that for the post-1950 period; however, armed with a few assumptions, the means for doing so exist, and in Chapter 8 we present estimates for economy-wide relative skill supply and demand shifts between 1890 and 2005. As an initial pass at this task, we examine here long-run changes in the demand for skill in the *manufacturing* sector to assess the degree to which skill-biased technical change was altered across the twentieth century.

The share of wages expended on nonproduction workers can be used as a proxy for the hiring of more-skilled workers generally. Fairly consistent data from censuses and surveys of manufacturing establishments exist. They are used here to compare the pace of labor market shifts toward more-skilled workers between two periods of comparable length at the start and end of the twentieth century: 1890 to 1929 and 1960 to 1999.

The wage bill share of nonproduction workers (in this case, officials and clerks) in the manufacturing sector increased from 17.2 percent in 1890 to 23.6 percent in 1929.[52] This means that the wage bill of the more skilled (nonproduction workers) relative to the less skilled (production workers) in manufacturing changed from 20.8 percent in 1890 to 30.9 percent in 1929. The nonproduction worker wage bill share for operating manufacturing establishments increased from 33.6 percent in 1960 to 40.9 percent in 1999.[53] Similarly, the wage bill of the more skilled to the less skilled increased from 50.6 percent in 1960 to 69.2 percent in 1999.

When the elasticity of substitution (σ) between the two skill groups is 1 (as in the case of a Cobb-Douglas production function), the ratio of the wage bills of the two skill groups cannot change unless the relative demand for skill shifts.[54] Thus, in the case of $\sigma = 1$, the growth rate in relative demand for nonproduction workers is given by the growth rate of the relative wage bill.

The simplifying assumption ($\sigma = 1$) implies that the relative demand for nonproduction workers in manufacturing increased more rapidly from 1890 to 1929 (0.0103 log points per year) than in the more recent period of 1960 to 1999 (0.0079 log points per year).[55] Even when there was generally acknowledged rapid skill-biased technical change during 1979 to 1999, the index of relative demand shifts for more-skilled manufacturing workers grew only slightly more than it did from 1890 to 1929 (0.0116 log points per year versus 0.0103 log points per year). In fact, the growth rate of the index of relative skill demand increased at exactly the same rate of 0.0103 log points per year for the most recent four decade period (1965 to 2004) as it did in the earlier four decade period with comparable data (1890 to 1929).

The technologically forward industries, which had a more highly educated workforce, grew faster than did others in the first half of the twentieth century, just as they have done since 1950.[56] The employment share of more-educated occupations also expanded at a similarly rapid pace in both the first and second halves of the twentieth century. Furthermore, the most rapid skill upgrading (as measured by the growth of the wage bill share of nonproduction workers) between 1909 and 1919 occurred in industries with the greatest increases in capital intensity and electrification, just as would be the case at the close of the century in industries with the greatest investment in computer technology. In fact, estimates of the degree of capital-skill complementarity within detailed industries are almost identical for the 1910s (1909 to 1919) and for the 1980s (1979 to 1989).[57]

The evidence both from the U.S. manufacturing sector and the broader economy suggests substantial continuity in the skill-bias of technological change and the rate of growth in the relative demand for more-skilled workers in the early and late parts of the twentieth century.[58] These findings raise the important but unanswered question of why the pace of skill-biased technical change has remained rather stable for over a century.

The Origins of Technology-Skill Complementarity

The Emergence of Technology-Skill Complementarity

Throughout the twentieth century, as we have just seen, physical capital and more advanced technologies have been the relative complements of human capital (by which we mean various types of skills including those gained in formal schooling). But if there was technology-skill and capital-skill complementarity throughout much of the twentieth century, did such complementarities exist during an earlier period, such as America's nineteenth- century industrial revolution? Can we extrapolate the relationships among skill, capital, and technology further back in time?

The answer we provide here is that the relationship was nonexistent further back in time. The origin of the complementarities can be found around the turn of the twentieth century and was associated with the shift to production methods such as continuous-process and batch operations and with the extensive use of electricity, all of which increased the relative demand for human capital in the industrial sector. The question we must address is precisely why these technological changes led to an increase in the complementarity among human skill, capital, and technology. We provide the answer through reference to a framework that we have developed more fully elsewhere.[59] We also use the findings of others on the evolution of manufacturing in the nineteenth century.

A wide-ranging literature has established that physical and human capital were *not* relative complements in the industrial past. Many of the major technological advances of the nineteenth century, according to this literature, substituted physical capital, raw materials, and unskilled labor, as a group, for highly skilled artisans.[60] Rather than being the relative complement to skill, physical capital was, for some time, a relative complement of raw materials and, together with unskilled labor, substituted for highly skilled individuals.[61] The prototypical example is gun making. Cheap lumber in America fostered the use of woodworking lathes and displaced hand fitting in the production of gun stocks by skilled woodworkers. The butcher, baker, glassblower, shoemaker, and smith were also skilled artisans whose occupations were profoundly altered by the factory system, machinery, and mechanization.[62]

Technological advance and human skill were not relative complements in the distant past but they are today. When did they become so? Although an extensive literature has established that today's relationships among skill, capital, and technology were not present in the more distant past, it does not address when the switchover occurred. We have argued above that technology-skill complementarity emerged in manufacturing early in the twentieth century as particular technologies, known as batch and continuous-process methods of production, spread. The switch to electricity from steam and water-power energy sources was reinforcing because it reduced the demand for unskilled manual workers in many hauling, conveying, and assembly tasks. Our evidence here takes the form of a framework as well as the empirical results offered in previous sections of this chapter.

A Framework for Understanding the Emergence of Technology-Skill Complementarity

We postulate that manufacturing production, for certain products, began in the artisanal shop then shifted to factories (1830s to 1880s), to assembly lines (early 1900s), and more recently to robotized assembly lines.[63] For other types of goods, however, the shift may have been from artisanal shops or factories to continuous-process and batch methods (1890s and beyond).[64] The production process shifts did not affect all goods similarly, and some were never manufactured by more than one method. But manufacturing as a whole progressed in the fashion we posit: from artisanal shops, to factories (also assembly lines), and then to continuous-process (also robotized assembly lines) or batch methods.

We have in mind rather distinct notions for each process, following a rich literature in the histories of technology and business. The distinction between the artisanal shop and factory is mainly in the degree of division of labor. Factories are larger, with more specialized workers and often more capital per worker. Batch operations, as we said before, are used for processing liquid, semi-solid, or gaseous matters, whereas continuous-process methods are used for products requiring little assembly and having few or no moving parts.

Few products went through all the stages we describe, but those that did are informative. Automobile production began in large artisanal shops. Like the carriages that preceded them, automobiles were first assembled by craftsmen who hand-fitted the various pieces.[65] Technological advances then led to standardized and completely interchangeable parts that were assembled in factories, later equipped with assembly lines as at Ford in 1913, by scores of less-skilled workers. Much later, the robotized assembly line appeared, using fewer unskilled operatives and more skilled machinists. In the history of automobile production, the first technological advances reduced the relative demand for skilled labor whereas later advances increased it.

How did these technological shifts affect the relative demand for skill? Let us sketch out the intuition of our framework.[66] The manufacturing process is assumed to contain two distinct segments. In the first segment, raw capital must be installed and maintained by skilled labor. We call this "machine maintenance" and it results in usable and workable capital. In the second segment, called "production," goods are assembled or created with the workable capital and unskilled labor. All workers in the production segment are assumed to be unskilled, whereas all workers in machine-maintenance segment are skilled.

Given this simple yet revealing framework, what effects did the successive production regimes—artisanal to factory to batch and continuous-process—have on the relative demand for skill? The answer depends on whether the increase in capital, by increasing the demand for skilled workers, outweighed the increase in the demand for unskilled workers to operate the usable and workable capital.

The transition from artisanal shop to factory production probably increased the capital-to-output ratio. More important is that the switch to the factory most likely *decreased* the demand for skilled relative to unskilled labor in manufacturing. That is, the industrial revolution with the shift into various types of factories was, overall, deskilling. The technological advances that later shifted production from the factory (or assembly line) to continuous-process or batch methods further raised the capital-to-output ratio. More important to our argument is that these advances served to *increase* the relative demand for skilled labor, because the workable capital did not require an abundance of laborers in the production segment. Reinforcing these technological shifts was electrification, the adoption of unit-drive systems, and the automation of hauling and conveying operations which decreased the demand for unskilled laborers.[67]

Our central point is that the technological shift from factories to continuous-process and batch methods, and from steam and water-power to electricity, were at the root of the increase in the relative demand for skilled labor in manufacturing in the early twentieth century. These technological changes provide the origins of the transition to technology-skill complementarity, which we believe to be in full blossom today.[68]

Industries adopting advanced technologies (e.g., continuous-process and batch methods) in the first part of the twentieth century, according to our framework, should have employed more skilled production workers on average and had a larger share of nonproduction (white collar) workers. These industries should have been more capital intensive and relied on purchased electricity for a larger share of their horsepower. These predictions are borne out by the empirical work we have presented in the previous parts of this chapter.

Overall, our evidence is consistent with the notion that the transition from the factory to continuous-process production, starting as early as 1890 and lasting through 1940, increased the relative demand for skilled workers. The previous transition, from the artisanal shop to the factory, appears to have involved an opposite force, although the evidence is less solid. We do know that many industries that remained artisanal (e.g., engraving, jewelry, clocks and watches) had far lower capital intensity but higher worker skill (education) than the majority of industries that shifted to factory production.

The role played by skilled labor in machine-maintenance means that capital and skilled labor are relative complements within any given manufacturing production process. But capital and skilled workers may be relative complements *or* substitutes in considering shifts among different manufacturing processes. For example, the movement from artisanal production to factories in the nineteenth century involved the substitution of capital and unskilled labor for skilled (artisanal) labor, while the adoption of continuous-process and unit drive methods in the twentieth century involved the substitution of capital and skilled (educated) labor for unskilled labor.

"It Isn't Just Technology—Stupid": A Summary

The primary purpose of this chapter has been to establish the continuity of the impact of technological change on the demand for labor across the twentieth century. Great technological advances in recent decades have increased the relative demand for skill; but, surprising as it may seem, the early part of the twentieth century also experienced great advances that increased the relative demand for skill, possibly to an equal degree. Technological changes, however, were not always skill biased. We located a turning-point in the late nineteenth century when technological changes became, on net, skill biased.

Technological changes are not, in themselves, responsible for the increase in inequality in the recent period, just as they are not responsible for the decrease in inequality during the earlier part of the twentieth century. Thus the central point of this chapter, to paraphrase a mantra of the 1992 presidential campaign, is that "it isn't just technology—stupid." Since "it isn't just technology," the reason for rising inequality cannot be located solely on the demand side. The other important part of the answer can be found on the supply side. To do this, we turn now to changes in the educational attainment of Americans.

II

EDUCATION FOR THE MASSES IN THREE TRANSFORMATIONS

4

Origins of the Virtues

Throughout much of the nineteenth century and most of the twentieth, the U.S. educational system worked admirably. It created an egalitarian system that put the elite systems of Europe to shame, as we saw in Chapter 1. The enormous U.S. advantage in education that developed in the twentieth century rested on a set of institutions that served to foster schooling in a variety of ways.

The key features of U.S. educational institutions—which we term "virtues"—that were present in 1900 had largely taken shape in the preceding century. Most had emerged in the period before the American Civil War. These virtues would determine U.S. educational development in the twentieth century and would enable the United States to lead the world in schooling, particularly in educating the masses. The subject of this chapter is the origin of these defining features during the republic's first hundred years.

The features include three of the most basic: public funding, public provision, and the separation of church and state. Three others were fundamental: a decentralized system containing thousands of fiscally independent districts, an open structure in which youthful transgressions were often forgiven, and the ability of girls to receive instruction in coeducational public schools, which we term "gender neutrality." These characteristics are so often taken for granted that it seems inconceivable that there was ever a time in the history of the United States when they did not exist. But, indeed, there was.

The Virtues of American Education

By *virtues*, we mean a set of characteristics that originated in basic democratic and egalitarian principles and that influenced the educational system. The virtues, to repeat, include public provision by small, fiscally independent districts; public funding; secular control; gender neutrality; open access; and a forgiving system. These virtuous features are summarized by the word "egalitarianism." They have held the promise (if not always the reality) of equality of opportunity and a common education for all U.S. children.[1]

An early outcome of the virtues was the rapid diffusion of schooling and educational institutions throughout much of the young nation. By the mid-nineteenth century, the enrollment rate among children and youth in the United States exceeded that of any other nation in the world.[2] The primary reason that we deem these features virtues is that they produced a relatively high level of schooling and educational attainment.[3] They did so from the mid-nineteenth century and continued to function well into the latter part of the twentieth century, when some of the virtues began to be questioned.

The U.S. lead in primary schooling narrowed toward the end of the nineteenth century as parts of Europe began to educate their masses, but then greatly expanded in the early part of the twentieth century as America was swept up in the high school movement and as the transformation to mass higher education began. The preeminence of the United States in mass education continued up to the 1980s; but this is getting ahead of our story.

The virtues of long ago need not be the virtues of today. They also need not have been virtuous in all places and at all times in the past. For example, the existence of numerous and small fiscally independent school districts was an important virtue of the past and it enabled educational progress, but today small fiscally independent districts, with their widely divergent levels of taxable land and real estate, are often seen as a source of inequity in school resources and a hindrance to educational advances for children from low-income communities.

The openness and forgiveness of the U.S. educational system is another example of changing virtues. These features were once viewed as enabling youths to make up for deficits in their backgrounds and to escape severe penalties for their past misdeeds. But these features are often seen today as an excuse for schools to lower academic standards and for teachers to avoid having to deal with problem students. Yet another example is local control in educational funding and decision making. Decentralization may have facilitated rapid educational progress across many communities, but local control also meant that *de jure* racial segregation persisted in many southern cities, even after *Brown v. Board of Education* (1954), and that *de facto* segregation existed in numerous districts outside the South.

The features we deem as virtues were often accompanied by corollary or accessory characteristics. The existence of large numbers of fiscally independent districts meant that small localities were required to raise funds for schools. Real estate taxation was the most effective means of funding education, in part because land cannot migrate whereas most other forms of capital can, and do, in response to taxation. The existence of small, local school districts meant that minor governmental units, such as townships, could compete for residents along many dimensions.

A forgiving system cannot be so severely tracked that students cannot make up for past wrongdoings. The educational systems of Germany and other nations of northern Europe, in which some students did an industrial curriculum and a far smaller group were groomed for the universities, were eschewed in favor of an academic, yet practical, curriculum in the United States.

The virtuous aspects of these features were not always intentional, as in the case of the small school districts. There were practical reasons to have large numbers of fiscally independent school districts in rural America, which derived as well from the American desire for community control and local taxation. Similarly, the more vicious aspects of these same features also may have been unintentional and their unintended consequences may have worsened over time. Some districts became considerably richer than others and had better schools. With rising inequality over the past several decades, geographic sorting on the basis of income has become more extreme.[4] Many have deemed this outcome a non-virtuous consequence of having small, fiscally independent districts.

But, for some time and in many places in America, the features we just described gave rise to extraordinarily beneficial outcomes. These outcomes were far better than were those of other countries, which often had educational systems with diametrically opposing features. If the decentralization of America led to the growth of mass secondary schooling, then the centralization of control that characterized most European school systems stifled it. If an open and forgiving system gave disadvantaged and errant youths a second chance, then the insistence on standards and accountability of many European systems reinforced a caste system. It is, in part, for these reasons that we deem the features of the U.S. educational system as virtues, at least for much of the period (in Chapter 6 we further defend the characterization of these features as virtues of the past). It is sufficient for now that they be seen simply as important aspects of the U.S. elementary and secondary educational systems that existed around the turn of the twentieth century.

The most important of the features of U.S. elementary and secondary education around 1900 were its public funding and public provision. These twin characteristics formed the essence of public education. Publicly funded education need not be publicly provided, as in the example of vouchers today; similarly, publicly provided education need not be publicly funded, as in the example of tuition payments and rate bills of the nineteenth century.

Schooling in 1900 was not only publicly funded, but it was also provided by tens of thousands of fiscally independent school districts. Some of these districts, mainly those of the nation's large cities, served enormous numbers of children and were geographically large as well. But the vast majority of districts in 1900 were small, both in terms of their geographic size and the number of children they served. The majority of children, moreover, resided in relatively small districts. The nation's school system, therefore, was a highly decentralized one at the start of the twentieth century and it has remained decentralized to the present day, although not nearly to the same degree that it once was.[5]

Education in terms of its funding, staffing, and curriculum, therefore, was largely a local affair at the turn of the twentieth century. In 1900, the U.S. federal government had virtually no involvement in primary and secondary education, although it had once played an important role in the granting to states of federal lands for educational funding. Far later in the twentieth century, the federal government would take on an increasingly important, yet still relatively modest, role. Most states around 1900 also played a relatively small fiscal role in education. In its decentralization, the U.S. system was, and continues to be, the polar opposite of many of its European counterparts.[6]

Decentralization is not the only contrast between the American and European elementary and secondary school systems. One of the most important differences around 1900 was that U.S. schooling was not an elite system in which only a small number of bright young boys could attain an upper secondary school education, and thus continue their studies in a college or university. Schools were, by and large, open to all and were forgiving to those who did not shine in the lower grades. U.S. programs trained young people to enter college or begin various jobs, including homemaking. Around 1910 most newly employed male high school graduates were in white-collar jobs.[7] Secondary schools, in the years to come, would place increasing emphasis on vocational and lifestyle courses and the change in curriculum would lead many to take a less academic track.

The extensive gender equality in U.S. education was another virtue of the American system. High school entering classes in 1900, for example, contained an almost equal number of boys and girls. Considerably more girls than boys were in attendance in the upper secondary school grades, and a larger proportion of females than males eventually graduated. Gender neutrality existed in the ability of girls to obtain more years of schooling at least through secondary school, but gender neutrality in the quality of schooling and the continuation to college were often other matters.

Equality in education was not ubiquitous across all groups, of course. In 1900, African American youth were educated in segregated schools throughout the South; they had once been educated in *de jure* segregated schools in parts of the North as well. The South's segregated schools were not equal in any manner, despite the 1896 *Plessy v. Ferguson* decision of the U.S. Supreme Court, which was interpreted to mean that schools could be separate only if they were equal. In the North, furthermore, many poor children, often of immigrant parents, were educated in inadequately funded schools even when other schools in the same district were better funded. Even so, educational advances touched more children in the United States than they did in Europe

and enabled considerably more youth to advance to secondary schools
and even to higher education well into the twentieth century.

Finally, public education in the United States was secular by 1900 in
the sense that organized religious groups generally did not receive state
and local funds to run schools. Nonsectarian education did not mean
that bibles and prayer were absent from the public school classroom,
for they remained part of it for a long time. In 1900, the separation of
church and state in the provision of public education was largely the
doing of the states, not the federal government. In contrast, recent interpretation of the Establishment Clause of the U.S. Constitution is an
important feature of more modern school debates.[8]

All the virtues just discussed were distinguishing features of U.S. elementary and secondary education long before the start of the twentieth century. Similarly, the U.S. higher education system contained
distinctively American features that had been formed prior to 1900.

Higher education in the early twentieth-century United States was
open, flexible, geographically close to its primarily rural constituency,
connected to state and local concerns, practical in its curriculum, and
diverse in many dimensions. It contained a wide-ranging, viable private
sector and a growing, multilayered public sector. Competition existed
both between and within the public and private sectors. In these many
ways the institutions of American higher education differed greatly
from those in most European nations. But, as in the case of secondary
education and probably to a far greater extent in higher education, further distinctive characteristics would take shape in the decades after
1900. The factors that drove these changes and their impact are addressed in Chapter 7.

The Origins of the Virtues: The First Hundred Years of the Republic

Our history of the virtues of U.S. education begins at the dawn of the
new republic as we explore the origins of six important features:
(1) public provision of education, thus the establishment of common
schools; (2) numerous and small fiscally independent districts, thus decentralization and competition; (3) public funding of schools, thus a
free education for all; (4) nonsectarian public schools, thus the separation between church and state in educational finance and control;

(5) gender neutrality in access to public education, thus a public education regardless of sex; and (6) an open and forgiving system, thus mass education.[9] Virtues (1) and (2) are intricately connected and our discussion aggregates them.

We focus here on common or elementary education in the nineteenth century and in later chapters shift to secondary and higher education in the twentieth century. Education in the lower grades diffused in the nineteenth century and secondary school and higher education followed in the twentieth century. Mass secondary schooling in the early twentieth century was possible only because universal elementary school education had already spread throughout most sections of the nation. Mass elementary school education, in turn, was attainable because common schools were free throughout the nation by the 1870s at the latest.

Founding Ideas

Many patriots of the new nation and signers of the Declaration of Independence, such as John Adams, Benjamin Franklin, Thomas Jefferson, and Benjamin Rush, wrote extensively about educational institutions. In some instances, the Massachusetts Constitution of 1780 drafted by John Adams for example, their ideas were widely shared and quickly adopted.[10] But that was not often the case and most of their plans never bore fruit. Although many of their statements about education were deeply held by themselves and most other Americans, some of their writings were propaganda tracts to help speed the development and integration of the new republic. These treatises, whatever their intent and immediate success, helped coalesce a growing sentiment in the new nation for a strong educational foundation. Perhaps most important is that their ideas spoke to the concerns of those who formed the union.[11]

Almost all these authors wrote compellingly of the critical importance of education in a democracy to enable Americans to perform their civic functions, such as voting, and to prepare them to run for office and lead the nation. Some revolutionary thinkers had more visionary ideas concerning the role of education and how the nation's educational system should be organized. Jefferson and Rush, for example, elaborated schemes for entire educational systems ranging from elementary schools to colleges and universities. Jefferson's plan for Virginia contained three levels of schooling: elementary schools, which formed the bottom level, would be open to all and publicly funded; academies, at the middle level, were to be publicly funded and available to able boys; and William and Mary College, the highest level, would award scholarships to needy, bright boys. Ironically, given his political views on the relative role of the federal and state governments, Jefferson's plan may have failed because it was insufficiently local in its funding.

To Benjamin Rush, the most prominent physician of his age and a renowned public figure, education mattered not only for the infant democracy, but also for its economy. Expenditures on education, according to Rush, would lessen taxes because they would increase "the profits of agriculture and [promote] manufacturing." He called for the study of practical subjects, a curriculum that would soon become part of the quintessentially American form of education. "Agriculture," he aptly noted, "is as much a science as hydraulics or optics."[12] "The study of Commerce and the principles of Money," he more idealistically suggested, would have an "effect as next to those of religion in humanizing mankind."[13]

Public Provision of Schooling by Small, Independent Localities (Virtues 1 and 2)

COMMON SCHOOLS IN THE EARLY REPUBLIC

Schools existed everywhere in the early republic for most free youths, in communities of every size, much as they had done prior to the revolution. The schools were often termed "common schools," a name that persisted into the twentieth century in rural areas. In most towns these schools were eventually called elementary schools and, in some cities, grammar schools existed for the older students. For simplicity of usage, we will call all schools in the pre-1870 era that educated children from about 5– to 14–years old common schools, even though many of them included older children and even though some were graded by age and would have been termed elementary or grammar schools.

The phrase "common school education" has meant many things and the term continued to be used for nearly a century and a half. In the early nineteenth century, a common school meant one that was publicly maintained and belonged to the community. The term was used in opposition to a private school, either secular or denominational. On a practical level it connoted an inclusive school, one of all the people and "common" to them. A common school bound the community together and taught commonly held principles and elementary subjects. As the graded elementary schools of the towns and cities increased in number and size, the common school came to mean the one-room schoolhouse of rural America, and the term "common school" became reserved for the ungraded elementary school.

DECENTRALIZATION OF CONTROL

The common schools of the early republic were organized at the local level and were funded in a variety of ways. "Local" did not always mean exactly the same type of governmental unit in all parts of the nation and its meaning has changed over time within regions. The New England township was for most of its history the smallest unit that governed school finance and curricula decisions. The township joined residents of a town with those of the rural and village communities that surrounded it. A smaller unit—the school district—had once been the locus of school governance in Massachusetts and existed until the 1840s.[14]

The New England township model migrated west with New Englanders and was replicated in much of the nascent Midwest. But school districts, of varying sizes, were the smallest levels of educational governance in other parts of the young nation, including certain parts of the Midwest. Thus the jurisdictional unit for educational decision making ranged in size and in organizing principle. There were township communities in New England, groups of farm families in the newer states, and religious parishes in Connecticut, to mention just a few.

FINANCING SCHOOLS: SCHOOL FUNDS AND THE PROPERTY TAX

Even though schooling was, by and large, a local concern in the nineteenth century, both state and federal governments played important enabling roles. That the states played a role in the nineteenth century should not be surprising, since education is one of the powers "reserved for the States" by the Tenth Amendment of the U.S. Constitution.

Most Americans in the period of the early republic, if we infer from their behavior, preferred that their local communities provide for their children's education and levy the taxes to pay for some or all of the expenses. But localities did not initially have taxing authority and states had to pass enabling legislation to allow localities to impose taxes. By the 1820s most of the states of the Northeast had passed legislation allowing their towns to impose taxes. For example, Connecticut did so in 1794, Rhode Island in 1828, and New Jersey in 1829.

States later passed legislation that forced localities to provide schooling free of charge for the full length of the school year. Massachusetts, in 1827, required that towns exceeding 50 families support their schools through taxation since it had in 1826 foreclosed the possibility of raising funds for teachers through a "user tax" (which we discuss in more detail later). Some states, such as New York, provided matching funds to localities for various school expenses, including teacher salaries and school buildings.[15]

The benefits of decentralized financial control are several. In Chapter 6 we discuss the advantages of smaller decision-making units in the spread of secondary school education, particularly in its early stages. Another benefit concerns efficiency gains in the production of education that may result through greater competition in the provision of schooling. Decentralization, however, is likely to increase inequality in funding across school districts.[16] Some districts will be richer than others and will be able to raise more tax revenue to support schools. Within this important trade-off of equity and efficiency in the provision of local public goods, the property tax looms large and has played an important role in the local finance of education.[17]

Under local control with a property tax, improvements in the production of school quality increase the market value of land, and thereby of housing, through a process known as capitalization. With a constant tax rate, the increased value of housing gets translated into greater revenue for the district.[18] The decentralization of fiscal control, the large number of fiscally independent districts, and the use of the property tax may all have been efficiency enhancing innovations; however, it is likely that each originated in the pragmatic concerns of the day and not from a recognized sense of the potential benefits.

The often forgotten player in early educational funding is the federal government. Its role in the nascent years of the republic was considerably greater than it was to be later in U.S. history.

The federal role in education was formalized at the start of the nation when the Land Ordinance of 1785 was passed under the Articles of Confederation. Written by Thomas Jefferson, it devised a way to deal with the lands in the northwest that were ceded to the federal government by the original 13 states. Jefferson's plan divided the lands into townships, six miles square, each consisting of 36 square-mile sections. Each township was to receive one section to finance the schooling of its children through the sale of the land. Two years later, in 1787, Congress adopted the Northwest Ordinance, which provided for the creation of from three to five states from these lands. Illinois, Indiana, Michigan, Ohio, and Wisconsin were eventually carved out of the Northwest Territories.

As the federal government acquired new territories the precepts of the Land Ordinance of 1785 were followed. Townships were surveyed with one section in each reserved as an endowment for schools. After California entered the Union in 1850, two sections, rather than one, were allotted to educational funds in each township, and four sections were allocated in the Southwest due to the low value of land.[19] The federal role in encouraging education through grants of land continued into the nineteenth century with land grants to the states to fund colleges, first in an *ad hoc* manner and later through the two Morrill land grant acts (1862 and 1890) discussed in Chapter 7.

Public Funding of Schools (Virtue 3)

The movement for free schools began in New England, where the fervor for a public role in education was the greatest. It spread with the migration of New Englanders to the new lands of the Midwest and West, diffused into the Middle Atlantic states, and was eventually imposed on the South after the Civil War.

The passage of state legislation in the pre–Civil War era requiring that school districts be fiscally responsible for the schooling of youths has been interpreted as a turning point in the history of education. These acts, according to many, were crucial ingredients in the increased education of Americans and the hallmark of the egalitarian nation. The actual history, however, is more complicated and the precise manner in which free schooling was accomplished is an involved tale.

The reason for the complexity is that many municipalities and school districts had free schools long before the states required that they did, and public funding existed for some part of the school year even when tuition was charged for the remainder. The real history of free schooling requires that we examine exactly how schools were funded.

THE RATE BILLS

At the start of the new nation, the funding of schools was accomplished through a combination of state funds accumulated through the sale of public lands, parental contributions (known as rate bills or tuition), and local taxes. The state funds were adequate at first, but population increase and expanding school enrollment meant that other fiscal means had to be found. The search for additional funds led first to fees called "rate bills" and then to the use of local taxes, mainly the property tax, and state taxes. Rate bills were imposed at different moments in each state's history.

Rate bills were tuition payments for the use of public schools and, at certain times and in particular areas, were levied on families whose children attended school. In most communities that had rate bills, these tuition payments were charged for days in attendance exceeding some number provided by the community free of direct charge. In others they were levied for the full term. Because the school teacher was the most important expense, and often the only one, families in rural areas and even in some cities often paid for services in-kind with the provision of room and board, in addition to a stipend.

In Massachusetts, a state at the forefront of education from the nineteenth century, towns with more than 50 families were required, to provide six months of elementary schooling out of public funds. The requirement did not rule out the private payment by parents of additional sums to extend the term. In New York State, funds from land sales and other state sources supported local schools until 1828, when the use of rate bills began. From 1828 to 1868 these private fees provided half the salaries for teachers whereas state funds and local taxation provided the rest.[20] In Connecticut, the sale of the state's Western Reserve lands in Ohio created a school fund of sufficient size to support local schools for some time. But, similar to New York, rate bills were eventually imposed to provide additional funds.

The campaign for free schools in most states was a battle to eliminate the rate bill and an appeal for alternative means to fund public schools. The crusade, known historically as the "common school revival," is among the best known and most studied episodes in American educational history. The campaign involved impressive figures such as Horace Mann, who was the leader of the movement between the Jacksonian era and the Civil War.[21]

The spread of free schooling in the antebellum period has rightly been hailed as fundamental to the diffusion of mass schooling in the United States. But what were the actual roles of the state laws that abolished rate bills and of the common school revival to the provision of free schooling and the increased education of youth? At first blush, the answers seem obvious. A campaign for free schooling was undertaken, state laws abolishing the rate bills were passed, and school enrollments, in consequence, increased. Not so fast. The real answer will take considerably more effort since the causal impacts of the state laws and the crusade may not be what they first appear.

The abolition of the rate bills occurred in a complicated manner in most states, even though it was relatively simple in some, such as Massachusetts and Maine, which abolished their rate bills before 1830, and New Hampshire, which never had one. Publicly provided common schools in most other northern states continued to be financed out of both public and private funds until the mid-nineteenth century; however, some major cities and other communities in these states abolished rate bills long before they were legislatively ended by the states. Another complication was that rate bills often covered just part of the total cost of schooling, and occasionally only a small part. The rest of the cost was a public charge long before the abolition of tuition fees.

The common school systems of the early republic formed a patchwork quilt of finance and provision. Most communities had publicly provided and publicly funded schools. Many allowed parents to pay for extended days beyond those provided by the district. Other communities had publicly provided but privately funded schools with rate bills covering the entire cost. In certain cities, most children were educated in private schools. Those without means were educated in pauper schools that were communally or philanthropically supported and often church-based. Because 80 percent of free Americans (84 percent of 5- to 14-year-olds) lived in rural places and small towns in 1850, not in larger cities, the system of quasi-public common schools with both taxes and rate bills must have predominated.[22]

The funding of the common schools in the early republic differed greatly from what would soon unfold. What developed would be the enduring legacy of a movement to fund schools entirely out of local and state revenues and rid the nation of rate bills and pauper schools.

In the years before the American Civil War an educational system evolved in which the community paid for the schooling of *all* its (free) children. Under this system the taxes of older property owners, whose children had already attended school, would pay for the schooling of other children, most likely those of younger members of the community. Communities would bond together in a system of "overlapping generations" in which the older members implicitly paid back to the community what they had received a generation before. Moreover, the schools would be common for all the children, no matter how poor the parents. With free schools for all, so the theory went, only the very wealthy would elect to send their children to private tuition-based schools and pauper schools would no longer exist.

Although full public funding for common schools spread throughout the northern and western states in the decades before the Civil War, the task was not fully complete in the North and West until 1871, and it was not until the 1870s that the South had free schools. The year that all rate bills were abolished for the states of the northeast and the "old northwest" is given in Table 4.1. Massachusetts, Maine, and New Hampshire did not have rate bills by 1826. In Connecticut, Rhode Island, New York, and New Jersey, statewide free schooling legislation was not passed until after the Civil War. In the 1870s, with Reconstruction, free schooling was imposed on the states of the Confederacy.

In many nations the impetus for state funding came from compulsory schooling laws. The state became obliged to provide sufficient schools and teachers when students were compelled to attend. But that does not appear to have been the case for the United States. For all states listed in Table 4.1, the first compulsory education law was passed *after* rate bills were eliminated, and in most cases long after.[23]

It should be clear from the inclusion of the South in the group of states with free schooling in the 1870s that the existence of public

Table 4.1. Free Schooling, Rate Bills, and Compulsory Education Laws in the North and Midwest

State	Year Rate Bill Abolished	Year of First Compulsory Education Law
New Hampshire	n.a.[a]	1871
Maine	1820[b]	1875
Massachusetts	1826	1852
Pennsylvania	1834	1895
Wisconsin	1848[c]	1879
Indiana	1852[d]	1897
Ohio	1853	1877
Illinois	1855	1883
Iowa	1858	1902
Vermont	1864	1867
New York	1867[e]	1874
Connecticut	1868	1872
Rhode Island	1868	1883
Michigan	1869	1871
New Jersey	1871	1875

Sources: Rate bill abolition: Cubberley (1947, orig. pub. 1919), p. 205, which is consistent with Adams (1969, orig. pub. 1875); Fishlow (1966a) for Indiana, Iowa, and Wisconsin. Cubberley makes no mention of the New Hampshire and Maine dates. Compulsory education laws: Deffenbaugh and Keesecker (1935), p. 8.

Notes: In addition to the New York State cities mentioned in the footnotes to this table, Providence, Baltimore, Charleston, Mobile, New Orleans, Louisville, Cincinnati, Chicago, and Detroit, according to Cubberley, had free schools for about 25 years before their respective states abolished the rate bill.

a. New Hampshire apparently never had a rate bill. See Bush (1898) and Bishop (1930).

b. Cubberley (1934, orig. pub. 1919) notes that Maine's constitution (1820) required towns to "provide suitable support for schools." See also Chadbourne (1928) and Nickerson (1970) who make no mention of a rate bill.

c. Fishlow (1966a) notes that Wisconsin abolished payments in its constitution, which passed in 1848.

d. Fishlow (1966a) gives the date as 1851.

e. Cubberley (1934, orig. pub. 1919) reports that many of the larger, and even some of the smaller, cities of New York provided for free schools long before the rate bill was abolished in the state. These include New York City (1832), Buffalo (1838), Hudson (1841), Rochester (1841), Brooklyn (1843), Syracuse (1848), Troy (1849), and Utica (1853).

funding did not mean that schools were well funded or that all children received equal amounts of funding. Even for the richer states, the quality of schooling varied across place and changed over time as the demand for better schooling increased. The average length of the Massachusetts public school session, for example, was about 165 days in the 1840s but about 192 days in the 1870s.[24]

The abolition of tuition payments in a state would seem to indicate that public schools became free overnight and that the cost to parents of educating their children changed suddenly. That may have been the case in some states and in some areas in these states, but in most states, the shift to free education was more gradual than is indicated by state law changes.

The significance of the demise of the rate bills depends on the fraction of the year that was paid for by districts prior to the establishment of free schooling. It also depends on the number of districts and municipalities that already had free schools before the advent of free schooling at the state level.

As we mentioned before, many municipalities, often the larger cities, abolished school fees earlier than did the entire state. The larger cities in New York State, for example, and even some of the smaller ones, abolished the rate bill long before the entire state did in 1867. New York City did in 1832, Brooklyn did in 1843, and many of the cities that lined the Erie Canal did in the 1840s.[25] Almost all the great port cities of the antebellum period, including those in the South, had free schools many years—even decades—before their states abolished the rate bill. Two-and-a-half decades before their respective states abolished the rate bill the southern cities of Baltimore, Charleston, Mobile, New Orleans, and Louisville had free schools, as did Cincinnati, Chicago, and Detroit in the "West," and Providence in the East. Even though Connecticut abolished the rate bill in 1868, the majority of school districts had established free schools about a decade earlier.[26] Dubuque and other Iowa school districts had free schools in 1856, although rate bills in Iowa City covered half the total expenditures in 1857.[27]

Not only did many districts abolish the rate bill in advance of state legislation, but the number of publicly provided free school days, even in districts that had rate bills, was substantial in many districts. We do not, however, know of any comprehensive study giving the fraction of the school year paid by districts prior to rate bill abolition and thus cannot fully assess the fiscal impact of the abolition of rate bills.

The shift in local finance from the levying of fees on families based on the number of weeks attended per child to taxation without any "user fees" should have increased enrollment and attendance if the rate bills really mattered. On the other hand, rate bills would have had less consequence if the publicly funded term was already long, if a substantial fraction of children already lived in free school areas, or if tuition payments mattered little in parental decisions about schooling. School enrollment and attendance data for the periods before and after the abolition of the rate bill can shed light on the issue. Even though the evidence is thin for the antebellum period, and must be interpreted with caution, revealing data exist for various states.

Two of the most studied states in antebellum educational history are Massachusetts and New York. Several researchers have extensively used the records of these states, as well as those compiled by the U.S. Census in 1830 and 1840 in its school censuses. Comparisons between the two states may be instructive because Massachusetts eliminated its rate bills in 1826 whereas New York State did not do so until 1867, some 41 years later.

In a careful study of the antebellum "common school revival," Albert Fishlow reported that 73 percent of all 5- to 19-year-olds in Massachusetts were enrolled in schools in 1830 and that 69 percent were in 1840. The comparable figures for enrollment in New York schools are nearly identical—74 percent in 1830 and 69 percent in 1840.[28] These figures suggest that the legislated abolition of rate bills had little impact on aggregate school enrollment.[29]

Other sources on antebellum educational statistics are the 1850 and 1860 U.S. population census manuscripts, which record whether an individual attended school for at least one day during the census year. The census school rate in 1850 for (white) children 5 to 19 years old in Massachusetts was 67 percent and 63 percent in New York. That in 1860 was 65 percent for Massachusetts and 62 percent for New York.[30] Although the levels implied for youths are high, probably too high using the enrollment or school rate data, there is no reason to believe that the data for New York and Massachusetts are biased *relative* to one another.

The comparison between New York and Massachusetts is telling. Around 1840 both New York and Massachusetts achieved almost equal enrollment rates even though Massachusetts had abolished the rate bill in 1826. Enrollment rates remained similar in 1850 and 1860, even though New York State did not abolish its rate bill until 1867.[31] One of the possible reasons for the lack of importance of the rate bill for New York State, as we have already seen, is the fact that many of its cities and some of its towns provided free schooling before the state mandated it.

Confirmatory evidence is provided by the states of the Midwest and reinforces the conclusion we draw from the New York–Massachusetts comparison. Schooling rates in the Midwest were substantial *before* the rate bills were abolished at the state level. Enrollment rates in the Midwest were around 55 percent in 1850 and they increased by another 10 percentage points to 1860. During that decade, or immediately before, four of the states (Illinois, Indiana, Ohio, Wisconsin) abolished their rate bills; yet much of the increase in enrollment in the area had already occurred.[32]

These findings do not indicate that the free-schooling campaign was unimportant. On the contrary, the provision of free schooling mattered considerably to mass education. Rather, the point we are making is that many localities were *already* providing a considerable number of community-funded school days prior to the state abolition of the rate bills.[33] In addition, many municipalities and school districts had adopted free schooling prior to the abolition of the rate bills at the state level. The evidence points to the importance of local, rather than state, regulations and to grassroots action rather than top-down campaigns.

The case of the rate bills is one of many instances in U.S. educational history demonstrating that local control fostered educational expansion and that local districts expanded educational access before state mandates.[34] Thus our findings, while not suggesting that the free schools for all crusade was unimportant, do point to the existence of and importance of grassroots educational movements.

HORACE MANN AND THE "SCHOOL MEN"

Historical interest in the abolition of the rate bills stems in part from the importance of free schooling, but it also derives from the individuals who mounted the campaign to spread public schooling throughout the young nation. Whether or not their personal feats affected the end result is a matter of some dispute. What is clear is that they were dedicated individuals whose appeals and arguments were often decades ahead of their time. Among the best known of these figures were Horace Mann, the secretary of the Massachusetts Board of Education from 1837 to 1848, and Henry Barnard, his counterpart in Connecticut and Rhode Island who served the former from 1839 to 1855 and the latter after his Connecticut position was temporarily abolished in the 1840s.[35] Both Mann and Barnard published widely read journals that disseminated their ideas about free and common schooling.

Horace Mann became the first secretary of the Massachusetts Board of Education in 1837, a decade after the state adopted legislation requiring localities to pay for all public schooling and constrained localities from imposing tuition charges. His goals, like those of other free school advocates throughout the nation, were to maintain the quality of the common schools, raise the standards for teachers, build more and better schools, and increase the length of the school year. To put it bluntly, Mann was a tireless advocate of mass education and he used every tactic he could muster to achieve his goal.

In response to a decreased interest in public education around 1840, possibly due to the depressed state of the national economy or to political events closer to home, Mann devised a way to convince the state legislators of the merits of an excellent public school system.[36] He constructed a questionnaire to demonstrate that education in the Commonwealth should be valued because it produced efficient workers. The educated worker, Mann wanted to show, was productive on the job, adapted more easily to new technologies and even added to them.

Mann's questionnaire was ahead of its time in its empirical sophistication but was probably more a tool of propaganda than of scientific inquiry. "My object," he wrote "has been to ascertain the difference in the productive ability—where natural capacities have been equal—between the educated and the uneducated." He chose to survey "manufacturers of all kinds . . . machinists, engineers, railroad contractors, officers in the army" who hired hundreds of people. He chose large employers to enable comparisons between workers with ample and with minimal common schooling. In his *Fifth Annual Report* (1841) he shared some of the completed questionnaires with his readers. Although the responses he offered were few in number (there were only five), his summary continued for 20 pages of exhortations regarding the enormous gains from having an educated work force.[37]

GRASSROOTS MOVEMENT OR TOP-DOWN CRUSADE?

A large and often contentious literature has emerged on the role of schools in socializing youth in the nineteenth century. Some historians

have interpreted the documents of the past as evidence that manufacturers and property owners wanted to fund schools to create a group of
docile workers out of immigrant children, often Catholics.[38] Horace
Mann and his counterparts were dedicated, motivated, energetic individuals who formulated clever, often valid, arguments for the public
provision of schooling and its increased funding. Yet some have interpreted increased schooling as a measure of the success of school bu-
reaucrats, such as the "school men" (dedicated champions of free
schooling), in expanding their sphere of operations. Was there more
behind the spread of public schooling than egalitarian virtues?

An important similarity between these arguments is the view that
educational expansion emanated mainly from those nominally in
power. This interpretation gives credit to the manufacturers, the large
property owners, and the so-called school men. According to this argument, each of these groups may have had a different reason for expanding education or making it free, although they were unified in
their desire to impose education on the masses.

A rather different view is that educational reform was largely a
grassroots movement. It may have been hastened by the efforts of the
school men, but educational reform according to this view was gen-
uinely demanded by parents and the community. Interestingly, these
two views of the spread of free common school education in the nineteenth century are repeated in the debates about the spread of secondary schooling in the twentieth century (we will return to these issues in the discussion of the high school movement of Chapter 6).

Each of the arguments contains certain elements of truth, some
more so than others. Across much of America, mass education was a
truly grassroots movement. Its popular base is clear from the referenda
in many states that led to the passage of constitutional amendments,
constitutions, and legislative statutes providing for taxation and free
public education.[39] It is also clear from the role of the migration of
New Englanders into the western lands and the institutions they
brought with them. Yet it is also the case that public education was
championed by energetic and persuasive school men such as Horace
Mann, and that some manufacturers and property owners, particularly
in the wake of the large migration of the Irish to New England, wanted
to create Protestant Americans out of newly arrived Catholics. Protestants, as the religious majority, wanted to prevent the growing minority

of Catholics from gaining control of school funds. In one important case—that of New York City—the battle over school funds led to the early separation of church and state in the provision of education. We will soon explore the case in some detail.

Separation of Church and State (Virtue 4)

NONSECTARIANISM

Mann's *Fifth Annual Report* of 1841, which we discussed above, may have been motivated by political attacks on him in the Massachusetts legislature around 1839 concerning what others saw as his creation of "Godless schools." Mann, a Unitarian, wanted schools to educate all children regardless of faith and he espoused nonsectarian schools. The old-line Puritans in Massachusetts rose up in protest. After much debate, precipitated by a request by Catholics for a share of the state's educational funds, Massachusetts amended its constitution in 1855 to prohibit state and local funds raised for education to be awarded to sectarian schools.

The separation of church and state for Mann and others, it should be noted, did not bar the teaching of religion. Nonsectarian schools in the nineteenth century did not imply secular, godless school rooms.[40] Although maintaining the Protestantism of the schools was an important ingredient in the separation of church and state, the issue of commonality within Protestantism loomed large as well. Religious fervor rose in early to mid-nineteenth-century America, as is apparent from the proliferation of Protestant religions and the disestablishment of the Congregationalist Church in Massachusetts in 1833.[41] A common education for all Protestant children required that public schools be nonsectarian but did not require that they be secular in their instruction. That change happened considerably later—in the mid-twentieth century.[42]

Six states preceded Massachusetts in banning the use of state and local schools funds by churches. New Hampshire had done so in its constitution in 1792 as had Connecticut in 1818. New Jersey, Michigan, Ohio, and Indiana followed suit from 1844 to 1851. Almost all the states that entered the union before 1876 amended their constitutions to provide for the restriction regarding the use of school funds by religious bodies. States that entered later were required by an act of Congress, passed in 1876, to include the same prohibition in their constitutions.[43]

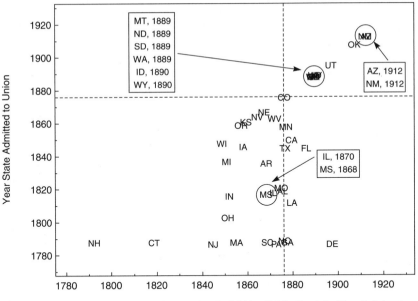

Figure 4.1. Year of Statehood and Prohibitions Regarding Use of Public Funds for Religious Schools. The dotted horizontal and vertical lines are given for 1876, the year Congress passed a law requiring new states to include the prohibition in their constitutions. Cubberley does not have information on the following states, some of which passed legislation to prohibit public funding of church schools: KY, MD, ME, NY, RI, TN, VA, and VT. For example, according to Torpey (1948, p. 234), the New York state legislature mandated in 1805 that public money for New York City's free schools could be used only by those institutions not controlled by a religious group. Sources: Cubberley (1934, orig. pub. 1919, p. 239); Torpey (1948).

A graph representing the year of statehood and the year the state amended its constitution (or adopted one) to prohibit the use of public funds to support sectarian schools is found in Figure 4.1.

CASE STUDY: NEW YORK CITY

Cities contained most, if not all, of the factions just described in the debates over free schooling. Nowhere were these factions more active and more involved in intrigue than in the story of the origins of free schools in early 1800s New York City. Religious charities, civic leaders, parents, local politicians, and state legislators all played a key role.[44]

In turn-of-the-nineteenth-century New York City, education was not a public concern. Private schools trained children whose parents could afford the tuition, and pauper or charity schools often run by church denominations dealt with the others. Many of the private schools, known as "common pay schools," were relatively inexpensive and served working-class families, such as those of the artisan or shopkeeper.[45]

But by the 1840s New York City schools were publicly funded for all who wanted to attend. The turnaround that led to free schooling arose not because of the common school crusaders of the period of Horace Mann. Rather, free schooling in New York City triumphed largely because of the hard labor and cunning of those who founded an organization to set up non-church-based charity schools.

The organization, founded in 1805, was originally called the Free School Society. At its inception it was a charitable, secular, private institution established to provide schooling to poor children who were not taken care of by particular church groups. It was overseen by well-known city fathers, such as its first president, De Witt Clinton. Beginning in 1806, the organization received state funds for its activities in serving the city's poor.

As the number of poor increased, the Society sought greater funding and it also endeavored to expand its mission. Catholics, who were just 2 percent of New York City's population in 1800, were 16 percent in 1810, 18 percent in 1830, and about 22 percent in 1840—all before the great Irish potato famine and the mass emigration that followed.[46] The Society began to advocate the separation of church and state not because of an ideological opposition to religion, but rather because of a desire to expand financially. Curtailing the funding of church-based schools meant increasing its own financial support.

In the 1820s the Free School Society began a campaign to put church-based charity schools out of business by denying them public funds and promising to provide free schools to *all* the children of New York City be they rich or poor, Protestant or otherwise. The campaign to deny funds to the denominational schools claimed success in 1825, with the passage of an ordinance by the Common Council. In 1835 the Free School Society group expanded its mission and changed its name to the Public School Society. The Public School Society functioned in much the same way current voucher or charter school systems do. The Society was a private organization that provided schooling for the city's children and was paid out of public funds. It offered, in essence, a publicly funded but privately provided education.[47]

The new Society opened schools in many parts of the city and could claim considerable success. But triumph led to unintended consequences and its eventual undoing. An elective New York City Board of Education was founded in 1842 and with its establishment came a group of free publicly provided schools. With limited resources the Public School Society could not compete with a governmental body that itself provided free public schooling. It disbanded in 1853 and turned over its schools to the city. The Society's early courting of Catholics led to a separation of church and state in the provision of publicly funded schooling in New York City. The free school movement in New York City, therefore, ultimately resulted in the establishment of publicly provided, not just publicly funded, schools and the secular control of education.

Gender Neutrality (Virtue 5)

By the 1850s, and probably before, girls and boys throughout America were educated to about the same degree, in terms of years of schooling, until approximately age 15 (see Figure 4.2). This aspect of gender neutrality was not always the case but arose during the first few decades of the nineteenth century.[48] To prove this point, we employed data from the U.S. decennial population censuses for 1850 to 1880, which asked whether an individual attended school (for at least one day) during the previous year.[49]

The data for 1850 indicate that from about age 5 to 14 girls attended school—most often common schools—to about the same degree as did boys.[50] Although boys began to lead girls in attendance at around age 14, the disparity was not substantial until age 16, when girls had a school participation rate about three-quarters that of boys.

By 1880 gender neutrality in school attendance had been extended by age. Nationwide, it was not until age 15 or 16 that boys had somewhat greater school participation rates than did girls, and the difference did not become substantial until age 17, when the ratio of boy-to-girl school participation was 0.81.

The data in Figure 4.2 are for the entire nation. Clearly, the expansion of gender neutrality from 1850 to 1880 varied by region. Larger

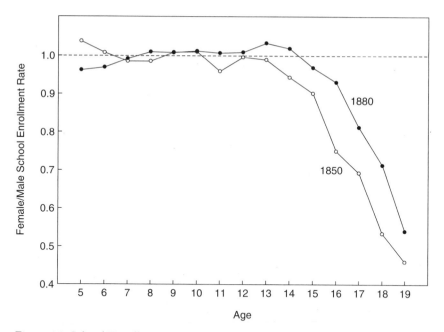

Figure 4.2. School Enrollment Rates by Sex (for Whites): Females/Males, 1850 and 1880. The census asked whether an individual had attended school during the previous year. We interpret these numbers as giving school enrollment (see text). Sources: 1850 and 1880 IPUMS of the U.S. Census of Population.

relative gains for girls were made in New England and the Middle Atlantic than in the states of the Midwest. The West, just beginning to be settled in the 1850s, emerged by 1880 as the most gender neutral of all. These relative gains for teenaged girls were rooted in the early extension of publicly provided and publicly funded schools to the secondary grades.

By the early 1900s, as we detail in Chapter 6, the fraction of teenage girls who attended secondary school was actually higher than it was for teenage boys, and these differences were considerable. In the nineteenth century, when schooling beyond age 15 generally took place in private schools and was often in residential schools, parents did not always treat their teenaged daughters as the equals of their brothers. Thus the expansion of publicly funded high schools increased the number of youths who went to school and also enticed parents to send their daughters to school.

Gender neutrality in attendance need not imply equality in the type of education, since some schools and courses of study were probably reserved for boys. Even though single-sex schools existed, the vast majority of public schools were coeducational by the end of the nineteenth century. In 1890, according to a national survey of public schools, 93 percent of all major U.S. cities (of which there were 628 in the survey) had adopted coeducation in all grades of their public schools.[51] Of the remaining 7 percent (42 cities), some schools and grades educated girls and boys separately. Certain cities that generally had coeducational instruction contained high schools that were reserved for boys (as in Boston, which had separate boy's and girl's Latin Schools) and these cities are included in the 42 mentioned above. Overall, by the end of the nineteenth century, coeducation existed in *all* schools and in *all* grades in most U.S. cities.[52] Rural schools were, according to the report, uniformly coeducational by 1890.

The gender neutrality that characterized American education by the mid-nineteenth century was in stark contrast to that existing in the leading European nations of the day. Many European educators who were delegates to the Chicago Educational Congress of 1893 were shocked at the degree to which America educated its girls together with its boys. "It seems strange," noted one of the German delegates from Prussia, "to see boys and girls not only of 13, but even of 16 years of age, sitting together or standing in mixed rows." The French were even more stunned. "Of all the features which characterize American education, perhaps the most striking," noted the female French minister of public instruction, "is the coeducation of young men and young women . . . at least it is most striking to a French observer, for it reveals to him a state of mind and of habits which is entirely strange to him."[53]

Open and Forgiving (Virtue 6)

The U.S. educational system has been open and forgiving in comparison with other educational systems. By "open" we mean that almost all children could attend school. By "forgiving" we mean that one could often advance to higher grades and institutions even if one failed to perform adequately in a lower grade. The U.S. educational system of the nineteenth and much of the twentieth century was not, we emphasize, open to all children in all parts of the country, and it relegated many to inferior and segregated schools. But it was a far more open and forgiving system than were those of other economically advanced nations of the day.

Because the U.S. system was highly decentralized, no national educational standards and generally few state standards existed, a subject to which we return in the discussion on secondary education. In England, France, and Germany, to the contrary, admission to publicly funded schools beyond the elementary years, or in a later period beyond the age of compulsion, was by examination, generally at the national level.

Nineteenth-century Prussia was a world leader in the education of its people at the lower grades.[54] However, as early as 1812, regulations were imposed on the granting of the secondary school degree (the Abitur), which conferred eligibility for entrance to the university. In 1834 alternative methods for gaining entrance to the university were abolished and the Abitur became the only avenue. Similarly in France, Napoleon established a centralized educational system in 1808 in which entrance to higher education was restricted to those who passed a national examination, the baccalaureate.[55] Because the "openness" and "forgiveness" of the U.S. system, in comparison with that of other nations, mainly concerns the transition to secondary and higher education, we will return to this subject in later chapters.

A Note on Nineteenth-Century Education Statistics

In our discussion of rate bills and the role of free schooling in increasing the attendance of youth, we touched on the subject of nineteenth-century educational statistics. The data derived from various federal census sources suggest high rates of school-going in the Northeast. We also noted that such data could be reasonably interpreted as enrollment rather than attendance figures.

We return here to a discussion of educational statistics using data from the U.S. decennial population censuses from 1850 to 1880.[56] These are the same data we used in the discussion of gender neutrality and are derived from the answers given to the census enumerators concerning whether an individual had attended school during the past year.[57] We term these "enrollment" data since there is no information on the fraction of the year attended. Also of interest was whether a youth who claimed to have enrolled in school also indicated labor force

A. Northeast

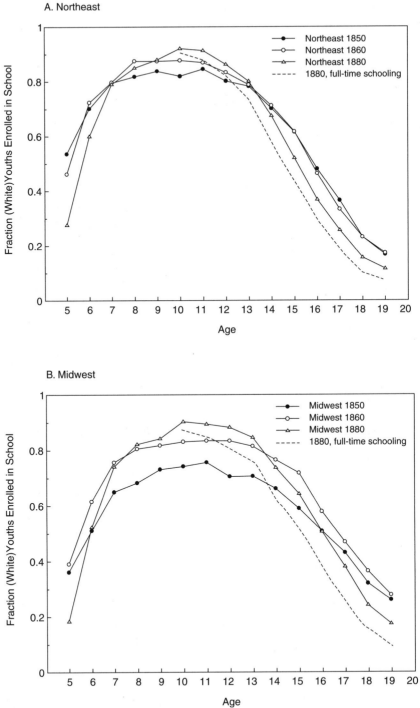

B. Midwest

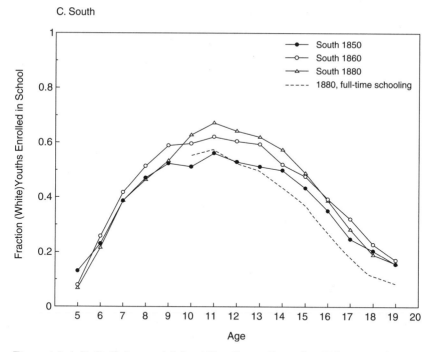

Figure 4.3 A, B, C. (facing page) School Enrollment Rates for (White) Youths in the Northeast, Midwest, and South: 1850, 1860, and 1880. Northeast is the states of New England and the Middle Atlantic, Midwest is the states of the East North Central, and South is the states of the South Atlantic. The school enrollment rate is the fraction of youth who reported attending school sometime in the past year. The "full-time schooling" rate is the fraction of youth who both reported attending school and reported no occupation in the census. Sources: IPUMS of the 1850, 1860, and 1880 population censuses.

attachment by declaring an occupation.[58] We focus on the school enrollment of white youths.

Enrollment levels for white youths were extremely high in the states of the Northeast in 1850, equally high in the Midwest by the 1860s, but trailed considerably in the South from 1850 to 1880 (see Figure 4.3 for a comparison among the Northeast, Midwest, and South). The overall enrollment rate of white youths 5 to 19 years old increased in all three regions from 1850 to 1860, declined (especially in the South) in the decade of the Civil War, showed little change in the 1870s in the Northeast and Midwest, and increased sharply to recover back to the pre–Civil War levels in the South from 1870 to 1880.[59] The changes in enrollment rates, however, differed substantially across age groups.

Enrollment rates increased across all three regions from 1850 to 1880 for children 7 to 13 years olds. Older and younger children, for the most part, experienced a decrease in enrollment rates during that timespan. Although the decrease might be interpreted as a backsliding in school participation, that for the older age group should more accurately be interpreted as a sign of the greater effectiveness of schools. Many children who enrolled in 1850 were in common schools that had limited days of operation and scant resources, particularly in rural areas. As annual days of schooling and attendance increased, these youths could finish the school program at younger ages.[60] The reason for the decrease at the lower ages is more complex and may concern state laws that allowed districts to restrict schooling to children above a certain age.[61]

The census data for 1870 and 1880 provide occupational information for individuals 10 years and older and thereby allow the calculation of a "full-time schooling" measure for youth. We assumed that individuals who attended school in the past year and did not list an occupation in the census were engaged in full-time schooling. Because the comparisons between the total statistics and those for full-time schooling apparently did not vary over time, we have graphed them for 1880 only. The full-time schooling fractions are lower than the totals, more so in the Midwest and the South, where agricultural employment was greater, than in the Northeast. But even though an adjustment for full-time schooling reduces the figures, they remain substantial for youth in their early teens in most parts of the nation.[62]

Although youths could have attended school for the regular term *and* have been employed on their family farm or in a town either before or after school, it is more likely that most of them did not attend school for much of the term. Thus the schooling figures derived from the population census and other education statistics are somewhat exaggerated as a measure of actual school attendance.

The Roots of the Public High School

Even during the era of the common school movement and the campaign for free education, schools that provided training beyond the elementary level were proliferating in the United States. The schools in this group were of many types. They began chronologically with the

Latin grammar school of the seventeenth century and were followed by academies and public high schools. By the mid-nineteenth century several, often competing, institutions provided schooling beyond the common school years. These included public high schools, private and quasi-private academies, preparatory divisions of colleges, and even some common schools.

The grammar or Latin schools were the earliest institutions in America to educate young men to enter the nation's colleges and the oldest in this group was the Boston Latin School (1635), which largely trained youths to enter Harvard College. There were never many of these schools and they were found in larger cities in both the colonial and early republic periods.

Even by the early nineteenth century some parents of ordinary means in towns and cities were demanding an education for their children that went beyond the common school years. They wanted their children to be trained for the counting house, mercantile establishment, and professional and business occupations. To fill this demand, many of the nation's larger cities established free public high schools.

The first of these high schools, the English Classical School, was established in Boston in 1821. During the next two decades about 30 public high schools were established in other Massachusetts towns.[63] Even though an 1827 act of the Massachusetts Commonwealth required towns with more than 500 families to support a public high school, it is likely that these schools were founded for other reasons. Of the 20 largest Massachusetts towns in 1830, 15 had a high school by 1841, but so did 11 towns that were smaller than the legal requirement. Furthermore, many towns that would have met the size requirement did not have a high school even by 1860.[64]

In 1838 Philadelphia established its first public high school, Central High School, which exists to this day.[65] The first public high school in New York City was not founded until 1848, although 25 years earlier a short-lived, privately funded high school had been formed.[66] Public high schools continued to expand in pre–Civil War America and by 1860 more than 320 were in existence.[67]

Throughout much of the nineteenth century the demand for public high schools in the sparsely settled parts of rural America and in its small towns was insufficient to warrant their provision. The public high school was a relatively expensive educational institution and was seen as one that would serve only a small fraction of the town or district youth. The external benefit that accrued from the common school, such as teaching children basic skills so they could function as good citizens and endowing them with republican virtues, was not the rationale for the high school. In rural communities, towns, and small cities, the common school often accommodated older youths, a practice that continued into the early twentieth century.[68]

In the mid-nineteenth century most communities outside the large cities could not yet support a public high school. Parents who wanted their children to continue their education beyond the common school years and had the funds to do so, supported private secondary schools. By the 1870s many of these schools, often called academies, filled the demand for post-elementary education. Academies grew so rapidly in the first half of the nineteenth century, and then disappeared even more quickly, that some have referred to the period in the history of American education as that of the "academy movement."[69] Academies appeared just as the common schools were increasingly supported by public funds and were expanding their enrollments.

Academies of all types existed. In their laissez faire proliferation and unregulated growth they became typically American institutions. Academies could be college preparatory schools, of which some of the best, such as Phillips Academy, still exist.[70] Most academies were not of this caliber but gave, nonetheless, adequate training in academic subjects. Academies often taught vocational subjects, such as bookkeeping, surveying, drafting, and navigation, that drew on the academic subjects they offered, such as mathematics, English, and history. Some academies taught music, dance, and other less scholarly subjects demanded by the parents of proper young women.

Because academies generally taught youths who resided outside the town, many of these schools boarded students. Although some academies were institutions of substantial size, many were established by a schoolmaster or schoolmistress who taught pupils at home. Even though academies charged tuition and were privately controlled schools run by trustees, many received subsidies from the locality or state in the forms of land and buildings and were, therefore, only quasi-privately supported.

Academies arose at the precise moment when publicly funded high schools began to diffuse across the nation, and in consequence academies were largely ephemeral institutions.[71] They folded as the publicly funded, "free" institutions arose and took their place. Few survived for long, and although some left records most left none at all. For that reason we do not know the extent of their spread nationally other than from limited county-level data in the 1850, 1860, and 1870 U.S. population decennial censuses. We leave further discussion of the academies to Chapter 5.

Since the mid-nineteenth century, reformers had argued that free public high schools were an integral part of the democratization of education,[72] but certain of their contemporaries questioned whether the provisions of the states that gave school districts the ability to raise taxes to support the common schools could also be used for the benefit of high schools. Without the ability to raise taxes, there could be no publicly funded high schools. The Michigan Supreme Court in *Kalamazoo* (1874) unambiguously answered that question.[73]

The *Kalamazoo* decision set down the ruling that local funds could legitimately be used to support high schools, Many districts and municipalities across America had already used the state laws to raise funds for high schools, but the *Kalamazoo* case marked an important turning point because various citizen groups had opposed using funds for high schools on the grounds that these schools did not educate the majority of the children. With *Kalamazoo*, the argument against the use of school funds to support a public high school no longer had legal validity.[74]

Egalitarianism: A Summary

This chapter traced the origins of the many virtues of American education—its public provision and public funding, existence of small fiscally independent units, separation of church and state, gender neutrality, openness, and forgiveness. These features can be summed up as egalitarian in nature. The democratic, republican vision of education triumphed over an elitist one in which private schools would exist for some and charity or pauper schools would serve the others.

The public provision and public funding of education were clearly part of the republican vision of an open and common system. The separation of church and state in the funding of schools was a logical extension of that view, as the case of New York City demonstrates. If denominational institutions received state funds, then children outside these denominations would be excluded, and children would not learn in a common, inclusive setting. But the history of the Massachusetts ban on the use of state funds by sectarian schools suggests an additional motive. The potential use of funds by Catholic churches prompted a prohibition on the use of funds by any denomination.

The creation of publicly funded common schools and their spread throughout much of America was the first great transformation of education in America. Even before mandated free schooling spread throughout the states, with the abolition of rate bills, years of education among Americans had surpassed that of the citizens of any other country.[75] Free public schooling, which had diffused nearly everywhere in the nation by the 1870s, set the stage for the next great educational expansion—the growth of public high schools.

America's commitment to publicly provided and, later, publicly funded schooling began with a desire to create educated and informed citizens who could vote and stand for election. By the end of the nineteenth century, and probably before if Horace Mann's *Fifth Annual Report* is any guide, education was increasingly looked upon as it is today: a means of acquiring skills for the world of work and for life in general.[76]

As we have seen in this chapter, the second great transformation—the high school movement—began slowly in the nineteenth century with the creation of public high schools in the nation's larger cities. Private academies increased the reach of secondary education through the mid-nineteenth century. The "academy movement" clearly reveals the strong demand by parents for public secondary schools. The second transformation picked up its greatest steam with the diffusion of public high schools, and in the first few decades of the twentieth century these schools reached even the smallest rural communities in America.

The second transformation was built on foundations that were set down in the nineteenth century. The virtues of the past, largely put in place before the American Civil War, fostered mass education at all levels during the twentieth century. Whether these virtues have continued to be beneficial and whether some have become hindrances to educational excellence are issues we take up in Chapter 9. But we must first explore the high school movement of the early twentieth century.

5

Economic Foundations of the High School Movement

By the middle of the nineteenth century the United States had the most educated youth in the world.[1] Mass elementary schooling had swept through much of America and came to many states even before it was fully funded by local governments. The citizens of other industrializing nations would have to wait another three to four decades to attain elementary school enrollment rates comparable to those in 1860 America. Mass elementary and publicly funded education eventually did come to Europe, often after the franchise was extended.[2] Elementary school enrollments began to soar in Austria, Denmark, France, Sweden, and especially Britain during the late nineteenth century.[3]

But just as Europe began to narrow the educational gap with America at the elementary school level, a second great educational transformation started to gather steam in the United States. The educational movement would serve to widen the gap between the educational attainment of youths in Europe and America and it would leave Europe in the educational dust for some time. The gap in educational attainment would not again begin to close until well into the latter part of the twentieth century.

The second educational transformation that catapulted the United States to another peak in mass education, and one that would last for much of the twentieth century, was known then and today as the "high school movement."[4] This chapter concerns the origins of that second transformation.

The Second Transformation of American Education

The high school movement rapidly changed the education of American youth. The typical young, native-born American in 1900 had a common school education, about the equivalent of six to eight grades. But the average young person in 1940 was a high school graduate (Figure 5.1). Outside of the South, the transition was more rapid: as early as 1930 the median youth in the New England states and parts of the West was a high school graduate.

The high school movement was swifter than the first educational transformation and enabled America to begin its third transformation, that to mass higher education, by the close of World War II. Mass higher education did not make an appearance in most of Western Europe until after the 1970s.

The central driving force behind the high school movement was the existence by 1910, if not earlier, of substantial pecuniary returns to schooling beyond the elementary grades (or common school years). We demonstrate that fact by referring to earnings by occupation and to the data on schooling and earnings presented in Chapter 2. But the existence of high returns alone was insufficient to bring forth the high school movement. The other factor propelling the movement was the increased demand for workers with education beyond that given in the common schools and the elementary grades. The financial returns to obtaining a secondary school degree appear to have been high even in the mid-nineteenth century; however, with educated workers spread out across the United States, demand was insufficiently concentrated to call forth a major public response to build and staff high schools until the early twentieth century.[5]

In parts of the nation that did not yet have accessible public high schools, youths often continued with their education in the common schools beyond the usual eight years. These additional years in a common or elementary school, our data show, were not nearly as valuable in the labor market as those spent in a high school. The secondary school was a distinct institution created to fill an educational void.

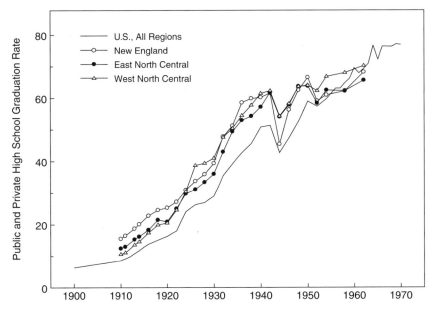

Figure 5.1. Public and Private High School Graduation Rates for the United States and Three Regions. Includes graduates of public and private schools where private schools include parochial high schools, academies, and the preparatory divisions of colleges and universities. Rates are produced by dividing by the number of 17-year-olds. The number of 17-year-olds for years between the censuses is obtained by extrapolation. Sources: See Appendix B.

Youths also attended quasi-private secondary schools, often called academies, and schools that trained youths in specific skills. Prior to the spread of public high schools, academies and related institutions emerged in towns and cities throughout America. The switch from academies to public high schools required more than an increased demand by some parents for the further education of their children. It required the support of a larger community. The public high school movement could take off only when a sufficiently large group of citizens was convinced that there would be personal gain to them from having a high school. They were mobilized when they realized that their children would benefit from a secondary school education and that paths to success were no longer the informal ones of the office, the shop floor, and the family farm.

Because the U.S. educational system was highly decentralized, groups that supported the expansion of secondary schools need not have been large within the nation, or even within a state, for educational growth to progress. All that was needed was that the support be substantial enough within certain small communities and that the demand be sufficiently concentrated geographically. Thus the extreme decentralization of U.S. education served to advance the public schooling of U.S. youth, at least for some time.

This chapter is mainly concerned with *why* the second great educational transformation in America took place *when* it did. We found that the financial returns of attending and graduating from secondary school were substantial even in the mid-nineteenth century. The high school movement, however, began slowly in the mid-nineteenth century and then moved with spectacular speed and urgency in the early twentieth century.

The demand for educated workers grew markedly toward the end of the nineteenth century. Because the cost of providing secondary school education was high, most communities would not support publicly funded secondary schools until demand was sufficiently great and widespread. Part of the reason concerns the need for a larger scale of operation to reduce the per pupil cost of high schools. A more important point, we think, concerns public support for an expensive public good. Only when a sufficiently large group was willing to tax themselves for the public good would it be provided.

Public high schools were established in the nation's larger cities early in the nineteenth century, but were rarely encountered in its sparsely settled communities before the high school movement began. Because the United States was predominantly rural, relatively few secondary schools existed until the early 1900s. Only when demand expanded for educated workers in the major sectors of the economy, such as manufacturing, could a widespread movement emerge to build and staff secondary schools for America's youth. One of the reasons for their slow growth is that the minimum efficient scale of operations for a high school was considerably larger than for a common school because the curriculum was broader and some courses required specialized equipment. Larger scale also meant greater distances and transportation costs for students in rural areas. The rapid spread of high schools in the early 1900s outside the larger cities may have been assisted by reductions in transportation costs with advent of the automobile, the school bus, and better roads.

Contrary to many historical accounts, the high school movement was not precipitated by a series of state laws that legislated against child and youthful labor and that compelled children to attend school (in Chapter 6 we show that the increase of high school enrollments from 1900 to 1940 was also not much affected by the expansions of these laws). Rather, the high school movement emerged from a grassroots desire for greater social mobility. The consequence of this deep-seated interest was the establishment of a multitude of private academies and training institutes. As the demand for educated workers grew, publicly funded high schools followed. We return in Chapter 6 to the high school movement itself.

Changing Jobs, Changing Skill Demands

For most of the nineteenth century, the vast majority of American workers were employed in jobs that did not require much formal education. As late as 1870, 53 percent of the labor force were in agricultural occupations and 10 percent were domestics and personal service workers. Manufacturing operatives and laborers made up 13 percent of the workers and another 7 percent were craftsmen and shop-floor supervisors. Most manufacturing workers were employed by the older industries of the first industrial revolution, such as cotton, silk, and wool textiles and boots and shoes. Finally, only a small fraction of the 4 percent who were in transportation and communication were employed in jobs requiring secondary schooling. Therefore, only about 10 percent of the labor force were employed in an occupation that typically required an education beyond the elementary school years, whereas the other 90 percent were employed in jobs that did not.[6]

In contrast, by 1920 more than a quarter of the nation's workers had jobs in occupations for which a high school or college education was expected. In the previous half-century several sectors of the economy had grown considerably and these sectors had substantial educational requirements. Among them were office and sales work, which increased its share of the labor force by more than four times from 1870 to 1920 (Table 5.1). Office workers, including clerks, bookkeepers, and managers, earned a hefty premium for their education in the mid-nineteenth century, if not earlier. They continued to earn a comparable premium even as their numbers increased, as we will soon show. Other occupational groups that demanded more highly educated workers were professionals, including teachers, lawyers, and doctors, and managers. Those groups grew 50 percent faster than did all workers in the 1870 to 1920 period (Table 5.1).

Although the increase in the educated occupations occurred for both males and females, that for women was considerably greater. In the clerical and sales trades, for example, women increased their proportion by more than 20 times from 1870 to 1920 and by more than three times from 1900 to 1920. For men the increase was 2.6 times from 1870 to 1920 and 1.3 times from 1900 to 1920. Women more than doubled their representation among all white-collar trades (professionals and managers plus clerical and sales) from 1900 to 1920, whereas the increase for men was just 1.25 times (Table 5.1). Another important point to gather from Table 5.1 is that a large fraction of the increase in the educated occupations after 1870 occurred from 1870 to 1920, and for women, the vast bulk of the increase also occurred during that half-century. For all workers, 38 percent of the growth in the white-collar share from 1870 to 1970 occurred to 1920; for women 59 percent did.

Table 5.1. Fraction in Various White-Collar Employments: 1870 to 1990

Year	Males and Females		Males		Females	
	Professional and Managerial	Clerical and Sales	Professional and Managerial	Clerical and Sales	Professional and Managerial	Clerical and Sales
1990	0.332	0.256	0.333	0.147	0.331	0.384
1980	0.278	0.260	0.299	0.140	0.250	0.420
1970	0.234	0.252	0.256	0.146	0.199	0.425
1960	0.197	0.216	0.211	0.137	0.168	0.377
1950	0.178	0.195	0.182	0.130	0.169	0.365
1940	0.151	0.166	0.147	0.127	0.163	0.285
1920[a]	0.124	0.131	0.119	0.099	0.141	0.256
1910	0.116	0.099	0.115	0.089	0.118	0.136
1900	0.100	0.075	0.099	0.075	0.105	0.074
1880[a]	0.085	0.042	0.084	0.047	0.092	0.017
1870	0.080	0.034	0.083	0.038	0.065	0.012

Sources: *Historical Statistics, Millennial Edition* (2006), table Ba 1033–1074.
Notes: Proprietors are included in the professional and managerial category.
a. No data points for 1890 and 1930 are reported in the source used.

In addition to changes in the white-collar sector, employment in certain manufacturing industries, from the nineteenth century to 1920, expanded relative to others. The growing industries were disproportionately the newer ones and those in the higher technology fields, such as chemicals, machinery, and vehicles. They were (as discussed in Chapter 3) industries that had the greatest demand for more educated workers, even among operatives and laborers.

More concrete evidence on the relationship between occupations and education can be gleaned from the 1915 Iowa State Census. That document is the earliest available containing data on both occupation and education for a large and representative population. The information from the census confirms the growth of occupations that required greater levels of education.

Among men 25 to 34 years old in the professional and managerial category, 72 percent had some high school training and, on average, they had 12 years of schooling (Table 5.2). In the sales and clerical group 62 percent had some high school. But among manual workers, including blue-collar workers, farmers, laborers, and service workers, fewer than 20 percent had attended any high school. The data for men 35 to 44 years old show similar trends. There were large differences in the education of individuals by occupation and most of the difference is found in high school attendance and graduation.

Because of the smaller sample of workingwomen, we used the combined age group for females and present data only for occupational groups that contained large groups of women. The educational level of women was even higher than of men, especially in the professional category, which includes mainly teachers, and in the blue-collar group.[7]

The important point demonstrated by the data is that there was a far higher level of education for both men and women in the professional, managerial, and ordinary white-collar groups than in the other categories. Education was the ticket to obtaining a white-collar position. Among native-born males 25 to 34 years old with some high school education, 53 percent had a white-collar position, but only 13 percent of those without any years of high school were similarly employed. Put another way, a high school educated male was four times more likely to obtain a white-collar job than was his counterpart who did not continue beyond the eighth grade or common school.

Those with more education were disproportionately found in the office, not on the shop floor. But some manual workers also had a

Table 5.2. Education and Occupation: Iowa 1915

Occupational Groupings	Native-Born Males, 25–34 Years			Native-Born Males, 35–44 Years		
	Highest Grade (years)	% with High School	% High School Grad.	Highest Grade (years)	% with High School	% High School Grad.
Professional and managerial	12.1	71.6	50.0	11.5	59.5	41.8
Ordinary white-collar	10.7	62.5	39.2	10.6	57.9	38.9
Blue-collar	8.4	20.2	8.3	8.2	17.5	9.2
Farmer	8.8	18.1	6.9	8.7	16.4	7.6
Farm laborer and servant	8.2	17.9	6.6	7.8	9.5	5.2
Total for all occupations	9.2	30.2	16.3	9.2	28.1	17.1

Occupational Groupings	Native-Born Females, 25–44 Yrs		
	Highest Grade (years)	% with High School	% High School Grad.
Professional and managerial	12.9	88.1	67.6
Ordinary white-collar	10.8	67.4	47.4
Blue-collar	9.0	33.0	13.5
Farm laborer and servant	8.5	21.0	9.7
Total for all occupations	10.3	48.5	36.1

Source: Iowa 1915 sample, see Appendix A.

Notes: The occupational groups listed do not exhaust the total. Semiprofessionals are omitted for both males and females; farmers are omitted for women. The "ordinary white-collar" group includes clerks and salespersons. "Blue-collar" includes craftsmen, operatives, and non-farm laborers. The education data truncate years for the total at 18 and for common school at nine years. Individuals with college but not high school are presumed to have received preparatory instruction in a college. The entries for highest grade are the average years of completed schooling for each group. The percentage with high school is the percentage of each group with any years of high school. High school graduates are those who completed four years of high school including those who continued to college.

secondary education. As we just noted, for the 25- to 34-year-old group of men, 20 percent of blue-collar workers had attended high school and 8 percent had graduated. These more-educated manual workers were found, disproportionately, in the newer industries of their day. They were electricians, linemen, plumbers, auto mechanics, and auto repairers. They were members of an elite set of tradesmen that included printers, compositors, railroad engineers, stationary engineers, watchmakers, engravers, jewelers, machinists, and mechanics, and they were factory supervisors and inspectors. The educated could advance even if they preferred industrial life to the office.

Among young women, however, those who had some years of high school and yet had not joined the ranks of professionals and other white-collar workers were far less privileged than were comparable men. To be sure, some were in the more elite printing trades and others were in supervisory factory positions, but most were confined to the usual remunerative jobs women often did in their homes as dressmakers, seamstresses, tailoresses, and milliners. Not surprisingly the vast majority of employed women who attended some high school were white-collar workers, whereas far fewer of the men were since they had reasonable opportunities in other sectors. Among the 25- to 44-year-old group of native-born women who were working and had attended some high school, 81 percent were white-collar workers, compared with just 54 percent of the men.

The 25- to 44-year-olds in 1915 would have been teenagers from the 1880s to the early 1900s. Thus the data to which we have just referred reflect the occupational advantages of those who received a high school education mainly in the last decades of the nineteenth century. Although some of the advantage could have been due to the influence of more prominent families or to other family background factors, it is likely that much of the difference was due to what was taught and learned in high school and not to who went to high school.

In 1915, the year of the Iowa census, the high school movement was just beginning and thus older individuals surveyed by the census would have gone to school *before* the enormous spread of public high schools throughout rural America. The average 45- to 54-year-old native-born male had 8.8 years of education and 18 percent stated that they had attended some high school. The average 25- to 34-year-old native-born male had 9.2 years of education and slightly more than a quarter said they had attended a secondary school.[8] Their high school education was a tribute to their eagerness to obtain an education beyond that offered in the common school since many who attended high school before the spread of public secondary schools had to board in a nearby town.

Ordinary White-Collar Occupations in the Early Twentieth Century

The enormous growth in white-collar occupations during the early twentieth century was the product of several factors. It was fueled by increases in the size of firms in the manufacturing, public utilities, communications, and transportation sectors as well as by the complexity of their activities. The appearance of larger firms with far-flung operations and an increasingly complex division of labor raised the demand for ordinary office personnel, sales workers, and managers to process information, coordinate production, and interact with customers.[9] Advances in information technology and office machinery (such as the typewriter, adding machines, and improved filing systems), as well as reductions in communication costs because of the telephone and improved telegraph networks, further stimulated the demand for white-collar workers. The increased demand was also induced by changes in the structure of the economy that increased employment in particular sectors, such as banking, insurance, real estate, communications, and retail trade. Employment in these sectors consisted almost entirely of white-collar workers. Retail stores became larger and this, too, heightened the demand for sales employees rather than owner-operators.

The largest single increase in office workers around the turn of the twentieth century was in the manufacturing sector. Even though employment increased greatly in sectors that hired office and other white-collar workers exclusively, manufacturing provided the largest increase in white-collar workers. Manufacturing was, quite simply, a very large economic sector and the employment share of its nonproduction workers expanded significantly during the early decades of the twentieth century. Nonproduction workers, as the name would imply, include primarily those in office work, sales, and management.

In 1899 nonproduction workers were just 7 percent of all manufacturing employees, but in 1909 they were 14 percent. In just one decade the fraction of nonproduction workers in manufacturing doubled. Almost a half century later, in 1954, the employment share of nonproduction workers in manufacturing was 21 percent, just 7 percentage points higher than in 1909.[10] The growth of nonproduction jobs around the turn of the twentieth century was spectacular both absolutely and also in comparison with that during the next half century.[11]

The increased demand for office personnel was met, at first, by hiring more workers into pre-existing positions. In an office setting, this entailed hiring more secretaries, bookkeepers, stenographers, typists, and clerks. These pre-existing positions often demanded highly skilled and educated personnel. The secretary of the late nineteenth century was the "keeper of the office secrets," the right-hand to the company president, and an office worker who had to master a variety of tasks. A nineteenth-century office bookkeeper, to give another example, was the firm's principal accountant rather than someone who kept the accounts of just one division and summed long columns without ever seeing the entire balance sheet.[12] These workers were to the office what the artisan had been to the shop floor—they were familiar with the *entire* business or product.

An exhaustive list of office occupations in a large group of manufacturing, trade, and insurance establishments in 1895 reveals fewer than 15 separate positions of any quantitative importance. The office jobs listed in manufacturing and trade were those of bookkeeper, cashier, clerk (including billing and shipping clerks), messenger, office boy, salesman, stenographer, and typewriter. Chief clerk, loan inspector, and policy writer are additional occupations listed for insurance companies.[13]

With the increased demand for office work, many of its jobs became rationalized, routinized, and subdivided. The producers of office machinery responded to the increased demand for office work by vastly expanding their product lines. The equipment required of an ordinary office rapidly proliferated from the 1890s to the 1920s.[14] Standard office equipment in the 1910s included typewriters, dictating equipment, adding machines, carbon paper, and filing systems. The typical office of 1924 had far more. There were dictating, bookkeeping, calculating, addressing, duplicating, and automatic typing machines, in addition to the typewriters, adding machines, filing systems, and business forms of the previous decade. Some offices even had tabulating machines, a forerunner of modern computing systems.[15] Office equipment catalogues from 1924 reveal that the mechanization of the office had advanced so far that it was virtually complete, at least for some time to come. Technological change in the office had taken off from around 1915 to the early 1920s and revolutionized white-collar jobs, particularly for young women. Although there were just 15 separate office jobs defined in 1895, there were 45 in 1922, and more than 100 by 1940.[16]

The industrial revolution had come to the office. The secretary of the mid-nineteenth century had been a corporate officer in the making and the bookkeeper had managed the books for the entire firm. The office worker of the early twentieth century, however, was afforded a narrower purview of the firm's entire operations.

Even though the division of labor in the office reduced the skill required in each of several occupations, the work of the office nevertheless required far greater skill and education than that demanded of most workers on the production floor, and probably more than that ordinarily required in agriculture. Therefore, the use of office machinery and the division of labor in the office greatly increased the demand for high school educated workers. Put another way, the transition from the all-purpose nineteenth-century office worker to the single-task twentieth-century clerical worker decreased the skill required for the average office worker, but the overall skill demand was more than made up for by the expansion in the numbers of office jobs of which all required some modicum of education.

The education level that most of these jobs demanded was that provided by secondary school. Office workers were required to have mastered the essentials of literacy, such as proper grammar and correct spelling. Dictated letters often had to be redrafted by the stenographer. A foreign language or two was often considered useful and, for many jobs, mathematical skills were beneficial.[17]

In the late nineteenth century the vast majority of office workers were men, even among those who were clerks, typists, stenographers, and secretaries, a group known collectively as ordinary office or clerical workers. In 1890, for example, 81 percent of ordinary office workers were male.[18] But women rapidly increased their numbers in each of these occupations and their fraction in the total group soared. Whereas 19 percent of ordinary office workers were female in 1890, 24 percent were in 1900, and 35 percent were in 1910. By 1920 almost half of all clerical workers (48 percent) were women.[19]

One might think that the increase of women in office occupations from 1890 to 1930 came about because of an increase in the women's overall labor force participation rate. But that was not the case.

Women were just 18 percent of the total labor force in 1900 and 22 percent in 1930. Rather, the increase in women's employment as white-collar workers was due to an enormous shift in white-collar occupations among all workingwomen. Put another way, just 4 percent of all working women were ordinary office workers in 1900 but almost 20 percent were in 1920. Women accounted for almost 60 percent of the fivefold increase in the number of clerical workers from 1900 to 1930 and they accounted for 41 percent of the almost threefold increase in all white-collar workers during the same period.

Even though the occupations that increased to the greatest extent in the early twentieth century were not among those requiring the most schooling, almost all required some years of high school if not a high school diploma. As the demand for office workers increased in the early twentieth century, young women flocked to high school. They went to secondary school in greater numbers than did young men and they graduated from high school to a far greater extent (detailed in Chapter 6). In the absence of a high school education, a young woman had rather bleak job prospects. She could become a manufacturing operative or a domestic, neither of which was an occupation for a respectable woman and both of which paid little. A young man, on the other hand, could become a manual worker and earn not much less than he could as an office worker. The 1915 Iowa State Census reveals that the pecuniary return to business school training was greater for a young woman than it was for a young man and years in high school, if not a high school diploma, was often a complement to business school training or business college, as these schools were often called.

In the popular literature of the early twentieth century, the white-collar employee was seen as a fresh new face and was afforded a prominent place. Office workers were frequent magazine story characters. The most distinguished authors of the early twentieth century incorporated the new middle-class clerical workers in their novels and stories. O. Henry wrote about an itinerant woman typist, John Dos Passos featured a stenographer in his novel *1919*, and Sinclair Lewis had a female stenographer ascend the corporate ladder in *The Job*.[20] These were not the pitiable factory women who were the heroines of a previous literature. Rather, the men and women of this literature formed a new middle class to which many young people of the day aspired. To attain their goal, they went to high school.[21]

Manual Jobs in the Early Twentieth Century

Formal Education and Manual Jobs

Most white-collar employees, certainly by the turn of the twentieth century, arrived at their first job with general skills taught in high school. At the same time, employers of *manual* workers in the newer industries of the day also began to demand the types of general skills provided in secondary schools. By the first decade of the twentieth century, certain blue-collar positions required a few years of high school and some even called for a diploma.

A contemporary account of the role of education in industries, such as machine building, noted in 1908 that: "The boy who goes into the shop in his early youth . . . should understand . . . mechanical drawing, algebra and geometry, and have a fair command of the English language."[22] Firms with more scientific content in their production processes had a strong inclination for production workers with general skills. By the late 1910s they had a stated preference for those with an ability to decipher manuals, knowledge of algebra, a mastery of mechanical drawing to read and create blueprints, and an understanding of the elements of chemistry and the fundamentals of electricity.[23]

These requirements were a far cry from what was being demanded of ordinary operatives in many industries that hired large numbers of newly arrived immigrants. Physical characteristics, such as muscle and might, not cognitive skills, were often all that mattered. The skill divide between the white-collar and the blue-collar groups had begun to fade, a bit.[24]

Technological change in the early twentieth century shifted the economy's industries from older ones, such as cotton textiles, to newer ones, such as chemicals. Older industries that demanded scores of operatives, most of them unskilled, decreased in importance relative to those that used continuous-process and batch technologies which had greater needs for more skilled personnel. The newer industries of the age, such as petroleum refining, photography, automobiles, and aluminum, as well as certain older ones that had undergone technological changes, such as steel, sugar refining, canning, soap, and paints and varnishes, had greater demands for more educated workers than did most others. Several firms, such as National Cash Register and Deere

Tractor Works, were in the forefront of a larger group that looked to the high school educated boy to provide the talent to use, maintain, and install machinery.[25]

The manager of the Deere Tractor Company in 1902 made it clear that he would "not take boys in the office unless they are at least high school graduates." The head of the employment department at National Cash Register Company in Dayton, Ohio, noted in 1902 that: "The office boys must . . . have high school education, or at least must have had two years of such." He added: "In the factory we like the boys to have a high school education if possible."[26] It had become customary to hire office workers with a high school education but it was, as yet, more extraordinary to hire production or manual workers with that level of education. However, that too, would soon become routine in the newer and more technologically advanced industries in the 1910s and 1920s.

According to a historian of the U.S. machine tool industry, "by the end of the nineteenth century the machine tool industry had developed to the point where a more reliable supply of skilled workers was needed who were able to read engineering drawings and written instructions."[27] Mechanical engineers, according to the premiere historian of the occupation, increasingly came from "the high school, not the shop, and they often had a disdain for the dirt and roughness of the shop."[28]

America never developed the extensive apprenticeship system in manufacturing and the crafts that Britain and certain continental countries did. The apprenticeship programs that had existed in the United States were, by the late nineteenth century, in considerable decline.[29] By the early twentieth century only a small fraction of manufacturing workers had ever been apprenticed, but a few apprenticeship programs did persist into the early twentieth century.

A few technologically advanced firms from the 1900s to 1920s required a high school diploma or some years of high school of their apprentices. The General Electric Company, for example, required a high school diploma for young men who were being trained as draftsmen, designers, electrical and steam turbine testers, technical clerks, and manufacturing and erecting engineers in their apprentice program. Youths who had not completed high school but had mechanical abilities could become apprentices, but not for the most lucrative and elite craft positions. They could, instead, train to be machinists, tool and die makers, pattern makers, steamfitters, blacksmiths, and iron, steel, and brass molders. The National Cash Register Company took only apprentices who had finished at least two years of high school. These young men were taught to be tool, model, cabinet, and pattern makers; electricians; pressmen; designers; compositors; and electrotypers.[30]

All of General Electric's apprenticeship programs had a schooling component. The young men who had only a grammar school education had to learn algebra, mechanical drawing, geometry, plane trigonometry, elementary physics, and practical electricity, among other courses that became standard in most of the nation's high schools by the 1910s. The apprenticeship program for those who had not attended high school, therefore, mimicked parts of the contemporary high school curriculum. For those who had graduated high school, the required courses in the apprenticeship school included advanced algebra, advanced electricity, thermodynamics, and machine design.

Farmers, too, began to recognize that a formal secondary-school education was of value in learning about new crops, animal health, fertilizers, machinery, and accounting techniques. Such knowledge was becoming essential to the running of a modern farm. *The Twentieth Century Farmer* noted, in 1905, that "it takes a better trained mind to be a successful farmer or business man today than it did even ten years ago. . . . The new conditions of life on the farm demand improvement in general education for the farm boy or girl."[31] Farmers with more education adopted technological advances at a faster rate than did others, as the example of hybrid corn demonstrated.[32] Although the diffusion of hybrid corn began in the late 1920s, more than a decade later than our other examples, it provides one of the finest illustrations of the benefits of education to farming.

General Training: The Case of Telegraphers

In the mid-nineteenth century the mastery of a specific skill, which could be acquired without a formal education, was often a steppingstone to success in the business world. One of those steppingstones was telegraphy.

Telegraphers were bright, energetic young men and women who learned the trade from others in the telegraphy business. From the

1840s to the 1860s, a telegrapher was a well-rounded worker who was often the "clerk, bookkeeper, battery man" in an office and who could even "fix the lines in a pinch."[33] Telegraphers often rose through the ranks of a business. Among the most famous of the self-made men of telegraphy was Andrew Carnegie, who began as a telegrapher in 1845 at age 14 and was promoted through the ranks of the Pennsylvania Railroad. Thomas Edison was another supremely successful individual who began his career "at the key." Although Edison, who became a telegrapher in 1863 at age 16, never rose through the ranks in the manner of Carnegie, he used his talents to improve telegraphy.[34]

But just 20 years later, by the 1880s, telegraphers were "mere manipulators of a key."[35] Some managed to continue the tradition of rising through the ranks but, for most, the job had become a dead end. No longer did telegraphers learn the trade primarily as apprentices. Rather, telegraphic day and evening schools spread rapidly throughout the United States and trained scores of young men and women to be dedicated manipulators of the key. Only the most extraordinary young telegraphers could amount to much more without further education.[36] The highly specific skills of the telegrapher had become less valuable relative to the general, portable, and flexible skills of the high school or college graduate.

The point we are making is that specific skills, such as telegraphy in the nineteenth century and computer programming more recently, were picked off by proprietary schools and taught as separate trades. When skills are used in separate tasks, the job can easily become a dead-end one and is no longer a gateway to higher positions. Only with a more general, flexible education could a youth prove himself and rise through the ranks.

Immigration as an Impetus to Formal Schooling

With the enormously increased immigration to the United States starting in the early 1900s and ending abruptly with the outbreak of World War I there was additional reason for parents to desire more schooling for their children. The new wave of immigrants from southern, eastern, and central Europe flocked to urban manufacturing jobs and depressed wages. Increased competition existed not only for the low-skill, muscle and brawn jobs but also for craft positions.[37] The new immigrants often arrived from Europe with skills in specific trades and crafts against which native-born youths could not easily compete.

The foreign-born had a stronghold on, and in some cases *de facto* monopolized, certain trades in which they may have served as apprentices in their native lands or worked at before they left for America. Whereas the foreign-born were 42 percent of all workers in manufacturing in the northeastern states, they were 80 percent of its bakers, 67 percent of its textile dyers, 65 percent of its cabinetmakers, 61 percent of its textile weavers, 55 percent of its blacksmiths, and 50 percent of its jewelers. Various countries furnished particular tradesmen. The British were 3.3 times more likely to be jewelers than they were to be in any other manufacturing occupation, Italians were 3.3 times more likely to be textile dyers, and Scandinavians were 3.5 times more likely to be cabinetmakers.[38]

Parents realized that a high school education could enable their children to become white-collar employees, often referred to by parents as "business work." Even if their children were not destined to be in business or be white-collar workers, parents recognized that higher education could lead to work as manual employees in occupations that used the formal skills acquired in school and that were, therefore, less subject to competition from immigrants, most of whom had spent only a few years in a European common school.

Compared with the foreign-born, native-born blue-collar workers in 1920 were found disproportionately in manufacturing industries that were "new" or "high tech," as we would call such industries today. They also worked in occupations that used their English language skills. The native-born, either of native-born or foreign-born parents, were disproportionately compositors, typesetters, and white-collar employees such as bookkeepers, clerks, and managers. None of these findings is unexpected. What is far more surprising is that the native-born were disproportionately electricians, plumbers, and pipe fitters in the building trades; rollers, tinsmiths, coppersmiths, tool makers, and structural metal workers in the metal trades; and mechanics and stationary engineers in various industries.[39]

Native-born manual workers were found disproportionately in certain industries such as metal trades and the more modern construction occupations that were an important part of the "new" economy of the early twentieth century. The foreign-born, on the other hand, were disproportionately in the older wood, cloth, and food processing industries. The native-born were not just in industries and occupations that favored their language skills and general knowledge of America, they were in occupations that used the formal skills they garnered in schools.[40]

Returns to Education before the High School Movement

Occupational Wage Premiums, 1820s to 1910s

The data on the earnings of clerical workers relative to production workers in manufacturing, presented in Chapter 2, Table 2.2, revealed a slight upward trend for women from 1890 to 1914 and general stability for men. The main conclusions we drew from those data were less concerned with the trends during the period and more focused on the levels. The point we drew was that there was a substantial occupational wage premium from the 1890s to the 1910s for both men and women and that the implied returns to a high school education were substantial. We now demonstrate that returns to schooling were probably just as high in the early nineteenth century.

Relative to female production workers, female clerks earned 85 percent more in 1890 and greater than 100 percent more in 1914. Male clerks earned 70 percent more than production workers in 1914. If the average production worker had an eighth grade education and the average clerk had a twelfth grade education, then the return to a year of schooling would be well over 20 percent. The implicit return to becoming a clerk around the turn of the twentieth century was extremely high for both males and females.[41]

How do these ratios compare with those for the rest of the nineteenth century? Did the return to education rise during the nineteenth century as the demand for educated workers increased? This is a difficult question since data for white-collar earnings in the nineteenth century are scarce. Fortunately, there is a series for 1820 to 1860 assembled by Robert Margo (2000).[42] The white-collar workers in the sample were civilian "clerks" working in U.S. forts. These clerks were a mixture of lower-skilled clerks and relatively high-skilled bookkeepers and managers who did the purchasing for the forts. The forts were often in settled parts of the United States, but some were closer to the frontier. The method for constructing the annual means from the data, called a hedonic regression, allows one to hold the observable differences constant, such as occupation and place. Forts also hired common laborers. Thus the data allow one to obtain a skill differential in earnings: the ratio of the earnings of ordinary clerks to the earning for common laborers. The ratio for the early period appears to be similar to that presented in Chapter 2 for the later period.

The ratio of clerk earnings to laborer earnings was 1.93 for the 1826 to 1830 period and 1.99 for 1856 to 1860.[43] To make a comparison with data from the period around 1900 involves one small change to the earnings ratio just presented. Clerical earnings for the 1890s and beyond were compared with production workers earnings, mainly because of their numerical importance; however, for the antebellum period, the clericals have been compared with ordinary laborers, mainly because of their predominance at the forts and in the general economy.

If the male clerk data for the later period are compared with laborers, the resulting ratios are 2.50 for 1895 and 1.96 for 1914.[44] Thus, the ratio of the earnings of ordinary clerks to laborers was somewhat higher around the turn of the twentieth century (in 1895) than in the antebellum period, but about the same in the early period as in 1914. The decrease in the ratio from 1895 to 1914 is due to a sharp increase in the relative earnings of the lowest paid workers in manufacturing. Since the quantitatively important comparison is between production workers and clerical workers and that ratio remained fairly constant from 1895 to 1914, it is likely that the returns to skill were somewhat higher around the turn of the twentieth century than in the antebellum period.

Because we do not know precisely how the skill ratio changed across the nineteenth century and into the early twentieth, we must remain somewhat agnostic. It seems clear that the ratio did not rise by much and even more obvious that it did not decline. The quantity of skilled workers, such as clerical, sales, managerial, and professional workers, increased greatly during the period (Table 5.1). Thus the relative quantity of skill expanded rapidly while relative wages were increasing slightly or were stable. Putting these facts together within the simple supply and demand framework of Chapter 3 implies that the demand for skills was growing at a rate greater than that of supply or about the same. The most relevant and robust evidence we have uncovered is that the return to skill was relatively large throughout the period examined.

Returns to Formal Schooling c. 1915

By the early twentieth century, according to our estimates, relative earnings for various white-collar workers in the clerical group were high and the number of workers in the more educated occupations had increased relative to all others. The evidence we presented to make these points concerned occupations, such as those in the clerical sector and in offices of all types, that were known to have hired more educated workers relative to those in many of the manual trades. We used such evidence because direct information on the educational attainment of workers and their earnings is not available for a large national sample until 1940, when they are included in the federal census of population. At the state level, as we previously mentioned, the Iowa State Census of 1915 provides similar evidence and affords a unique snapshot of the difference in earnings by educational level for men and women of various ages. The data also reveal differences in the occupations of the more and less educated as well as the special role of the high school. We presented information in Chapter 2 on the returns to education both between and within occupations. We will now review the evidence that is germane to the issue addressed here concerning why the high school movement began in the early twentieth century and what its impact was on the structure of earnings.

The private, pecuniary return to a year of secondary school in 1915 was 10.3 percent for employed men 18 to 65 years old (see Chapter 2, Table 2.5). It was 12.0 for younger men, 18 to 34 years old, and 10.1 percent for younger unmarried women. Not only was the return to a year of high school substantial for men across all occupations, it was also high *within* various occupational groups. The return was slightly larger within all blue-collar occupations (0.091 for 18- to 34-year-olds) than it was within all white-collar occupations (0.083). Therefore, even though there were pecuniary benefits to education in enabling individuals to shift to white-collar occupations, the benefits of more education also accrued within occupational groups. The return to a year of high school, in the data just offered, dropped minimally from 11.4 percent for all non-farm occupations to 9.1 percent for blue-collar occupations. A high school education may have enabled some young men to become bookkeepers and office managers, but it also imparted useful skills that enabled others to become electricians and automobile mechanics.

Youths in 1915 had enormous reason to continue with their education and attend high school. But even in a high-education state like Iowa, most adults in 1915 would not have had a public high school in their district or even in their township when they were of school age. Many who wanted to continue with their education would have been forced to advance their education in their local common school. Such attendance was legally ended in 1912 in Iowa when the state legislature passed a law against the provision of secondary school education within the common schools. Only stand-alone high schools having more than one teacher could instruct in the upper grades.[45]

Many older youths, prior to 1912, remained in common school after eight years in an attempt to learn the material of the upper grades. The information in the Iowa State Census reveals that the additional years youths spent in the common schools produced a rather low rate of return. The return to a year of common school education yielded a 4.5 percent return up to nine years. After nine years the return fell to 2.9 percent (see Chapter 2, Table 2.5, col. 2). A year of high school, on the other hand, produced a return of 11 percent per year or three times what would be garnered from remaining in the common school.

The instruction Iowa youths received in high school was considerably more valuable than that received in the common schools even though many of their high schools were small—some having no more than 40 to 60 students in total. The high school curriculum, even in the smaller high schools of Iowa's tiny towns, differed from that offered in the elementary grades and common schools, and the teachers had considerably more advanced qualifications.

What we have tried to demonstrate here is that the returns to secondary schooling were substantial in the nineteenth century but that the demand for educated workers was insufficient to establish public high schools in all but the most populous cities. Private academies, institutes, and seminaries took the place of public secondary schools. Although these schools were generally short-lived institutions, and their histories have been shrouded in mystery, we now delve into this important moment in U.S. educational history. These schools were the precursors to public high schools and their emergence reveals the grassroots origins of the high school movement.

Academies: Precursors to Public High Schools

Varieties of Academies

Even as the common school was diffusing throughout antebellum America and spreading in the Northeast and Midwest, the demand for the subsequent step in education was emerging. The next level took different forms in the nineteenth century and as it evolved its name changed. It is now called high school or secondary school, but it was once termed grammar school, preparatory school, academy, institute, or seminary, especially for females.

College and university attendance, before the early twentieth century, required a particular type of preparation that was provided by grammar schools such as Boston Latin, academies such as Phillips and Andover, and the preparatory divisions of universities. But the burgeoning demand for secondary schooling in the mid-nineteenth century was often for a largely different purpose. It was a way of training young men and women for an occupation, not necessarily for college, and it was seen as a means of preparing young people for life in general.

In the nation's larger cities, the demand for more years of schooling led to the establishment of public high schools, but in the less populated sections of the nation, demand was insufficient in the nineteenth century to support public secondary schools. Rather, the increased demand for more education led to the growth of private secondary schools, often called academies. The precise number of academies that existed in the past cannot be determined with great accuracy even though the U.S. federal population censuses from 1850 to 1870 inquired about the schools at the county level. Only classical academies, those training students for college, were recorded in the 1870 census. In the two prior censuses both classical and nonclassical academies, it appears, were included. More vexing is that many private elementary schools were included in the academy figures for 1850. Despite these data problems we do have some sense of the relative importance of these institutions.

Because the nation was sparsely settled, classical academies in the mid-nineteenth century outnumbered public high schools, even though the average urban public high school had somewhat more students than did the average academy.[46] The number of public high schools probably did not equal the number of private academies until the 1880s, and the number of public high school students probably did not exceed those in all types of private academies and preparatory departments of universities and colleges until the 1890s.[47]

Private secondary schools were of many types. At one end of the spectrum were institutions offering a classical course that would prepare young men and women for college. Some were elite institutions that were loosely affiliated with a college and prepared students for the college's entrance exam. Others were less prestigious schools that offered a range of classical courses, occasionally in a coeducational setting but more generally separately by sex. At the other end of the spectrum were schools that offered a combination of academic and commercial training and those that gave courses mainly in vocational fields or in music and the fine arts. Small schools often consisted of a schoolmaster and his students, who were taught in his residence.

The information that exists on academies is thin and much is elusive. For one, the nonclassical academies were small affairs and most survived for just a few years. They were transitory institutions that left almost no permanent records. Even the classical academies are difficult to track. For New York State, however, the data on the classical academies are reasonably complete even for the early nineteenth century. The New York State Regents, which oversaw the state's educational institutions, provided public funds on a per student basis to classical academies and had reason, therefore, to collect information about them, including their curriculum.[48]

The courses offered by academies speak to the training that youths wanted to receive, or at least to the training their parents wanted them to have. Academies offered academic courses, such as English, history, mathematics, geography, foreign languages, natural sciences, physics, zoology, and astronomy. The classical academies by the 1840s even provided vocational courses in bookkeeping, surveying, stenography, drafting, optics, law, civil engineering, and navigation that often built on the more academic subjects and offered a range of nonacademic courses including dance, calisthenics, and music.[49]

Relative Importance of Academies

Much of the historical evidence that has been cited on academies comes from a special survey of the U.S. Census.[50] In the three census years—1850, 1860, and 1870—the Census had its agents record a host of "social statistics" at the county level, in addition to collecting information from individuals, farms, and firms for the censuses of population, agriculture, and manufactures.[51] The data they were asked to obtain concerned wealth, churches, libraries, poverty, wages for various occupations, and schools. With regard to the latter, agents were asked to obtain data about the numbers of teachers and pupils, and financial information for all public schools, private schools, and colleges.

The data were summarized by state in the printed volumes of the U.S. Census. Because of the paucity of other evidence, these published summary statistics have been widely used for historical research on academies. But the summaries were often incorrect and must be interpreted, as we do here, by consulting the original manuscripts.[52] The errors in interpretation have led to a substantial overstatement of the number of students attending academies in the mid-nineteenth century. We correct these data and find that even though the amended data are far lower than those often cited, they indicate, nonetheless, a substantial demand for secondary schools at a time when public high schools were few in number and were found only in the nation's larger cities.

The most widely cited data on academies is that for 1850.[53] The summary in the 1850 census of population reported that there were 6,032 "academies *and other private schools*" nationwide with 12,297 teachers and 261,362 pupils.[54] The data, if correct, would imply that the fraction of 15- to 18-year-olds attending academies in 1850 was more than 15 percent if academies were four-year schools (double the number if they were two-year schools). But the figures are *not* correct, largely because they were not restricted to academy students. Inspection of the extant manuscripts from the Census of Social Statistics for 1850 indicates that the vast majority of the students listed were in the lower grades and in the South virtually all were.[55] Most, it is clear, were attending "other private schools," not academies.

Rather than using either the 1850 or comparable 1860 data, both of which compound students in academies with those in the lower grades in private schools, we use the 1870 data which do not. In 1870 the

Social Statistics asked census takers to list private schools in various categories. "Classical academies" were listed separately, as were commercial, music and art, technical, parochial, and "day and boarding" schools. It is probable that the number of private secondary school students is understated by using only the academy data, a point we will address shortly.[56] The bias, however, will reinforce our conclusion that enrollments of post-elementary school youths in public high schools were small relative to those in private schools and that there was a substantial demand for public secondary schools as shown by the enrollments in private academies.

Nationwide in 1870, according to the census data, about 6.5 percent of 15- to 18-year-olds were attending either a public high school or a classical academy. That figure is considerably smaller than the one mistakenly derived from the 1850 and 1860 data.[57] But it is large for the mid-nineteenth century and is considerable since the populations of the South and the West are included. The Northeast, for example, had an enrollment rate in public high schools and classical academies of almost 8 percent.

Of equal interest is that the fraction of all secondary school students attending public high schools, according to the 1870 census, was just 38 percent for the entire United States and less if all nonclassical private schools were included. It was 55 percent in Iowa, a state that already had many public high schools and would soon be a leader in high school education. Our point is that even though a substantial fraction of young people were attending secondary schools, less than two-fifths were in public high schools. Private schools, such as classical academies, were serving an important function in educating young people and demonstrating the demand for *public* high schools.

Classical academies and public high schools alike, in 1870, were relatively small affairs by later standards. There were larger institutions in the bigger cities, to be sure. But in the Northeast, where we have the best data from the census manuscripts, there were on average three to five instructors per school and 60 to 90 students in attendance.

In addition to the larger and more prestigious schools reported in the 1870 Census, smaller ones existed. Some were unincorporated classical academies and others were institutes that taught a combination of academic, commercial, and vocational subjects. We will never know their exact number, but we do have ample evidence of their presence. In New

York State, for example, the 1870 Census listed only the academies that were recognized by the Regents. But advertisements in mid-nineteenth century newspapers reveal the existence of smaller institutions that were apparently not included in the census count.

Many of the smaller schools held classes in the teacher's home, as was the case of the "Family School for Boys" in Prattsville, New York, which announced that the "Rev. Wright, Rector of Grace Church will receive into his family a few boys to be educated with his own sons" and "fitted for college or business." The same was the case of a Miss Havens, who was reported to "reopen her school for young ladies in the home of her father."[58] Mr. Bingham's school, it was claimed, regularly "fitted pupils for Harvard and Yale Colleges" as well as prepared them for business. At George S. Parker and John McMullan's "pupils are prepared for college or the counting room."[59] Most of the schools that advertised were academic, although some were purely commercial and others specialized in music or drawing. Clearly academies and other private post-elementary schools were widespread and, we presume, the number of youths attending them in the mid-nineteenth century was substantial, even though our ability to estimate their numbers is severely hampered. It seems reasonable, however, to treat the 1870 Census numbers as a lower bound on the number of youths attending private secondary schools.

One important difference between academies and public high schools was that academies almost always charged tuition and public high schools generally did not.[60] Although academies often obtained public funds and had income from endowments, these grants were relatively small. Private schools depended almost entirely on tuition, which varied from school to school.

Elite preparatory schools and military academies charged in the hundreds but most were considerably less expensive. In New York State the median tuition for academies was $35 per student-year.[61] In Massachusetts the median academy charged $54 per student-year and in Iowa it was just $25. Income per capita in 1870 was around $800 for a skilled worker and about $900 for an office worker, such as a bookkeeper.[62] Thus tuition for one child would have been about 5 percent of gross earnings for even a skilled worker, and tuition was only half, or less, of the total expense of sending a child to an academy since most youth were boarded. Board generally raised the total expense to around $150 to $200 for the full year, probably less in the smaller schools where pupils boarded in the master's home.[63] The cost of an academy, therefore, was substantial in relation to the average income of Americans in the late nineteenth century.

The transition from private to public schools was a change from schools that charged fairly steep fees per pupil to those that charged nothing to the student's family. It shifted students from a school building that was sufficiently distant that most youths had to be boarded to one that was part of a community, with boarding for some students. What became of the academies? As public high schools spread, local school districts leased or bought some of the academy buildings and transformed them into local high schools.[64] In other cases, the academies disappeared without a trace.[65]

Academies emerged because of a grassroots movement that was generally uncoordinated by the state and largely free of the influence of school propagandists, such as the "schoolmen" of the common school revival period. The public high schools that coexisted with the academies in the larger cities, and that eventually displaced them nearly everywhere, were also largely grassroots institutions. Both the academies and the public high schools were called forth by the demands of parents. Shopkeepers, merchants, and professionals of various types wanted their children to have skills—such as greater literacy and numeracy—to continue in their businesses or embark on some other.[66] Parents who were manual workers wanted their children to be freed of physical toil and those who were farmers recognized that some of their children would be unable or unwilling to work the family's land.

Thus academies by the mid-nineteenth century were quantitatively important educational institutions. By sending their children to academies, parents affirmed their desire for secondary schools. The size and extent of the academy movement demonstrates the enormous latent demand for public secondary schools. While the impetus for the public high school was individual and distinctly grassroots in origin, more than just grassroots action was needed. What was required was a legitimate basis for raising public revenue.

The Role of Compulsion

We have demonstrated that a series of fundamental changes beginning in the nineteenth century increased the demand for more educated workers and thus heightened the desire by youth for higher levels of formal education. The increase in education was initially supplied mainly by the private sector through academies and other private schools. The economic return to a secondary school education was substantial in the nineteenth century, but publicly provided high schools had to await a thickening in demand. Demand for more educated workers eventually soared and communities, one by one, built high schools. The high school movement was the consequence.

A different view of the early spread of public secondary schooling holds that the state had an interest in legislating against youthful delinquents and wanted, as well, to control the children of the burgeoning immigrant population in the nation's larger cities. Compulsory schooling laws and child labor laws were passed in many states beginning in the mid-nineteenth century. Some view the high school movement as a direct result. Massachusetts, in 1852, was the earliest state to have a compulsory schooling law. By 1890, 27 states had passed a compulsory schooling law and in 1910 41 had one.[67] Child labor laws of varying type were in force in 40 states by 1910.

But the laws were not stringent enough to have had much impact in the period up to about 1910.[68] Enforcement was weak and the laws often exempted youths who had only minimal education. Compulsory schooling and child labor laws were not a major impetus to the rise of public secondary school education around the turn of the twentieth century. The impetus was provided by the middle classes that sent their children to private secondary schools and switched them into the public high schools when the schools became available.

Even though the laws did not spark the movement, they may have subsequently increased secondary school attendance and allowed the movement to flourish. After all, it was not until the early twentieth century that compulsory education and child labor laws could have had a large impact on secondary schooling. After 1910 the ages that the laws were intended to constrain increased, the bureaucracy to enforce them expanded, and the education and labor portions of the laws were better coordinated. All of these changes should have given the laws new bite to constrain the behavior of youth of secondary school age. We demonstrate in Chapter 6, however, that the actual impact of compulsory schooling laws was small in increasing high school enrollments from 1910 to 1940.[69]

From Economic Imperatives to Educational Outcomes: A Summary

There were abundant reasons why parents wanted their children to continue to high school in the first decade of the twentieth century and why youths should have remained in school. The demand for educated workers had increased, the earnings of educated workers such as those in offices were relatively high, and the newer and more technologically advanced industries were demanding greater levels of education from production workers. Even farmers recognized that schooling mattered to the livelihoods of their children. The Iowa data for 1915 reveal that the pecuniary return to a year of secondary school was substantial and that years in secondary school endowed youths with more marketable skills than did additional years in the local common and elementary schools.

Success meant more than the ability to earn a greater income. By enabling their children to obtain more education, parents could free them from various hardships in life. If they became blue-collar or production workers, more education would help them escape the harsh conditions and long hours of the older industries. If they, instead, became office and clerical employees, more education would enable them to work shorter hours and have better working conditions regardless of their industry. Education was also a form of insurance, allowing the more educated to respond faster to economic change and thereby providing some unemployment protection.

The increased demand for more education could be satisfied in the nation's cities by enrolling more youths in existing high schools. In 1915, virtually all cities with populations in excess of 10,000 had at least one public high school and the vast majority of cities with populations above 5,000 did.[70] But the majority of the nation's youth in 1910 did not have a high school in their school district or township and African American youths in the segregated South had virtually no secondary schools. Even in cities that contained public high schools, many youths lived too far from it.

The next several decades would witness a veritable explosion in the number of secondary schools and students. In thousands of small communities across America the demand for public secondary school became sufficiently great that people were willing to tax themselves to provide it. Those who did not want to take part in the educational advance could choose to move to another community, but many found that property values were enhanced when the community had a fine secondary school and so chose to stay and support the school.

The method of gaining public support for an expensive quasi-private good is well summarized by the following pro-education statement at the start of the high school movement:

> The landlord who lives in town . . . may well be reminded that when he offers his farm for sale it will be to his advantage to advertise, "free transportation to a good graded school." Those who have no children to attend school . . . should be interested in securing to the children of the whole community the best educational advantages possible . . . if they live out their years with no children to depend upon in old age, they must of necessity rely upon someone, they know not whom, who is today in the public schools. Their only safeguard lies in giving the best advantages possible to all.[71]

Urban property owners often wanted high schools as an attraction for newcomers who would increase demand for their land, homes, and businesses. "Somerville," remarks Reed Ueda, "began its high school rather early [c.1850], in part to attract families to this relatively wealthy suburb."[72]

Secondary schools were called into existence by parents, youths, schoolteachers, district administrators, and state legislators. "The idea appeals to the people," noted a California school report, "and they respond to it with a promptness and generosity."[73] An Iowa school report of the 1890s noted: "Many a boy unable to go away to college, obtains at his home high school an education to serve him well in the affairs of life and to increase his capacity for happiness."[74] The report termed the schools "the colleges of the common people."

With schools that were "free" to the user, although costly to the taxpayer, the pent-up demand for education beyond the common school was fulfilled and multitudes entered the schools. The rapid increase in secondary schools and student enrollments would continue for the three decades from 1910 to 1940. We now turn to the remarkable second transformation of education in America—the high school movement—that was an institutional response to wide ranging economic changes.

6

America's Graduation from High School

"The rise of the high school is one of the most remarkable educational movements of modern times."
California, Office of Superintendent of Public Instruction (1913/14)

During the first half of the twentieth century the educational attainment of American youth greatly increased; almost 60 percent of that increase was due to the rise of high school education. Mass secondary schooling was indeed a "remarkable educational movement" and set America far ahead of other nations for decades to come, even the rich European ones. Greater levels of education enhanced economic growth and also led to a more even distribution of its benefits.

In this chapter we examine how the increase in education occurred. The growth of secondary schooling varied significantly across the nation and we explore the factors that caused some places to lead and others to lag. Although economic factors had increased the value of education to individuals, as discussed in Chapters 2 and 5, for secondary education to spread communities had to support public high schools.

Americans pioneered the modern secondary school in the early twentieth century, tailored it for the masses, and rejected more elitist European institutions. We assess the impact of U.S. high schools and the changes in their curriculum across the twentieth century. America's graduation from high school in the first half of the twentieth century set the stage for the third great educational transformation, that to college. But we must first explore the high school movement. In this chapter we cover the period to 1950 (occasionally 1970); we review recent high school graduation trends in Chapter 9.

The High School Movement

In 1910, barely 9 percent of all American 18-year-olds graduated from secondary school and 19 percent of 15- to 18-year-olds were enrolled in a public or private high school.[1] By 1940 the median youth across the entire nation had a high school diploma and 73 percent of American youth were enrolled in high school (see Figure 6.1).

The increase in secondary schooling from 1910 to 1940 was extremely rapid, even more so for regions outside the South. In each of the non-South regions of the nation, with the exception of the industrial Middle Atlantic, the median youth was a high school graduate by 1935. Just 12 percent had been 25 years before. Enrollment rates in many states reached so high a level by 1940 that these rates would not again be exceeded until the mid-1950s.

The extraordinary increase in the education of the nation's youth is known today as the "high school movement."[2] It is dated as beginning in 1910, when it gathered steam in many rural parts of the country. It is conventionally deemed to have ended by 1940, when the median 18-year-old was a high school graduate, even though enrollment and graduation rates continued to increase until the 1970s in almost every part of the nation.

The high school movement is not just a latter-day historian's term. Rather, contemporaries used it to describe the educational changes of their day. "The rise of the High School," reported the California Superintendent of Public Instruction in 1910, "is undoubtedly the newest and most striking of recent educational movements. . . . The idea appeals to the people, and they respond to it with a promptness and generosity."[3] The educational commissioner of North Carolina, perhaps the most educationally progressive state in the South, noted in 1910 that: "The new public high school movement in the South did not become general nor well-defined until about four or five years ago."[4]

The significance of the zealous activity that built schools, hired teachers, enrolled youth, raised tax dollars, and revamped the high school curriculum was as clear to Americans who experienced it as it is to us with the benefit of hindsight. Americans were keenly aware that they were involved in an historic achievement and knew, as well, that they were setting the United States on a course far different from that being followed elsewhere in the world. They were embarking on an experi-

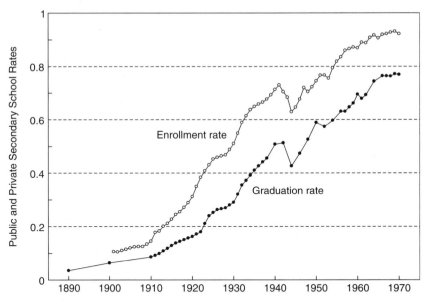

Figure 6.1. Secondary School Enrollment and Graduation Rates: Entire United States, 1890 to 1970. Enrollment numbers are divided by the number of 14- to 17-year-olds; graduation figures are divided by the number of 17-year-olds. Males and females in public and private schools are included. Year given is end of school year. Sources: U.S. Department of Education (1993) and Goldin (1998) for 1910 to 1930 graduation rates.

ment as grand as any in American history. For decades to come, no other nation would come close to putting as large a fraction of their citizens through secondary school.

High school education diffused rapidly across America in the early twentieth century. Even youth in rural and isolated places in America's heartland were, by the 1920s, within reach of a high school. Enrollment rates in the more sparsely populated parts soared once schools were established. But secondary schooling did not advance uniformly across the nation. In some parts—such as the Northeast, the West, and much of the central part of the nation—high schools spread like wildfire and youth flocked to them in droves, but in other parts the movement was delayed. The unevenness was geographic, racial, and ethnic. Oddly, any unevenness that was gender specific favored girls, who went to and graduated from high schools at considerably higher rates than did boys, at least until the Depression of the 1930s.

Geographically, there were two lagging areas in the high school movement: the industrial North and most of the South. Within some of the larger American cities and elsewhere in the industrial North, high school enrollment rates were comparatively low and advanced slowly until the 1930s. These low rates are curious because public secondary schools had been founded in most large U.S. cities in the nineteenth century. Unlike the small towns of rural America, these large cities already had high schools. The large flow of European immigrants to America's cities in the early twentieth century is partly responsible for the lower enrollment rates; but there are other reasons, including the job opportunities for youth, particularly boys, in cities. Whether their school decisions were due to capital market constraints, a curriculum that did not keep their interest, or impatience are other matters to consider.

The South was the other laggard in the high school movement. Youth in most parts of the South had the lowest rates of high school enrollment in the nation and this is true even if one excludes the African American population, whose schooling rates were abysmally low. The low rates for most southern youth were due, in part, to their more rural setting, but sparsely settled farm areas in the North and West had much higher rates of secondary school enrollment. The low enrollment rate for African Americans in the South is less difficult to explain. Until school integration was achieved, African Americans went to segregated schools and there were few public secondary schools for blacks until the 1930s. What schools did exist were in the larger cities, not in the rural places where most blacks lived.

The diffusion of secondary schooling was influenced by various factors, among them the taxable wealth of residents, the distribution of income, and the opportunity cost of youth employment. Because homogeneity of voters increased the probability that citizens would agree to establish a secondary school, the decentralization of schooling decisions helped propel the spread of high schools. In fact, the existence of thousands of relatively small, fiscally independent, and competing school districts increased educational expenditures as the high school movement got underway.[5] The decentralization of school decision making in America—even within states—is to be contrasted with the far more centralized decision making in Europe.

Despite the wide differences that existed in secondary schooling rates across America, the high school movement was nonetheless an extraordinary educational transformation. The transformation would soon make the United States the first nation in the world to have mass secondary school education. The process of institution-building did not appear *de novo*. Rather, states and communities built on prior commitments to mass common school education.[6]

The second transformation of American education was primarily a grassroots movement, just as the common school movement had been in the early nineteenth century and the academy movement had been in the mid-nineteenth century. Given the highly decentralized nature of American education, the spread of the high school was, by and large, an uncoordinated activity almost completely outside the realm of the federal government's control and not much affected by state compulsory schooling laws and by state child labor acts. However, particular state laws that governed a school district's fiscal responsibilities may have been of more importance in the increase of enrollments and the building of schools. States often provided poorer school districts with grants to build schools and supplemented teacher salaries, although these incentives were small in comparison with local funds. Additionally, a somewhat obscure set of state laws, known as "free tuition laws," were probably far more important in the spread of high schools than the better known compulsory schooling laws. These laws mandated that school districts without high schools be fiscally responsible to pay tuition to other school districts.

Rather than stemming from compulsions, restraints, and subsidies, the roots of the increase in high school enrollments can be found in more fundamental factors that affected the demand for educated workers and for schooling in general. These factors encouraged young people to continue with their education and drove parents to demand that communities provide more education for their children.

In the 1890s, when the high school movement was first emerging in various parts of the country, secondary schools were highly varied institutions differing from place to place. By the 1920s, the American high school was a fairly uniform institution. A high school in rural Indiana taught subjects similar to those offered in urban Chicago and a high school in northern Maine was not unlike its counterpart in Los Angeles. Americans today would find a high school in the 1920s to be a familiar institution but would be less at home in a secondary school of the late nineteenth century.

The modern high school that emerged by the 1920s served a multiplicity of functions including college preparation, general education, vocational instruction, and commercial training. As the public high school broadened its curriculum, it expanded its reach and appealed to a greater fraction of teenagers and a larger group of tax-payers.

The period of the high school movement is judged to have ended around 1940, since the greatest increases had already occurred. The completion of the great expansion in the education of America's youth came at a decisive moment in the history of the nation. The United States was drawn into World War II at the moment the median American youth had become a high school graduate. The increased education of the labor force meant that the nation was better prepared for war. It also meant that when World War II ended the nation was poised to begin the transformation to mass higher education.

As national data reveal, World War II cut deeply into high school graduation and enrollment. In all regions of the country and in every state, graduation rates plummeted and enrollment rates for older students were markedly reduced. The military draft was not the only reason, as evidenced by the fact that graduation and enrollment rates decreased for young women of high school age.[7] Notably, the decrease in schooling rates for older youths was greatest in states that received the largest defense contracts on a per capita basis, such as those in the New England and the Pacific regions. The three states of the Pacific were among the top five states having the largest absolute decline in the male graduation rate and were also in the top 12 in total defense contract spending per capita.[8] Independent of geography, the earnings of the unskilled and the young soared in the 1940s relative to those for the skilled and those with more labor market experience. The youths who left high school for the labor market may have thought they would eventually return, but most did not. Unlike their slightly older peers who went into the military and received the GI Bill at the end of the war, these youth became a lost generation.

Enrollment and Graduation Rates by Region and State

Measurement Issues

To understand the educational changes brought about by the high school movement, we present data on enrollment and graduation rates. The schooling data include *all* students in grades nine through twelve in public and private secondary schools, as well as those in the preparatory departments of colleges and universities, an important component until the 1920s.[9] We have used a host of sources and methods, given in Appendix B. In brief, the data are derived mainly from the administrative records of the U.S. Office of Education.[10] These records were published on an annual basis until 1916 in the *Report of the Commissioner of Education* (the *Annuals*), then every other year in the *Biennial Survey of Education* (the *Biennials*), which was superseded in 1962 by the *Digest of Education Statistics.*[11]

The two main indicators are high school enrollment and high school graduation, neither of which may reveal much about educational quality. Because those who enroll need not attend, graduation is a better indicator since students often have to complete courses and pass exams to be promoted and thus graduate. We will use the graduation rate as our main indicator of educational achievement. Graduation and enrollment measures are highly correlated across the states. In the past, as today, each state had somewhat different standards for high school graduation and in some cases the criteria were determined by localities.[12] Because the graduation and enrollment data are counts of students, we have produced rates by dividing by the number of youths who could have attended or graduated from high school.[13]

The nationwide trends in enrollment and graduation rates are depicted in Figure 6.1, in which the U.S. aggregate public and private secondary school series from 1890 for graduation and 1900 for enrollment to 1970 are graphed.[14] The series reveal several aspects of the high school movement nationally. Most important is the impressive increase from 1910 to the 1930s, the large decrease with World War II, and the resumption of the increase in the late 1940s. What cannot be seen in the graph, but will be apparent in Chapter 9, is that the rates achieved since 1970 have been relatively stagnant.

The United States is a land of enormous diversity and there is no reason to believe that every part of the country underwent change at the same time. In some states, high school graduation and enrollment rates increased sharply in the 1920s, but in others, even those outside the South, the high school movement came somewhat later. For ease of presentation, we aggregate the state data to the regional (census division) level. Even within these regions, certain states were far ahead of the pack. We employ maps to show variation within region. We first address the change by region and state, and then examine differences by size of city, race, gender, and ethnicity.

We divide the data into the nine census divisions—New England, Middle Atlantic, South Atlantic, East South Central, West South Central, East North Central, West North Central, Mountain states, and Pacific states.[15] Graduation data by region for 1910 to 1970 are given in Figure 6.2. Panel A of the figure contains the Middle Atlantic, New England, Pacific, and West North Central regions and Panel B has the South Atlantic (also for whites only from 1930 to 1954), the East South Central, and the East North Central for comparison. (All the enrollment and graduation data by region are in Appendix B, Tables B.1 and B.2.) State variation can be observed in maps spanning the beginning (1910), middle (1928), and end (1938) of the high school movement (Figure 6.3).

Leaders and Laggards in the High School Movement

The increase in graduation rates from 1910 to 1940 in the four regions of the North and West was considerably more impressive than for the entire nation. At the beginning of the high school movement the graduation rate in the New England region exceeded that in other parts of the country. Another high education region at this early date was the East North Central, to which many New Englanders had migrated. Even though New England remained in the educational forefront for some time, other regions closed the gap with the leader during the 1920s. By 1924 several non-southern regions of the country had graduation rates that exceeded that of New England. New England youth had not fallen behind; rather, various parts of the rest of the nation invested heavily in education and had rapidly caught up to the leader. Some regions of the country, such as the Pacific and West North Central, had enormous increases in high school graduation rates that resulted from substantial investments in school building and staffing.

The parts of the nation that began in the 1920s to invest heavily in education include states that, at first glance, would appear to defy easy categorization. Included in the group are the Prairie states of America's heartland in the West North Central division, such as Iowa, Kansas, and Nebraska, and the Pacific states of California, Oregon, and Washington. As different as these states may appear, they were similar in

Panel A: Four regions in the North and West

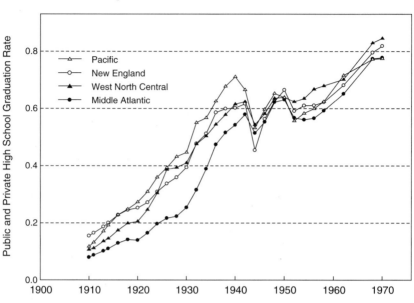

Panel B: Two regions in the South and the East North Central

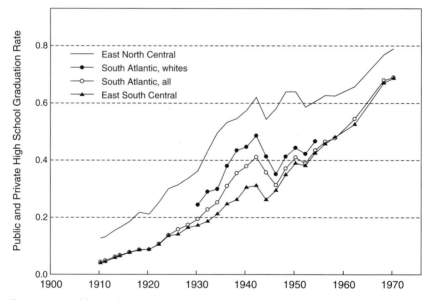

Figure 6.2. Public and Private High School Graduation Rates, 1910 to 1970.
Includes both males and females in public and private schools (including
preparatory departments of colleges and universities). Graduates are divided by
the approximate number of 17-year-olds in the state. Constant growth rate
interpolations of population data are made between census years. Sources: See
Appendix B.

various ways. Most important among their similarities was their high value of per capita taxable wealth. High school education was considerably more expensive than was elementary or common school education—in fact, it cost about twice as much per pupil-year. More wealth meant that taxpayers could more easily afford expensive education. Of additional importance was the low level of manufacturing activity in these states. Less industry meant that myopic youth were not lured from school by the attractions of a job.

The best way to see the differences across the states is from the maps in Figure 6.3. In each of the maps the darker the shade, the higher rate for enrollment or graduation. In 1910 the darker states are almost all in New England, although there are some high education states in the center of the country and on the West coast. By 1928 a group of states in the central part of the country and on the Pacific had emerged as the education leaders.

In 1928 the states with the highest rates appear to form an "education belt" across the midsection of the nation. The belt contains all the states of the Pacific, some of the Mountain states, various prairie states of the East and West North Central, and much of New England. What began in New England jumped clear across the nation to the rich states of the far West, then spread to the nation's heartland in the Middle West, skirting the Middle Atlantic states—New York, New Jersey, and Pennsylvania—for the time being. The belt clearly excludes the South, which is an obvious outlier as a region. It also omits some of the more industrial states in the North, although to a lesser degree.

One may wonder why some of the northern states lagged in high school enrollment and graduation relative to the leaders. Several are states in the East North Central region (e.g., Michigan, Illinois), one is a New England state (Rhode Island), and three of the most obvious comprise the Middle Atlantic region (New Jersey, New York, and Pennsylvania). The difference between the graduation rates among the northern laggard states and those of the northern educational leaders was substantial. The graduation rate in Iowa (44 percent), Kansas (48 percent), or Nebraska (46 percent) in 1928—to take three contiguous high-education states—was more than twice that in New Jersey (23 percent), New York (20 percent), or Pennsylvania (24 percent)—to take three contiguous ones at the other extreme.

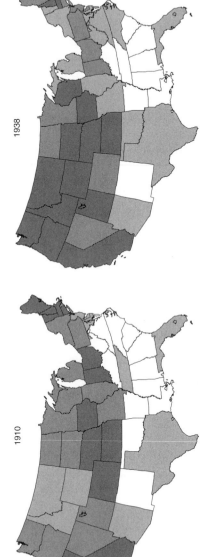

1938

1910

1928

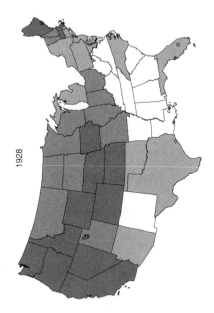

Legend for Maps

Shading	Graduation Rate Ranges for States		
	1910	1928	1938
	2.6 < 4.3	11.8 < 18.4	29.7 < 33.7
	4.3 < 7.5	18.4 < 31.5	33.7 < 50.4
	7.5 < 11.9	31.5 < 38.8	50.4 < 60.0
	11.9 < 22.0	38.8 < 55.6	60.0 < 76.9

The northern states with lower high school rates had one feature in common: their cities were mainly dependent on manufacturing and some were heavily industrial. Many of the industrial areas of the nation joined the high school bandwagon late. The states of the Middle Atlantic region are the most obvious latecomers in the North. Michigan was also a laggard in the otherwise high education region of the East North Central. But each of these states narrowed the gap considerably in the 1930s, when the Great Depression caused enormous unemployment in industrial areas. A large fraction of the population of these states lived in industrial cities where youths in the 1910s and 1920s had often dropped out of school between ages 14 and 16 in favor of work; that changed with high unemployment rates in many industries in the 1930s.

The recession that would soon become the Great Depression was but a minor blip on the macro-economic radar screen in 1929 when the graduation rate for the states of the Middle Atlantic was 22 percent. But by 1932, when the Great Depression had resulted in an unemployment rate of almost 24 percent nationwide, the graduation rate in the three states of the Middle Atlantic region had soared to 32 percent.[16] In 1936, when the downturn was in its seventh year, the graduation rate increased to 47 percent. From 1929 to 1936 the graduation rate in the states of the Middle Atlantic had more than doubled and the region had been catapulted into the high school movement.

The reason for the increased enrollment of high school students extends somewhat beyond the general inability of workers to secure productive employment during the 1930s. It also concerns particular legislation of the first New Deal. The National Industrial Recovery Act (NIRA) prohibited the employment of youths less than 16 years old in manufacturing work. But because schooling rates increased before the NIRA was passed in 1933, the act must have merely reinforced the response to the immediate effects of the unemployment of the Great Depression. Bad economic times are often good for education and that appears to have been true of the 1930s, although the

Figure 6.3. (opposite page) Public and Private High School Graduation Rates by State: 1910, 1928, and 1938. States are shaded by graduation rates so that darker states have higher rates. Approximately equal numbers of states are in each group in each year. Sources: See Appendix B.

high school movement was clearly well underway before the 1930s in many parts of the country.

The South was the major schooling laggard during the high school movement, and the region continued to rank far behind the North and the West until the 1970s. Within the South, the East South Central had the lowest schooling rates during the high school movement, the South Atlantic was next, and the West South Central, dominated in population by Texas and also including Oklahoma, was the highest.

The high school movement in the South was held back by many factors. Racially segregated schools existed in all southern states, and the overwhelming majority of African Americans lived in the South—in 1910 almost 90 percent did and in 1930, 79 percent did.[17] Secondary school enrollment rates for African Americans in that region were barely above zero until the 1930s, when they reached about 6 percent.[18] Until the 1920s most African American youth who finished eighth grade and entered secondary school attended a private school, often founded by northern philanthropies.[19] By 1929 the situation was somewhat less bleak and more than 1,000 public high schools for blacks had been established in the South, together with about 110 private high schools, even though many of them offered primarily industrial and vocational courses.[20]

But schooling rates for white youths in the South were also far below those in the rest of the country (see Figure 6.2, Panel B, which shows the schooling rates for whites in the South Atlantic region from 1930 to 1954). The fact that African Americans had extremely low rates of high school enrollment and graduation is not a complete answer for why the South had low levels of secondary school education.

Another reason proffered for why the South fell behind is that it was an agricultural region and was sparsely settled. Yet other parts of the nation were about as agricultural and just as sparsely settled and they were educational leaders. Compare Iowa and Georgia, for example. In 1930 both Iowa and Georgia had low and comparable population densities and both had substantial and nearly identical fractions of their adult male labor forces involved in agricultural occupations.[21] Yet Iowa had a public and private high school graduation rate of 44 percent in 1928, while Georgia's overall graduation rate was just 14 percent, or one-third the rate in Iowa. The rate for just the white youth of Georgia was about 17 percent. Thus even the white secondary schooling rate in Georgia was far behind that for Iowa's youth.

For all states in the West North Central region the fraction of the male labor force employed in agriculture was 40 percent in 1930 and the population density was 29.3 persons per square mile. In a seven-state portion of the U.S. South, where cotton was the dominant crop and its fertile land lent it the moniker of the "black belt," 52 percent of the male labor force was employed in agriculture and 52.8 persons, on average, resided in each square mile.[22] The graduation rate in the West North Central region was 41 percent in 1930 whereas that among white youths in the seven states of the cotton South was 24 percent.

Even though the South was a more densely populated region than were the farming areas of the Midwest, it had a considerably smaller fraction of its population living in cities and towns. In the West North Central, for example, 42 percent of the population lived in cities and towns having 2,500 or more people (the official census definition), whereas just 27 percent did in the seven states of the cotton south. However, a substantial fraction of the population in the West North Central region resided in towns with fewer than 2,500 people, less than the census required for the place to be deemed urban. Take the case of Iowa.

In 1910, Iowa contained 2.225 million individuals, 680,000 of whom lived in urban areas (those with populations greater than 2,500). But nearly 460,000 Iowans resided in incorporated areas below the census cut-off of 2,500 persons. The vast majority (82 percent) of these individuals lived in Iowa's 830 incorporated tiny towns, each of which had a population of less than 1,500.[23] If we use an augmented definition of urban and include those in incorporated cities, towns, and villages regardless of size, 51 percent of the population would have been deemed urban in 1910, whereas that number was 31 percent using the official census definition.[24]

Agriculture in the South did not place the same demands on the processing of the crop as it did in many other parts of the country. Cotton and tobacco, for example, required minor processing on or near the farm. After ginning or curing, the crops were brought to commercial cities on major waterways, such as the Atlantic Ocean, the Mississippi River, and the Ohio River, and sent on their way to Europe or the North. In contrast, the tiny towns dotting the Midwest added value to the crops and were a part of a major distribution system. The towns contained the grain elevator's and were stops along the railroad. Most important for the development of educational institutions, these small towns were incorporated places with taxing authority and they became centers of the local school districts that built and staffed the high schools. In the South, on the other hand, counties were the fiscal units that would have built and staffed high schools. The towns of the Midwest, especially the tiny towns, were wellsprings of social capital and thus of schools and human capital.

Explaining Differences in High School Graduation Rates

Reasons for High School and Its Expansion

To understand why some states led and others lagged, we must analyze the determinants of (public and private) high school graduation rates by state. To do this, we first analyze graduation rates at the dawn of the high school movement in 1910 and then traverse the initial surge in high school by exploring the transformation in education from 1910 to 1928. Next, we explore changes from the eve of the Great Depression to just before World War II, from 1928 to 1938.

The motivation for all our estimations is a standard model of human capital investment in which the educational return, opportunity cost, and capital constraints affect private decisions. This simple, and well-known, formulation of the educational investment decision does not, however, speak to the public nature of most schooling. Public support for secondary school was rarely justified on the basis of the creation of a literate citizenry, the way that the publicly supported primary school had been in the nineteenth century. Rather, it was rationalized, often implicitly, on the grounds of capital-market imperfections. Under this theory, communities are groups of families at different stages of their lifecycle, and publicly funded education serves as an intergenerational loan, a means of consumption smoothing.[25]

Under many reasonable scenarios, the wider the distribution of income, given its mean, the less support there will be for public education, since the rich can opt out and the poor will have a lower demand.[26] However, the extent to which individuals consider themselves members of the same community should provide an extension of the public choice framework. Greater social cohesion, intergenerational propinquity, and community stability should increase support for publicly funded education.[27]

In the next section, we analyze the determinants or correlates of the public and private high school graduation rates at the state level in two years (1910, 1928) and across two eras (1910 to 1928; 1928 to 1938). The variables we use approximate the key determinants of family-level education decisions and the factors relevant in a public choice framework.

To do the analysis, all youth were assumed to face the same (national) market for white-collar employment conditional on receiving a high school degree.[28] Because remunerative employment opportunities for older youth in the period were likely to be found in manufacturing, we used the fraction of the workforce in manufacturing and the manufacturing wage as the opportunity cost of high school education. Various estimates of income and wealth (state income per capita, taxable wealth per capita, and agricultural income per agricultural worker) were used to measure household capital-constraints and the consumption demand for education. The distribution of income or wealth was more difficult to obtain for the period. We will make the case that automobile registrations per capita is a reasonable substitute for the share of voters wealthy enough to favor financing an expensive public good, such as a high school.

Because the return to high school was probably greater where publicly supported colleges were available, we included (in the change regression for 1910 to 1928) the public university enrollment rate in the base year. The social stability of communities was inferred from the proportion 65 years and older in the state. Social distance or propinquity was proxied by variables relating to the fraction foreign-born (or Catholic) since both, within bounds, increased social heterogeneity.[29]

The determinants we identified above are precisely those contemporaries recognized as important during the high school movement. Take the following explanation of why Portland, OR, had an increased high school enrollment in the 1910s:

First, Portland is not a manufacturing city, and consequently, does not offer the attraction for boys and girls to drop out of school. Second, the increased wealth enables parents to keep their children in school longer. Third, the nature of the population of the city of Portland tends to keep children in school. By this we mean that the people of Portland . . . demand high standards of education . . . and take advantage of the opportunity. The fourth reason . . . is found in the nature of the high schools. Instead of offering a narrow college preparatory course . . . the high school now offers many courses of a more general and industrial nature.[30]

Estimation of a Framework to Explain State-Level High School Graduation Rates

Our estimations using high school graduation rates are suggestive of the forces that both encouraged and impeded secondary-school education, although they are, admittedly, reduced form models. In Table 6.1 we summarize the main results of our analysis: three of the columns (cols. 1, 2, and 3) give regressions in levels (one for 1910 and two for 1928) and three (cols. 4, 5, and 6) give the regressions in first-difference form.[31] The last two columns give the means.

With just 48 states in each year we have been judicious in our inclusion of variables, and a further constraint is that some variables are highly correlated.[32] Where only one of the many variables mentioned is included, the results are robust to the inclusion of the others.

The association between the key factors and high school graduation rates at the start of the high school movement in 1910 is summarized in column 1. Per capita wealth (in 1912), the proportion older than 64 years (in 1910), the percentage of the labor force in manufacturing (in 1910), the percentage Catholic (in 1910), and dummy variables for the South and New England are strong predictors of high school graduation and together they account for almost 90 percent of the cross-state variation.

Wealth per capita (or state income per capita, or agricultural income per capita), not surprisingly, is positively related to the high school graduation rate and the impact is reasonably large—a shift from the state at the 25th percentile to that at the 75th percentile increases the graduation rate by about 1.5 percentage points in 1910 (or by 16 percent around the mean). Having more manufacturing, on the other hand, was a drag on education; a move from the 25th to the 75th percentile reduces the graduation rate by 1 percentage point in 1910 (or by 12 percent around the mean). The greater the proportion older than 64 years, the higher is the graduation rate. This strikingly strong positive relationship at the dawn of the high school movement between high school graduation rates and fraction of older persons (a raw correlation of 0.79) is illustrated in Figure 6.4, Panel A. We attribute the effect to the stability of community and not to differential fertility or immigration, for neither of those variables reduced the positive impact.

That educational attainment in 1910 was positively related to the fraction of older persons in the state is opposite to a finding from a more current period (e.g., Poterba 1997). More recent estimates show that the elderly are less supportive of educational expenditures. There is good reason for the difference. Older citizens today are highly mobile as a group. A large fraction live far away from their community of origin and as a political unit they appear to have far less interest in the use of public resources to enhance education than did those early in the twentieth century who continued to reside in their communities.[33]

We examined the determinants of high school graduation rates in 1928 and found results similar to those for 1910, when translated into elasticities (see Table 6.1, col. 2). For 1928, we included variables that we could not for 1910 and they added significantly to the story. The most interesting of the new variables is automobile registrations per capita (in 1930).

In the 1920s automobile ownership required a higher level of income or wealth, relative to the mean, than it does today. Consider, for example, two symmetric income distributions each having the same mean but different variance and for which the cutoff point for automobile ownership is somewhere below the mean. The narrower distribution will have a higher fraction of car owners among the population. Thus, under certain conditions and given the mean of income (or wealth), the variable "automobile registrations per capita" is a good proxy for the variance of income (or wealth).

The auto registrations variable exhibits a substantial positive relationship to the high school graduation rate, even when a direct measure of per capita wealth is included. The specification in column 2 implies that increasing auto registrations per capita in the state at the 25th percentile to that at the 75th percentile increases the graduation rate by 5 percentage points (or by 17 percent of the mean level in 1928).[34] Automobile registrations per capita is a strong explanatory variable and speaks to the importance of a more equal distribution of wealth, given its mean, and a large share of middle class voters in the provision of education as a public good.[35] The states with the greatest

Table 6.1. Explaining (Public and Private) Secondary-School Graduation Rates across States

	(1)	(2)	(3)	(4)	(5)	(6)	(7)	(8)
	Levels			Differences			Means (s.d.)	
	1910	1928	1928	Δ1928–1910	Δ1938–1928	Δ1938–1928	1910	1928
Log per capita taxable wealth, 1912 or 1922	0.0236 (0.00901)	0.0852 (0.0368)		0.0857 (0.0260)	0.125 (0.0345)		7.471 (0.451)	7.926 (0.386)
% ≥65 years, 1910 or 1930	2.13 (0.260)	1.423 (0.788)	1.846 (0.774)	−1.749 (0.737)	−0.527 (0.866)		0.0414 (0.0143)	0.0547 (0.0142)
% of labor force in manufacturing, 1910 or 1930	−0.0673 (0.0335)	−0.144 (0.0972)	0.989 (0.481)	−0.0495 (0.0947)	0.126 (0.0934)	0.203 (0.0723)	0.248 (0.124)	0.255 (0.103)
% Catholic, 1910 or 1926	−0.0913 (0.0305)	−0.377 (0.0867)	−0.274 (0.0849)	−0.265 (0.0900)	0.0595 (0.0841)		0.150 (0.121)	0.151 (0.123)
South	−0.0449 (0.00932)	−0.0935 (0.0272)	−0.131 (0.0294)	−0.0735 (0.0267)	0.0375 (0.0306)			
New England	0.0444 (0.0121)	0.100 (0.0310)		0.0811 (0.0333)				
Middle Atlantic			−0.0635 (0.0338)		0.0620 (0.0188)			
Males in public colleges/17-year olds, 1910				1.09 (0.384)				
Wage in manufacturing, 1929			0.241 (0.0974)					
Wage × % in manufacturing			−0.827 (0.375)					
Auto registrations per capita, 1930		0.568 (0.230)	0.449 (0.218)					0.224 (0.0648)

Log agricultural income per agricultural worker, 1920						0.0985
						(0.0174)
Change in unemployment rate, 1930 to 1940						0.900
						(0.306)
Constant	−0.136	−0.468	−0.0962	−0.324	−0.814	−0.541
	(0.0709)	(0.273)	(0.115)	(0.199)	(0.276)	(0.104)
R^2	0.895	0.874	0.864	0.758	0.679	0.708
Root MSE	0.172	0.0451	0.0476	0.0474	0.0400	0.0368
Mean (unweighted) of dependent variable	0.0882	0.291	0.291	0.212	0.204	0.204

Sources: For complete notes regarding the sources see Goldin and Katz (2005), from which this table derives.

Notes: Standard errors are in parentheses; ordinary least squares regressions, unweighted except for the 1928 to 1938 change regressions (cols. 5, 6). Weight for state i is $(S_{i,28} \cdot S_{i,38})/(S_{i,28} + S_{i,38})$ where $S_{i,t}$ =share of state i 17-year-olds in U.S. total in year t. Weighting does not affect results in cols. 1 to 4. The 1928 to 1938 regressions are weighted due to two outliers (DE and NV). Number of observations is 48 in all columns; DC is excluded. AZ and NM were territories until 1912 but are included with the 1910 states.

Dependent variable: Public and private graduation rate by state: See Appendix B.

Independent variables:

Variables listed as % are entered as fractions. Note that in the change equations of columns 4, 5, and 6 the explanatory variables are those at the beginning of the period and reflect starting conditions.

Per capita taxable wealth, 1912 or 1922: Taxable wealth/population, U.S. Department of Commerce (1926), *Statistical Abstract.*

% ≥65 years, 1910 or 1930: Historical Statistics, series A 195–209.

% in manufacturing, 1910 or 1930: U.S. Bureau of the Census (1912, 1932).

% Catholic, 1910 or 1926: U.S. Department of Commerce (1930), *Religious Bodies: 1926,* Vol. I, table 29. The 1910 numbers are extrapolated from those for 1906 and 1916. All are expressed per state resident.

South: South includes the census divisions South Atlantic, East South Central, and West South Central.

Males in public colleges/17-year-olds, 1910: U.S. Bureau of Education (1910), table 31, p. 850. Military academies receiving public support are excluded. The denominator contains both males and females.

Wage in manufacturing in $000, 1929, Kuznets et al. (1960), table A 3.5, p. 129.

Auto registrations per capita, 1930: U.S. Department of Commerce (1940), *Statistical Abstract,* table 467.

Agricultural income per agricultural worker, 1920 (mean = $943): Kuznets et al. (1960), table A 4.3, p. 187. The variable is agricultural service income per agricultural worker. *Agricultural income per agricultural worker, 1930* (mean = 0.0574), *1940* (mean = 0.0883): U.S. Bureau of Commerce, *Statistical Abstract,* (1932) table 341, (1948) table 203. *Unemployment rate, 1930* (mean = 0.0574), *1940* (mean = 0.0883): U.S. Bureau of Commerce, *Statistical Abstract,* (1932) table 341, (1948) table 203. Unemployment for 1930 refers to April 1930 and is the sum of Class A (non-layoff) and Class B (layoff).

Panel A

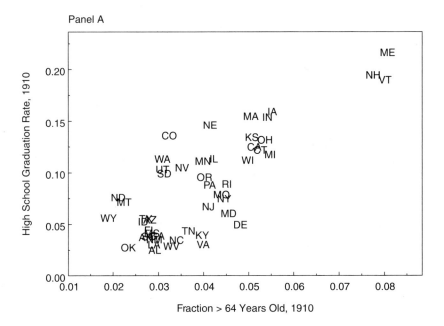

Fraction > 64 Years Old, 1910

Panel B

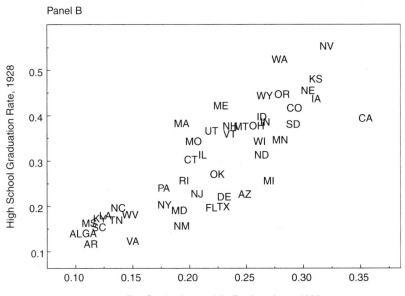

Per Capita Automobile Registrations, 1930

Panel C

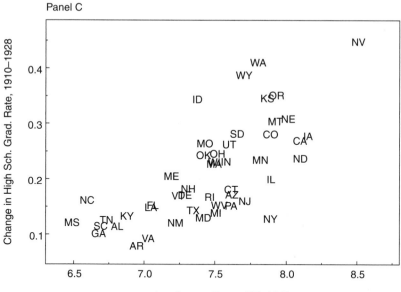

Panel D

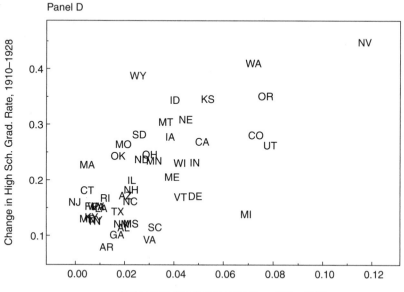

Figure 6.4 A–D. (facing pages) High School Graduation Rates and State
Characteristics: 1910 to 1930. Notes and Sources: See Table 6.1.

number of automobile registrations per capita in 1930—California, Nevada, Kansas, Iowa, and Nebraska—were all at the very highest end of the educational distribution in 1928 as can be seen in Figure 6.4, Panel B.[36]

Also of interest are the roles of manufacturing as a share of employment, the manufacturing wage, and their interaction (Table 6.1, col. 3). Having a greater percentage of the labor force in manufacturing, given the manufacturing wage, was a drag on education, as we found in the analysis of the 1910 data. But in the analysis of the 1928 data the interaction with the manufacturing wage revealed that the relationship held only when the wage was above the mean. Similarly, a higher manufacturing wage was not an impediment to education until the percentage of the labor force in manufacturing exceeded its mean. The lowest graduation rates outside the South were found in industrial states with relatively high manufacturing wages, such as the Middle Atlantic states. The opportunity cost of education in these states was high and the availability of manufacturing jobs was substantial enough to deter education.[37]

The difference regression for 1910 to 1928 (given in Table 6.1, col. 4) reinforces the findings in the levels regression and makes a causal interpretation more plausible.[38] The independent variables for the difference regressions capture the *initial* conditions in a state. For example, more wealth in 1910 hastened the growth of high schools from 1910 to 1928, but a greater share of the labor force in manufacturing in 1910 slowed it. The positive relationship between (log) per capita wealth at the start of the high school movement and the expansion of high schools from 1910 to 1928 is displayed in Figure 6.4, Panel C. The fraction of youth in the state who attended public colleges and universities in 1910 had a strong positive effect on the high school graduation rate (see Figure 6.4, Panel D), probably because the returns to high school were greater in states having amply funded public institutions of higher education.

Lastly, we analyzed the change during the 1930s. The estimation in Table 6.1, column 5 is configured similarly to that in column 4. Much appears to have been altered by the 1930s. Wealth remained an important determinant, but the fraction of the labor force in manufacturing no longer had a strong negative effect. In fact, manufacturing had a weak positive effect and the Middle Atlantic states were residually positive. The sparser specification in column (6) focused on factors unique to the Great Depression and added the change in the unemployment rate from 1930 to 1940.[39] High school graduation rates during the 1930s increased most in states that had the largest increases in unemployment, for given initial income, and for those with the largest manufacturing sectors.

In sum, we found that several factors can explain differences in the diffusion of high school graduation across the various states. The factors include wealth, income, relative homogeneity of the population, the distribution of income, the opportunity cost of youthful employment, state support for higher education, and the stability of the community through the retention of older members. But what was the role of state compulsion in the expansion of U.S. secondary schooling from 1910 to 1940? Such laws were apparently effective in countries such as Great Britain, where they were accompanied by large increases in educational access and spending.[40]

The Role of Compulsory Schooling and Child Labor Laws

As we mentioned in Chapter 5, compulsory education and child labor laws were first passed in the United States in the mid-nineteenth century.[41] But it was not until the early twentieth century, with the ramping up of compulsory education and child labor laws, that they could have had much impact on secondary schooling. The laws became more stringent just as youths were entering and graduating from high schools in considerably greater numbers. This coincidence has led many to assert that state legal changes spurred the increase in high school enrollments.[42] But did they?

From 1910 to 1940 the fraction of youths enrolled in public and private U.S. secondary schools increased from 18 to 71 percent. The fraction graduating nationwide soared from 9 to 51 percent. To find the effect of state compulsory schooling and child labor laws on secondary schooling we used contemporaneous evidence we compiled on high school enrollments together with information on the laws. Our estimation approach exploits cross-state differences in the timing of changes in state laws and controls for state fixed effects, year fixed effects, and other time-varying state level determinants of secondary schooling.

What Were the Laws?

The typical compulsory schooling law set down the ages during which youths had to be in school. The laws became more complicated in the early twentieth century when maximum ages increased in many states. The typical law was then altered to include a level of education that exempted a youth from the maximum age, and that factor became the binding constraint in most states.

Child labor laws modified compulsory schooling laws and generally exempted older working youths who were constrained by the compulsory schooling law. They set down the way youths could obtain a work permit and often a minimum level of schooling required. The child labor law was almost always the binding constraint. Statutes mandating continuation (or part-time) school attendance were added during the Progressive Era. Continuation schools educated youths who were below the maximum age of compulsory schooling and had left school to work.[43]

We have compiled information on these laws for all 48 states by year from 1910 and 1939 (see Goldin and Katz 2003, data appendix). The most important factors are: minimum age of compulsory schooling, maximum age of compulsory schooling, education for exemption from the maximum age rule, work permit age, education required to receive a work permit, and whether the state had mandatory continuation schools. From these data we constructed two variables, where s is state and t is year to approximate the state laws applicable to youth of high school entrance age (14 years old) in year t:

$$\text{Child Labor School Years}_{st} = max \ \{(\text{education required for work permit})_{st}, (\text{work permit age}_{st} - \text{school entrance age}_{s,\,t-8})\},$$

which gives the mandated time a youth had to be in school before taking a job according to state child labor laws, and

$$\text{Compulsory School Years}_{st} = min \ \{(\text{education for exemption})_{st}, (\text{maximum age of compulsory schooling}_{st} - \text{school entrance age}_{s,\,t-8})\},$$

which gives the minimum time a youth had to attend school according to state compulsory education laws.[44]

Legislation became more stringent over the 1910 to 1940 period in terms of the number of states having laws and the stringency of existing legislation. Most of the changes were fairly continuous over the period. The maximum age continued to increase until to around 1930 when 42 states set their maximum age at 14 years or higher, while the minimum age decreased throughout that period. The level of schooling required to be exempt from the maximum age of compulsory education rose over time. The age at which youths could get a work permit increased significantly around World War I and after 1935. The education required to obtain a work permit rose so that by 1930, 18 states had a requirement of eight grades and 31 had at least a sixth grade requirement. The vast majority of states joined the continuation school bandwagon around World War I and passed laws requiring that employed youths below the maximum age attend school during their workday.

Effects of Law Changes

How much of the increase in high school enrollments can be explained by the increased stringency of the laws? To find the answer, we examined the effects of state child labor and compulsory schooling laws on contemporaneous high school enrollment using the data we earlier presented on public and private enrollments in grades 9 to 12. A standard panel data model was estimated, which included state and year fixed effects, state law variables (including a dummy variable for a state continuation school law), and other state time-varying economic and demographic controls. Our identification of the effects of state child labor and compulsory schooling laws comes from differential law changes across states, conditional on the time varying state controls also in Table 6.1. We included, as well, a full set of census division linear time trends.[45]

The regression estimates reported in column 1 of Table 6.2 show small but statistically significant effects of both compulsory education and child labor laws. A one-year increase in "compulsory school years" is associated with a 0.45 percentage point rise in the high school enrollment rate and a one-year increase in "child labor school years" is associated with a 0.78 percentage point rise.[46] Having a continuation school law is connected to a 2.5 percentage point increase in high school enrollments.

How large was the contribution of child labor and compulsory schooling laws to the 50.4 percentage point increase in the high school enrollment rate from 1910 to 1938? The combined effects of changes

Table 6.2. Impact of State Compulsory Schooling and Child Labor Laws on Secondary School Enrollment Rates, 1910 to 1938

Dependent variable: Fraction of state's 14- to 17-year-olds enrolled in public and private secondary schools (mean = 0.441; s.d. = 0.204)	(1) Coefficient (s.e.)	(2) Mean (s.d.)
Continuation school law	0.0249	0.530
	(0.00940)	(0.499)
Child labor school years[a]	0.00777	6.51
	(0.00265)	(2.00)
Compulsory school years[b]	0.00453	6.91
	(0.00209)	(2.67)
No child labor law	0.0217	0.299
	(0.0185)	(0.170)
No compulsory schooling law	0.0563	0.102
	(0.0167)	(0.303)
Autos per capita	0.865	0.136
	(0.187)	(0.093)
Manufacturing employment per capita	0.134	0.0662
	(0.409)	(0.0389)
Fraction ≥ 65 years	3.06	0.0511
	(1.39)	(0.0128)
Fraction ≥ 14 years	−2.10	0.305
	(0.583)	(0.0474)
Other state demographic controls[c]	Yes	
State dummies	Yes	
Year dummies	Yes	
Census division trends	Yes	
R^2	0.978	
Standard error	0.0321	
Number of observations	720	720

Sources: Biennial secondary school enrollments by state, Appendix B; compulsory school and child labor laws, Goldin and Katz (2003, data appendix); various demographic data were provided by Adriana Lleras-Muney and are from the 1910, 1920, 1930, and 1940 Censuses of Population (linearly imputed in intervening years); other variables, see Table 6.1.

Notes: The regression and sample means have been weighted by the number of 14-year-olds in the state. Robust standard errors (s.e.) clustered by state are reported for col. 1.

a. Child labor school years$_t$ = max\{(education required for work permit$_t$), (work permit age$_t$ − school entrance age$_{t-8}$)\}

b. Compulsory school years$_t$ = min\{(education for exemption$_t$), (maximum age of compulsory schooling$_t$ − entry age$_{t-8}$)\}

c. Includes fractions black, foreign-born, and urban.

in child labor and compulsory schooling laws (both the years and existence of the laws) add 1.8 percentage points. There was a 60 percentage point increase in the share of students in states with continuation school laws and that can explain a 1.5 percentage point increase in the high school enrollment rate. About a 3.3 percentage points (or 5 to 6 percent) of the overall increase in the high school enrollment rate from 1910 to 1938 can therefore be accounted for by changes in child labor and compulsory schooling laws. In contrast, among the state control variables, just the increase in automobiles per capita from under 0.01 in 1910 to 0.22 in 1938 can explain a 19 percentage point rise in the high school enrollment rate. Changes in state child labor and compulsory schooling laws appear to have had some impact on high school enrollment rates from 1910 to 1938, but the impacts were modest relative to the rapid rise in secondary schooling rates during the era of the high school movement.[47]

Cities and the High School Movement

Public high schools, as previously mentioned, were established in many of the nation's larger cities in the early nineteenth century. By 1903, according to a survey done by the U.S. Office of Education, there were more than 7,200 public high schools in the nation's cities and towns. Almost all of the nation's cities with a population exceeding 3,000 persons had one or more public high schools at the turn of the twentieth century and many of the smaller towns did as well.[48] In Iowa, for example, all cities with a population greater than 3,000 persons had at least one public high school by 1903 and of the 63 towns having a population between 1,500 and 3,000 persons in 1910, 57 (90 percent) did by 1903. Even some of the smallest towns in Iowa—villages would be a better descriptor—had a public high school. Iowa had 830 incorporated places with a population below 1,500 in 1910 and 230 of these (or 28 percent) had a public high school in 1903. Although most rural youth in Iowa were far removed from a public high school in 1900, secondary schools were within reach of most youth who resided in cities, towns, and even many villages.

Throughout the United States in the early years of the high school movement, many of the towns that enumerated public high school students did not actually have a separate high school building.[49] Rather, many of the smaller cities and towns that offered high school instruction did so in a portion of the town's elementary school, often the upper floor. As the high school movement progressed, these smaller towns built dedicated high schools. As noted in Chapter 5, some states passed legislation prohibiting the provision of high school classes within the elementary school building and set down strict guidelines for the number of rooms and teachers an accredited high school had to have.[50]

As public secondary schooling spread in the early twentieth century, an interesting and meaningful relationship arose between high school enrollment and the population of cities and towns. Rather than increasing with the size of the city or town as one might have expected if there was a minimum scale needed for a high school, the fraction of youth attending secondary school actually *decreased* with city size. The decrease with the size of city or town, moreover, was monotonic, extending throughout the size distribution of cities.

The smallest towns—those with populations between 1,000 and 2,500—had higher secondary schooling rates than did towns with populations ranging from 2,500 to 10,000 persons. These larger towns had schooling rates for youths that were, in turn, higher than were those for small cities. And, finally, the small cities had school rates that were higher than were those for the larger cities and, within the larger cities, the largest had the lowest rates of all. The range from the smallest of the towns to the largest of the cities for 16- and 17-year-old youth was 46 percent to 21 percent—a difference of 25 percentage points—in 1910. Incredible as it may seem, rates of school attendance were higher still among incorporated rural places with fewer than 1,000 persons— "tiny" towns, veritable hamlets.[51] The reasons for the inverse relationship between the size of a place and the rate of school attendance concern greater educational benefits in the smaller areas, higher opportunity costs in the larger cities, and the greater homogeneity of demand for education in the smaller places.

Towns were hotbeds of secondary school education. Smaller places were fairly homogeneous by income, religion, and ethnicity. Town citizens may have found it easier to agree upon the type of education and the amounts to be spent.

At the other extreme were the large cities—those with more than 500,000 persons—which had the lowest rates of enrollment. They contained the greatest fraction of immigrants, the widest disparity between the incomes of the rich and the poor, and the greatest demand for youth in relatively well-paying manufacturing jobs. Yet the negative relationship between city size and enrollment persists in a regression framework even after controlling for many of these factors.

Across about 220 cities in 1923 and 1927, as shown in the regressions of Table 6.3, the negative relationship between city size and the attendance rate of public secondary school pupils exists even though we have included the fraction of the population that was of immigrant parentage, the fraction Catholic, per capita taxable wealth of the city, and the fraction of the population employed in manufacturing jobs as well as that in various skill categories within manufacturing employment.[52] The relationship is obvious, even though the smallest cities that could be included had a population of 30,000—far larger than the towns of the previous discussion that had even larger high school attendance rates.

In 1923 the smallest cities in the sample (with populations from 30,000 to 35,000) had a public secondary school attendance rate 17 percentage points higher than the largest cities (with populations of over 500,000) and 10 percentage points higher than the second largest group (with populations from 100,000 to 500,000). Because the average secondary school attendance rate for the sample cities in 1923 was about 43 percent, the differences across city size are large. The small-city advantage in attendance rates is modestly expanded once city-level economic, demographic, and regional variables are taken into account. The results are similar, albeit slightly smaller relative to the mean, in 1927 with the (regression-adjusted) secondary school attendance rate monotonically declining across city size categories in both years.

The urban education data also reveal how municipalities accommodated the enormously increased demand for secondary school education during the height of the high school movement. Various margins could have been expanded. More schools could have been built and more teachers could have been hired. That is, the "extensive" margin could have been pushed out. The "intensive" margin could also have been extended by adding more students to each classroom or more teachers to each existing school.[53]

The evidence from our large sample of U.S. cities, collected from Commissioner of Education documents, indicates that the extensive margin was greatly expanded from 1915 to the late 1920s, during the

Table 6.3. Public Secondary School Attendance Rates and City Size

| | Public Secondary School Attendance Rate | | | | |
| | 1923 | | | 1927 | |
	Coeff.	s.e.	Means	Coeff.	s.e.
City population variables					
30,000 to 35,000 persons	0.221	0.0398	0.123	0.209	0.0365
>35,000 to 50,000	0.169	0.0361	0.283	0.157	0.0331
>50,000 to 100,000	0.103	0.0356	0.320	0.0980	0.0328
>100,000 to 500,000	0.0798	0.0358	0.224	0.0674	0.0329
City economic variables					
Log per capita taxable wealth (1926)	0.0621	0.0245	7.396	0.0404	0.0225
% production workers in population	−0.239	0.143	0.153	−0.110	0.131
% female workers in manufacturing	−0.267	0.101	0.210	−0.240	0.0927
% semiskilled in manufacturing	0.106	0.116	0.230	0.163	0.107
% craft workers in manufacturing	0.171	0.108	0.473	0.230	0.0990
% managers in manufacturing	0.597	0.279	0.0710	0.440	0.257
City demographic variables					
% Catholic (1926)	−0.248	0.0785	0.256	−0.323	0.0722
% native-born of foreign parents	−0.135	0.132	0.194	−0.0380	0.122
Constant	0.104	0.210		0.269	0.193
R^2	0.604			0.632	
Number of observations	219			220	
Mean of dependent variable	0.425			0.469	

Sources: Appendix C.

Notes: The dependent variable is average daily attendance in the high school grades of public schools divided by the number of youths 14 to 17 years old, for each of the cities in 1923 and 1927. Only cities with populations exceeding 30,000 in 1920 plus Elgin, Illinois (which had a population just below 30,000) are included because of limitations on per capita taxable wealth data.

City population: City population in 1920, U.S. Bureau of the Census (1923).

Log (per capita taxable wealth), 1926: U.S. Bureau of the Census (1927).

% production workers: Production workers in manufacturing as a fraction of the city population. Production worker data are from the U.S. Bureau of the Census (1923a).

% female, semiskilled, craft, or managers in manufacturing: fraction of all manufacturing workers who are female, semiskilled, craft, or managers; data obtained from Robert Whaples.

% native-born of foreign parentage: fraction of city population that is native-born of foreign parentage in 1920.

% Catholic: fraction of the city population (average of 1920 and 1930) who were members of the Roman Catholic Church in 1926. U.S. Department of Commerce (1930).

A full set of region dummies is included. All variables are for 1920, unless otherwise specified. The excluded categories are cities with more than 500,000 persons and the Pacific region.

early phase of the high school movement. At that time, the intensive margins—students per classroom and teachers per school—remained much the same. Municipalities met increased demand during the 1910s and 1920s by building schools, adding classrooms, and hiring teachers.

As additional schools opened, more children lived closer to a school and the cost of attending school decreased. Even though all of the nation's larger cities had secondary schools by the turn of the twentieth century, many youths had to travel long distances to attend one. In the 1910s the entire Bronx, for example, contained just one high school—the Morris High School which had opened in 1897. The high school was large (more than 4,000 students in 1911) but the enrollment rate among teenagers in the Bronx was far below that in the other boroughs of New York City, which had smaller and more numerous schools closer to the students' homes.[54] "Quite probably," noted the New York City Superintendent of Education, "this is due to the fact that The Bronx has but one high school; even though it is one of our best high schools" (New York City 1911).

The expansion of secondary schooling was rapidly altered at the start of the Great Depression. During the economically difficult 1930s the intensive margins were greatly extended. Increased funding for schools and teachers was halted or delayed at the same time the demand for more spaces soared as jobs disappeared. Increased classroom crowding was inevitable, as our data indicate.

We present, in Table 6.4, some of the statistics relevant to these issues by using our urban panel data set (see Appendix C). The data set includes all U.S. cities with a population exceeding 20,000 in 1920 and covers five years—1915, 1923, 1927, 1933, and 1937—thereby spanning the early, middle, and late periods of the high school movement. The data set includes a large number of variables for public, although not private, schools such as number of students enrolled and in attendance; teachers by type of school; and mean salaries of teachers, principals, and superintendents.

About 290 cities are included in the full sample which, by 1915, were almost all unified school districts. About 35 percent of those who attended public secondary schools in the 1920s and 1930s nationwide are included.[55] The number of cities with complete reporting of the relevant information changed by year. The cities meeting the minimum population criterion for inclusion in the federal data rose in the 1910s and educational

Table 6.4. Public High School and Student Characteristics of Cities: 1915 to 1937

High School Characteristics	Balanced Panel of 215 Cities				
	1915	1923	1927	1933	1937
1. Mean student enrollment	2500	4403	5436	8189	8550
2. Mean student attendance[a]	2022	3683	4695	6988	7561
3. Attendance/enrollment	0.809	0.837	0.864	0.853	0.884
4. Mean number of high school teachers	96.1	162.8	206.7	265.0	285.5
5. (Students/high school teacher)	26.0	27.0	26.3	30.9	29.9
6. Mean number of high schools[b]	1.89	3.09	3.71	4.24	4.48
7. (Teachers/high school)	50.9	52.7	55.7	62.5	63.6
8. Average term length (days)	184.8	184.7	184.4	182.0	182.1
9. Dollars/pupil, constant 2000 dollars[c] [current dollars]	325.9 [55.4]	368.7 [106.1]	390.7 [114.4]	387.6 [84.8]	397.7 [96.4]
10. Mean public high school attendance rate[d]	0.289	0.437	0.476	0.625	0.674

Sources: See Appendix C.

Notes: The ratios (attendance/enrollment), (teachers/school), and (students/teacher) have been computed from the sample aggregate means. The other rows are the city-level means of each variable for the 215 cities in our balanced panel sample.

a. Student attendance data in 1915 are only available for 207 cities. We impute student attendance in 1915 for the eight cities with missing data by assuming they have the same attendance/enrollment ratio of 0.809 as for 207 cities with complete data.

b. The number of high schools is not given for 1915 and is estimated by using the number of secondary school principals per city multiplied by the average number of principals per school in 1923, for those cities included in the 1915 sample. This procedure reduced the number of cities in this cell to 206.

c. Current values are deflated by *Historical Statistics, Millennial Edition*, table Cc1. The number of observations for this variable is: 1915: 124; 1923: 214; 1927: 215; 1933: 214; 1937: 211.

d. Average daily attendance in the high school grades of public schools divided by the number of youths 14 to 17 years old.

reporting increased in the 1920s, both of which increased the sample. But many southern cities in the 1930s failed to report education information to the federal government and the sample size falls in consequence.[56] We present results from a balanced panel of 215 cities, but the conclusions are similar for the unbalanced panel and hold up, as well, for each of the regions although they are not presented here.

Of the more than twofold (0.78 log point) increase in enrollments from 1915 to 1927 (Table 6.4, row 1), 87 percent was accommodated by an increase in the number of schools (row 6), 12 percent through an increased number of teachers per school (row 7), and just 1 percent from an increase in classroom size (row 5). The sharp rise in enrollments partially reflected population growth but was also due to the 60 percent increase in average daily attendance rates—from 29 percent in 1915 to 48 percent in 1927—in the typical city (row 10).

As the country moved from economic exuberance to the depths of the contraction in 1933, the situation radically changed. Enrollments increased by 1.5 times from 1927 to 1933, more on an annual basis than from 1915 to 1927. The increase in enrollments had accelerated in these cities but the cities were strapped for cash. Although schools continued to be built, the extensive margin accommodated only 33 percent of the increase. Whereas the intensive margin had been of trivial importance in the earlier period, the increase in students per classroom now contributed 39 percent of the total and the increased number of teachers per school accounted for 28 percent. At this point, schools were seriously constrained financially.[57]

Even though enrollments soared from 1915 to 1937, the ratio of attendance to enrollment (row 3) did not decrease and, in fact, actually increased. That is, the enrollments were not coming from marginal students whose attendance rates were lower than the previous group. Average term length (row 8) was almost 185 days from 1915 to the 1920s and although it decreased at the start of the Depression, the decrease was of trivial proportion. Finally, the average amount (in real dollars) spent per student enrolled (row 9) increased by 20 percent from 1915 to 1927. Although spending per pupil stagnated at the beginning of the 1930s, it managed to increase slightly from 1933 to 1937.

Public and Private Schools

As we saw in Chapter 5, the high school movement was preceded by another grassroots action. Known as the academy movement, it was motivated by a desire to increase schooling beyond the common school year and the elementary grades. Academies were almost always private, tuition-charging institutions, although they often received some public funding. Most were rapidly supplanted in the late nineteenth century by publicly provided and funded secondary schools. Yet a substantial fraction of youth, in various parts of the nation, attended private secondary schools in 1910 when our state-level data begin.

Some of the students we classify as attending private schools were in the preparatory divisions of colleges and universities. That group comprised 31 percent of all private school students in 1910 but diminished in importance to 22 percent in 1920, and to 10 percent in 1930 (see Appendix B, Table B.6).[58] About half of the students in preparatory departments in 1880 were training to continue to college.[59]

Many colleges and universities prepared their students prior to entering college, and this role was more important when high schools in the area were few in number. The University of Nebraska, for example, was founded in 1871 when there were few public high schools in the state. Even in 1900, the university enrolled more students in its preparatory division than in its collegiate department. In the more settled parts of the nation, and especially in the East, colleges and universities could rely on private and some public high schools to train students to undertake the rigors of college. Nevertheless, many colleges in 1910, even in the older settled areas, had numerous students in their preparatory departments and this was particularly true among those with strong religious ties. In the nation as a whole, there were only 2.4 students in the collegiate departments of universities and colleges for each preparatory student, although the figure was 4.6 in the states of the northeast.

The fraction of youth enrolled in and graduating from all private secondary schools, among high school students and graduates, was highest in places where public high schools had not yet made inroads, such as in the less settled parts of the country and the South. Private school enrollments and graduates were also relatively higher in parts of New England where Catholic schools had long prospered. As public secondary schools spread in the 1920s, the fraction of high school youth enrolled in and graduating from private schools greatly declined and this trend was reinforced by the relative increase in public school enrollments during the Great Depression. However, by the 1950s, the fraction of high school graduates coming from private schools had so greatly increased in the eastern part of the nation that it resumed the level from the 1910s. (See Appendix B, Table B.3 for the fraction of all high school graduates from private schools by census region.)

In the nation as a whole the fraction enrolled in private high schools decreased from 16 percent in 1910 to just 6.5 percent in 1940. Some of the decline was due to the dual impact of the Great Depression in increasing all enrollments while decreasing the absolute number in private schools.[60] The fraction of youth in private schools was just below 9 percent on the eve of the stock market crash. But most of the decrease over the long period from 1910 to 1940 was due to the simple fact that the spread of public secondary schools expanded the fraction of youth who attended public high schools. Other than during the early part of the 1930s, the absolute number of youths attending private high schools did not decline in any region. Rather, it almost always increased, but simply could not keep up with the increase in the number of youths attending public schools.

After World War II the fraction of youth in private secondary schools began to increase again. Even as early as the 1950s the fraction in private schools increased to 11 percent, about the level achieved nationwide in the 1920s. In New England and the Middle Atlantic, the level achieved in the post war era was actually higher than in 1910, although in the West and the South it was considerably lower.

Differences by Sex

As the high school movement took off, graduation rates for girls exceeded those for boys in every state. Their advantage was greatest in relative terms in the early years of the movement but girls retained their advantage well after the revolution in high school was completed. Enrollment rates for girls were also generally higher than those for boys. Only during the early 1930s was the gap substantially reduced. In the Middle Atlantic region it disappeared altogether in the early years of the Great Depression when jobs for youths, particularly young men, virtually evaporated.

Girls had a 5.6 percentage point advantage nationwide in secondary school graduation rates during the first two decades of the high school movement, other than in the World War I years when the gap was a bit greater.[61] The sex difference translates into a 39 percent graduation rate advantage of girls over boys. The difference varied by region. For example, during the years from 1910 to 1928 the graduation rate of girls exceeded that for boys by 6.8 percentage points (35 percent) in New England and by 7.2 points (31 percent) in the Pacific region. In the Middle Atlantic region the female excess was 3.3 percentage points (30 percent) and in the South Atlantic it was 3.9 percentage points (52 percent).[62] In Figure 6.5 we provide the graduation rate by sex for the entire United States from 1910 to 1970.

That girls remained in secondary school longer than did boys and thus attained higher graduation rates is not surprising. The difference between having a high school diploma and not having one was as large for a young woman as it was for a young man in terms of remuneration. But it was even larger for a woman in terms of occupational status. A woman who could secure a position as a clerk, stenographer, or, better yet, secretary was freed from the drudgery of piece-rate work in a factory or, worse yet, employment as a domestic servant. Most important at the time, perhaps, was that as a white-collar worker a woman would have a better chance of securing a higher income husband.

Creation of the Modern High School

In 1900, high schools in Vermont were termed "vague, nondescript, individual, and independent," even though Vermont was one of the New England leaders in nineteenth-century education.[63] "What constitutes a high school," noted an Iowa school report in 1893, "has never been defined."[64]

According to a turn-of-the-twentieth century U.S. Office of Education survey, the nation's public high schools varied widely by size, curriculum, and type of building.[65] Out of the 7,200 public high schools in the nation in 1904, 72 percent had fewer than four teachers and 30 percent had just one. Almost 70 percent of public high school students were taught in schools with 10 or fewer teachers and 37 percent were in schools with fewer than four. To Edward Thorndike, the early twentieth century educational psychologist and authority on testing, the high school around 1905 was "an institution of enormous variability as regards its capacity for educational work and its administrative and educational arrangements."[66] Even by 1930, I. M. Kandel, the prolific writer on education in the early twentieth century, noted: "The public high school of the United States . . . in its present form . . . is of recent origin, and it is still in a stage of transition" (1930, p. 496). Yet, the public high school had made decisive steps in the previous three decades.

In those 30 years the public high school had been rapidly transformed from an institution whose chief, but not sole, function was to prepare young people to enter college to one that primarily educated those who would end their education somewhere between ninth and

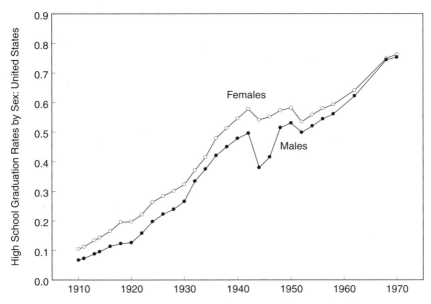

Figure 6.5. High School Graduation Rates by Sex. Graduation data by sex are provided in the Office of Education reports only for public secondary schools. We divide by the number of 17-year-old youths. We have scaled these rates by the ratio of the total graduation rate to that for public institutions. Sources: See Appendix B.

twelfth grades and not continue to college, immediately or ever.[67] As the high school movement got underway not only were the number of schools, teachers, and students vastly expanded, but the entire institution was altered.

The public high school was recreated in the early 1900s to be a quintessentially American institution: open, forgiving, gender neutral, practical but academic, universal, and often egalitarian. It was reinvented in a manner that moved it away from its nineteenth-century elitist European origins.[68] Many of the changes in the curriculum of the high school were the inevitable result of the proliferation of academic disciplines more generally. As we will see in Chapter 7, the number of separate academic disciplines practiced in the nation as a whole exploded in the 1890s. The social sciences, for example, were largely created around the 1890s, as were many of the modern sciences. But not all of the changes in the American high school were rooted in more basic changes in the structure of knowledge. Rather, whether intended or otherwise, the reinvention of the high school brought secondary education to the youth who would otherwise have gone to work. The American high school in the early twentieth century became a "modern" school and secondary education became mass education.

Changing Mission

An important consequence of the expansion of high school enrollments was that the fraction of youth who continued to college decreased while the fraction who graduated from high school increased. Surveys of graduating seniors in the early part of the twentieth century (see Table 6.5) indicate that the fraction of high school graduates who continued to college or some professional training was greater in the 1910s than it was to be until around the 1970s.[69] Odd as this may seem, it is a logical result of the creation of mass high school education.

Until college became a form of mass education, increasing the fraction who graduated from secondary school decreased the fraction continuing to college. In the nineteenth century one of the most compelling reasons to graduate from a secondary school was to attend a college or university.

The fact that the fraction of high school graduates who continued with their education decreased after the 1910s may seem like a small point to make, but it is actually quite important in the history of American education. The change in the immediate destination of high school graduates is a key aspect of the transition to the modern high school. It is also an important difference between secondary education in the United States and in Europe during the early part of the twentieth century.

The data in Table 6.5 appear to contradict the notion that college attendance involved but a small fraction of young people before the mid-1940s,[70] but they are an accurate assessment of the fraction of high school graduates who intended to continue to college. The fraction of Americans who attended college before the 1940s *was* small—10 percent for males born around 1900. But the fraction who graduated from high school prior to the 1920s was also small. Therefore, the fraction of high school graduates who went on to college was fairly high around the 1920s and then decreased. Although some of the youths who were reported by high school principals to have intended to continue to col-

Table 6.5. High School Graduates Intending to Continue Their Education: 1901 to 1937

	Percentage of High School Graduates Intending to Continue Their Education					
	1901	1910	1914	1923	1933	1937
Continuing to college from:						
Public high schools	31	34	35	31	21	24
Public high schools, males only	40	45	45	37	23	26
Public and private high schools	33	35	35	32	23[a]	—
Continuing to college and other institutions[b] from:						
Public high schools	—	49	50	44	25	29
Public high schools, males only	—	55	55	49	25	30
Public and private high schools	—	49	50	46	25[a]	—

Sources: U.S. Office of Education, *Annuals* (various years); U.S. Office [Bureau] of Education, *Biennials* (various years).

Notes: "Percentage continuing" was reported by the Commissioner of Education who gathered the information from the reports of school principals. It is the percentage who intended to continue, probably in the immediate future. Some did not eventually do so, and others went back to school at a later date having never indicated any intent to continue their education. These figures probably do not include students in the preparatory departments of colleges and universities. Their inclusion would increase the continuation percentages in the earlier years.

a. Private school graduation numbers are for 1932.

b. "Other institutions" probably include normal, nursing, and library schools.

lege may not have done so, immediately or ever, the vast majority actually did.

The most informative evidence we have on this point comes from the Iowa State Census of 1915, which we have used in previous chapters (see Appendix A).[71] The census asked all residents of Iowa how many years they had attended schools of various types. We have selected males 20 to 29 years old in 1915 who completed four years of high school or who reported that they attended school for more than twelve years and went to college. These men would have graduated high school between around 1904 and 1913. The only complication in these data is that some individuals who went to college do not indicate that they ever attended a school called a high school. Some would have gone to an academy, some would have gone to the preparatory department of a college or university, and others could have been home schooled. Depending on the assumptions we employ, we find that from 43 to 57 percent of the group continued to college either immediately following high school or preparatory school completion or within a decade after. If, instead, we restrict the sample to those who were living in the larger cities of Iowa in 1915, the range is 49 to 61 percent.

Iowa contained a far lower fraction of its population living in large cities than did the rest of the United States and even the largest city in Iowa—Des Moines—was just the 62nd largest in 1910.[72] Thus the Iowa data are likely to understate the fraction of high school graduates who continued to college since, as the Iowa data show, a larger fraction of those continuing lived in large cities. But even with this qualification, the point we are making is that the Iowa data are in the range of those from the surveys of high school principals given in Table 6.5.

The data in Table 6.5 are crucial to understanding the forces behind the creation of the modern high school. Important changes occurred in the curriculum of the high school beginning in the 1920s, although there were precedents extending back to the mid-nineteenth century. These curriculum changes have been debated ever since. To some, curriculum change occurred because of a shift in the composition of high school students. To others, the changes were due to misguided progressive reformers who diluted the curriculum and set the nation's schools on a wrongful course.[73] The curriculum, as we will see, was altered for most high school students. But the motivation for change was often to attract to the high schools youths who would otherwise have gone directly to work. By vastly increasing the enrollments, attendance, and graduation from high schools, the reinvention of the high school led, eventually, to mass high school education.

Curriculum change was both cause and effect of the rise in enrollments in U.S. public high schools. The creation of the modern high school made secondary school more appealing to youths. But it was also the case that changes in the demand for education above that given in the lower grades, as we detailed in Chapter 5, did not often involve knowledge of the subjects that had been an integral part of the usual college preparation course of study.

To make high school appealing to a mass market, it was altered in a variety of ways. These changes were not the same in all parts of the country. Urban youth had different needs and desires from those in rural areas and on farms. What unified the transition was that the high school would now serve a variety of functions. It would instruct youth in a larger group of academic courses and many non-academic subjects.

The "vague, nondescript" school of 1900 was transformed into an institution that was well-defined only in its eclectic and "anything goes" character.[74]

Curriculum Change

The increased demand for secondary school education that was voiced by parents and youth in the late nineteenth and early twentieth centuries was for instruction in the subjects already part of the high school curriculum as well as for a broadening of its teachings. Many parents wanted their children to be educated in subjects that were part of the traditional courses offered in secondary schools in what was known either as the Classical or the English courses of study.[75] They wanted their children to learn more history and English, to begin foreign language study, and to advance their knowledge of mathematics to include algebra and geometry. Others, however, wanted a broadening of the curriculum to make it more relevant to the positions their children would eventually undertake. Many who lived in the nation's larger cities were already paying tuition to give their children expertise in commerce, real estate, stenography, and a variety of other commercial and vocational skills. During the very early twentieth century the high school curriculum in both the large city and tiny town was broadened to include a wider set of academic courses and a far larger group of non-academic courses.

The non-academic subjects included commercial courses in typing, stenography, bookkeeping, and various business applications of English, law, and geography. Vocational instruction was added, including shop classes in woodworking, electricity, metals, and a variety of manual training courses for the boys and cooking, sewing, and other household arts for girls. Courses for "life" were added in music, dance, theater, arts, hygiene, and exercise. In rural areas the high school curriculum was expanded to provide instruction in the care of farm animals, farm machinery, botany, and accounting techniques. Although most changes in the public high school curriculum began as early as the late nineteenth century, many of the non-academic subjects were offered by academies in the mid-nineteenth century, including those that provided instruction in classical courses for the majority of their students.[76]

Academic courses increased through the addition of new areas as well as by subdividing the older ones. Language offerings increased to include modern languages. Science courses were expanded to include laboratory sciences. In small towns and villages the greater offerings often accompanied an increase in the scale of schools. When a high school had just 60 students and 3 teachers, the range of courses was necessarily limited. Yet even in tiny high schools a wide range of courses emerged in the 1920s. The larger schools of the big cities had always been able to give an expanded set of offerings, but among these schools the curriculum also expanded with the high school movement.

The World War I years marked a turning point in the proliferation of high school courses. "The modern high school program is indeed more comprehensive than was that of the old-time college," reported the Washington State high school inspector in 1922.[77] The transition occurred in most schools, save the very largest in which many of the courses were already offered. Whereas in the years before 1910 small to moderately sized schools offered courses in only the more traditional disciplines, the same schools in the 1920s offered courses in more than twice the number of subjects, which now included nontraditional academic subjects as well as a host of non-academic ones. The schools were, admittedly, mostly larger in size. But cross-section evidence we have located for the 1920s confirms that even tiny schools stretched their resources in the 1920s to offer a wide variety of courses.[78]

Some of the new courses were variants on an older theme. Until around 1920 English was a generic subject. But in the 1920s a teacher who previously had offered only English also gave courses in public speaking, journalism, and debate. Similarly, mathematics became a more differentiated subject that included geometry, algebra, and commercial mathematics. In other cases entirely new areas were added, including commercial and vocational subjects and a wide range of "life" courses.

For example, in 1900 the high school in Ottumwa, IA, a city of almost 20,000 people, had ten teachers and offered courses in ten subjects. In 1917, just before the Ottumwa junior high school opened, the high school had 29 teachers who offered 18 subjects. Of the eight additional subjects, two were spin-offs of the original academic courses but the remaining six were a mixed bag of commercial, vocational (e.g., mechanical drawing), and life (e.g., music, drawing) courses. In 1925 the high school had 52 teachers (plus there was a junior high school) and offered 35 separate types of courses (29 if the five commercial, two home economics, and two music courses are grouped).[79]

A cross-section of 20 Iowa towns and small cities that we selected for 1924 includes places ranging in population from 200 (Laurel) to 18,000 (Muscatine) and in high school enrollments from about 60 to 600.[80] High school enrollments in the smaller towns, it should be noted, came from both the town and the surrounding rural areas. In 1924 the median number of separate subjects offered was 13 (the mean was 13.6), and even the smallest town in the sample offered 13 subjects with just 4 teachers. The school's science teacher taught physics, general science, geometry, algebra, and even domestic science. The number of subjects offered per teacher decreased with the size of the school, and teachers taught one subject only when enrollments exceeded about 200 students. Just two decades earlier most teachers would have offered but a single subject regardless of the size of the school.

Courses offered do not necessarily indicate what students actually took. We now turn to complementary data on curriculum collected by the federal government. Beginning around 1890 the U.S. Commissioner of Education asked public high school principals to report the number of students taking particular subjects during the school year. These data span a long time period and include a large cross-section of public schools and their students. Principals were not asked the duration of each course. It is likely that the academic courses met every day for about one hour and that the non-academic courses met more infrequently. Nor were principals asked the school grade or year of the student. Another ambiguity is that certain courses were enumerated separately in some reports but were grouped together in others.[81] The earliest year of fairly complete reporting is 1915.

Enrollment in commercial courses was not requested until 1915, when 15 percent of all students were taking one or more commercial subjects. The last pre–World War II year for the curriculum survey is 1934.[82] We can, therefore, examine the change in courses across four years during the high school movement: 1915, 1922, 1928, and 1934. The fraction of all youth attending high school soared across those years, but of those in attendance the fraction intending to continue their education beyond public high school fell (it will be recalled from

Table 6.5 that it decreased from 50 percent in 1914 to 44 percent in 1923 and then to 25 percent in 1933). It should not be a surprise therefore, that the average number of academic courses fell and the average number of non-academic courses rose during this period.

From 1915 to 1934 the average number of academic courses taken each year per student declined from 3.83 to 3.04, and from 1922 to 1934 the decline was 3.47 to 3.04.[83] The decrease in academic subjects occurred across the board, although the greatest decline was in mathematics. Of the 1915 to 1934 decline, about half was due to the decrease in math enrollment, and of the 1922 to 1934 decline, all was due to the decrease in math courses.

Part of the decrease in academic subjects came from a change in the distribution of students across grades. As the high school movement spread and youths spent more years in school, a larger fraction were in the upper grades. A higher fraction of the students were in the lower grades in 1915 than 1934 and academic subjects were taken somewhat more frequently, at that time, by students in lower than upper grades of high school.[84] In 1915, 71 percent of high school students were in the lower grades (9 and 10), whereas 60 percent were in the lower grades in 1934.[85] The change in the distribution of students can explain part of the decline, but does not explain the decline altogether. Although the decrease in the fraction of all students taking math from 1915 to 1934 was 27 percent, when the decrease is weighted by each grade it is less. Although the precise number cannot be calculated using the Bureau of Education data, it ranges from just 9 percent to the actual figure of 27 percent.[86]

Since the number of academic subjects taken by students decreased, something else must have increased. The subjects that expanded to fill out the school year were primarily those of a commercial nature. A wide range of non-academic courses surfaced but, without knowing how many times they met each week, they are difficult to aggregate. Even with these problems of interpretation, it seems clear that a range of non-academic subjects increased.[87]

The data on curriculum suggest broad changes affecting schools of all sizes and in all types of places. Courses that had direct applicability to office employment increased greatly, as did those that were useful for professional development, such as teaching. These changes, however, were not confined to public high schools. Even academies that taught a classical curriculum emphasized commercial skills. Proprietary schools that charged fees for commercial courses, such as shorthand, typing, and accounting mushroomed in the late nineteenth century in the nation's larger cities. The public high school rapidly supplanted them.

In 1894, the first year for which we have data on enrollments in proprietary schools and commercial courses in public high schools, the latter accounted for just 12 percent of the total. By 1910 public high school students were 38 percent of the total and by 1924 they were almost 70 percent.[88] The public high school provided services similar to those of the proprietary schools, but at a much reduced cost. As the public high school supplanted many of the proprietary commercial schools, a greater fraction of all commercial students were female. In 1875 just 9 percent of commercial students were female, in 1900 36 percent were, and in 1930 67 percent were, an almost linear increase.[89]

The changes observed in the high school curriculum were not entirely motivated by a top-down campaign to dilute academic subjects. Rather, a grassroots movement altered the curriculum to include subjects of relevance to the lives of youths in a modern economy.

One of the many reasons offered for expanding the high school curriculum in the 1910s and 1920s was to make a secondary school education more appealing to youngsters who were dropping out of school around age 14 or 15. Another way to achieve that goal, according to some Progressive reformers, was to create a new educational institution—the junior high school—that would entice youth to remain in school at least for an additional year, to the end of ninth grade. The first junior high schools were founded in 1909 in two university towns (Berkeley, CA, and Columbus, OH), and the concept rapidly spread. By 1923 48 percent of all U.S. cities with populations greater than 25,000 had at least one junior high school and in 1927, 69 percent did.[90]

Teacher Quality

The soaring enrollment of high school students from 1910 to 1930 greatly increased the demand for secondary school teachers across the nation, especially so in the states that were the educational leaders of the high school movement. In the nation as a whole the increase in the number of teachers averaged 7.6 percent annually, whereas the increase in the number of primary school teachers averaged only 1.6 percent annually.[91] The number of secondary school teachers increased by 4.6 times from 1910 to 1930 while the number of primary school teachers increased by just 1.4 times.

In California, public school teachers of the high school grades increased at 9 percent average annually from 1911 to 1930; the number of elementary school teachers increased at 4 percent. In Kansas the growth of high school teachers was more than 6 percent annually over the same period. In the nation's larger cities, the growth averaged 6.6 percent annually from 1915 to 1927.[92] In all cases, the increase for teachers in the high school grades greatly exceeded that for elementary school teachers.

Because the number of secondary school teachers increased so rapidly in the entire nation and truly exploded in many parts, one might have expected a reduction in the quality of teaching personnel, at least for a while. One would expect that the earnings, credentials, and fraction female among secondary school teachers to have been greatly altered. Oddly, the enormous increase in the demand for secondary school teachers does not appear to have had these expected effects.[93] The increased demand for secondary school teachers did not lead to an increase in the earnings of secondary school teachers relative to elementary school teachers, did not produce a reduction in their credentials, and did not immediately lead to an increase in the fraction female among high school teachers. It appears that the supply of quality teachers was extremely elastic and the large increase in the demand for teachers did not lead to a decrease in the quality of teaching personnel at the secondary school level. Perhaps one of the reasons that the high school movement could proceed so smoothly is that the supply of high school teachers was actually quite elastic. As the high school movement proceeded, more females were able to continue to teachers colleges and universities making the supply even more elastic.

We have been able to obtain information on the fraction female among public secondary school teachers and their earnings relative to elementary school teachers for all reporting school districts in two states (California and Kansas) from around 1910 to 1940, and for a large sample of U.S. cities during the period of the high school movement. We also have the fraction female among high school and junior high school teachers for the nation as a whole. Teacher credentials by sex are available for Kansas and Oregon in the 1920s.

In the U.S. data and for both California and Kansas separately the fraction female among high school teachers was about 0.60 to 0.65 during the 1910s, before U.S. entry into World War I (see Table 6.6). The fraction increased to around 0.70 during the war, as would be expected during a period of military draft. But the fraction female then proceeded to decrease almost immediately in all of the series except the larger cities of Kansas.[94] In all cases, however, on the eve of the Great Depression the fraction female among public secondary school teachers was at about the same level as it had been in the early 1910s. The series having data on the fraction female during the Depression show a considerable decrease, consistent with the increase of men in teacher training institutes in the 1930s (see Chapter 7). Many school districts in the Depression had instituted "marriage bars," or enforced existing regulations concerning the hiring of married women and the retention of single women who married while in the employ of the district.[95] So it is not surprising that the fraction of high school teachers who were female decreased during the 1930s, but what is surprising is that it did not increase much before.

Similarly, the earnings of high school teachers did not advance greatly relative to those for elementary school teachers during the high school movement (see Table 6.7). Relative earnings for both male and female high school teachers fell somewhat from the 1910s to the 1920s and then far more rapidly during the 1930s. It seems clear that the earnings of high school teachers did not increase relative to those for elementary school teachers, even within sex, despite the far greater increase in the demand for the former. If relative earnings did not increase then perhaps credentials for high school teachers declined, allowing elementary school teachers to be used to teach the upper classes. But the evidence indicates that teaching credentials for high school teachers did not decline. In fact, credentials increased as more districts passed and enforced rules governing who could teach in the upper grades.

The data for Kansas cities, which cover 1912 to 1922, show an increase in the fraction of teachers who graduated from a four-year college or university. In 1912, 68 percent of high school teachers in the

Table 6.6. Fraction Female among Public Secondary School Teachers: 1910 to 1940

	Fraction Female among Public Secondary School Teachers				
	(1)	(2)	(3) Kansas		
Year	United States Entire Nation	California Entire State	City	Town	Village
1910	0.535	—			
1911	—	—	0.659	0.777	0.549
1912	—	—	0.644	0.766	0.632
1913	—	—	0.625	0.661	0.725
1914	—	—	0.622	0.656	0.651
1915	—	—	0.640	0.708	0.632
1916	—	0.665	0.645	0.711	0.600
1917	—	0.659	0.658	0.669	0.586
1918	—	0.693	0.678	0.724	—
1919	—	0.718	0.711	0.755	—
1920	0.685	0.708	0.682	0.726	—
1921	—	0.672	0.696	0.708	—
1922	—	0.673	0.711	0.730	0.567
1923	—	0.669	0.727	0.702	0.586
1924	—	0.668	0.720	—	—
1925	—	0.670	0.713	0.706	—
1926	—	0.664	0.720	0.691	—
1927	—	0.673	—	—	—
1928	—	0.667	0.703	—	—
1929	—	0.656	0.689	0.663	0.621
1930	0.651	0.654	0.680	0.653	0.628
1931	—	0.614	0.674	0.648	—
1932	—	0.615	0.666	0.644	—
1933	—	—	—	—	0.564
1934	—	—	—	—	0.549
1935	—	—	0.660	0.593	0.549
1936	—	—	0.642	0.577	0.632
1937	—	—	—	—	0.725
1938	—	—	—	—	0.651
1939	—	—	0.615	0.544	0.632
1940	0.578	—	0.600	0.550	0.600

Sources: Col. 1: U.S. Office of Education, *Biennials* 1938–40, p. 35, table 31 for rows 1 and 2. Col. 2: California, Office of Superintendent of Public Instruction, *Biennial Reports* (1915/16 to 1931/32). Col. 3: Kansas, State Superintendent of Public Instruction, *Biennial Reports* (1911/12 to 1939/40). See also Frydman (2001) on the Kansas data.

Notes: United States: Data include junior high school teachers. The ratios for the last two years given are almost the same as data for teachers in only grades 9 through 12. Kansas: Starting in 1924 information on junior high schools was reported separately. Prior to that the high school numbers include the junior high schools. In calculating the ratios, high school teachers are assumed to include one-third of junior high school teachers by sex after 1923. Cities ("cities of the first class") had populations exceeding 15,000; towns ("cities of the second class") had populations exceeding 2,500; villages are all other incorporated areas and include county high schools.

Table 6.7. High School Relative to Elementary School Teacher Salaries by Sex: Kansas, California, and U.S. Cities

| | (High School Teachers/Elementary School Teachers) Annual Salaries | | | | |
| | Kansas Cities | | California | | 180 U.S. Cities |
Year	Females	Males	Females	Males	Balanced Panel
1911	1.376	1.286	—	—	—
1912	1.420	1.210	—	—	—
1913	1.443	1.305	—	—	—
1914	1.393	1.295	—	—	—
1915	1.395	1.268	—	—	1.560
1916	1.339	1.325	1.557	1.729	—
1917	1.377	1.362	1.555	1.739	—
1918	1.364	1.299	1.492	1.603	—
1919	1.307	1.416	1.505	1.443	—
1920	1.333	1.225	1.407	1.527	—
1921	1.246	1.277	1.382	1.597	—
1922	1.275	1.265	1.353	1.395	—
1923	1.221	1.309	1.339	1.402	1.398
1924	—	—	1.365	1.376	—
1925	1.259	1.343	1.343	1.373	—
1926	1.251	1.294	1.304	1.380	—
1927	—	—	1.348	1.469	1.453
1928	—	—	1.354	1.458	—
1929	1.238	1.318	1.348	1.463	—
1930	1.048	1.138	1.348	1.435	—
1931	1.200	1.229	1.357	1.388	—
1932	1.171	1.240	1.342	1.394	—
1933	—	—	—	—	1.222
1934	—	—	—	—	—
1935	1.130	1.145	—	—	—
1936	1.129	1.135	—	—	—
1937	—	—	—	—	1.187
1938	—	—	—	—	—
1939	1.145	1.159	—	—	—
1940	1.136	1.166	—	—	—

Sources: California, Office of Superintendent of Public Instruction, *Biennial Reports* (1915/16 to 1931/32). Kansas, State Superintendent of Public Instruction, *Biennial Reports* (1911/12 to 1939/40). City data set see Appendix C and Table 6.3.

Notes: Kansas: Data include teachers in cities of the "first class," generally those with more than 15,000 residents in 1920. The Kansas data listed junior high schools beginning with 1924. Data for high schools are adjusted to include teachers in the ninth grade of junior high schools and those for elementary schools include those in the seventh and eighth grades. When salaries are listed as monthly, the annual figure is nine times.

California: Data include teachers in the entire state.

U.S. cities: All cities with more than 20,000 people in 1920 are included.

state's 11 largest cities graduated from a four-year college or university. In 1922, about 71 percent did.[96]

The information from Oregon reinforces the finding that secondary school teachers had increasing qualifications during the high school movement. The data for Oregon's high school teachers in 1923 contain detailed information on sex, college attended, year graduated, high school courses taught, and years of teaching experience.[97] These data reveal that a larger fraction of the less-experienced group of high school teachers had a college degree than the more-experienced group.[98] The finding holds even though the teaching of vocational courses had greatly expanded over time and those who taught the courses did not have to have a college degree. When the vocational teachers are excluded, the increase in college degrees is even greater. Among males with less than ten years of teaching experience, 82 percent graduated from college; 92 percent excluding vocational teachers. Among males with ten or more years of experience, 68 percent had graduated from college; 85 percent excluding vocational teachers. The figures for female teachers are 94 percent for the younger group (97 percent excluding vocational teachers) and 74 percent for the older group (80 percent excluding vocational teachers).

There is no suggestion, therefore, that the quality of high school teachers was compromised in any discernible way during the peak expansion of high schools. Although we do not have detailed information for all parts of the nation, the data we have collected suggest that the supply of high school teachers was quite elastic. As demand increased, so did the quantity of teachers, without much impact on their earnings.[99] When the Great Depression hit, high school teachers experienced unemployment, but not to the same extent as the entire nation's labor force, although that may have been partially mitigated by the fact that many female teachers had been forced to leave the labor force through the extension of marriage bars.

Why the United States Led: A Summary

The high school movement in America, a transformation that began around 1910, was fairly complete by 1940. This transition to mass secondary school education was rapid and it was remarkably so in certain parts of the nation, such as the Midwest, New England, and the Pacific states. Years of educational attainment in the United States soared in the middle part of the twentieth century. The increase was largely due to the high school movement.

Equally noteworthy is that during these decades the United States led the world in mass secondary school education. No other nation would come close to putting as large a fraction of its youth through secondary school until the latter part of the twentieth century. Why did America lead? We can learn much from examining why certain parts of the United States led and why others lagged.

The leading areas in the early part of the high school movement had various characteristics in common. They were places with large numbers of competing school districts, considerable homogeneity among their populations, high levels of (taxable) wealth per capita, and a low degree of inequality. They were places that managed to retain their older citizens and thereby maintain a sense of community. We have found that smaller towns had far higher enrollment rates than did large urban places. Larger cities had greater job opportunities for young people and even though the rate of return to additional schooling was high, myopic youth may have been misled by immediate opportunities. Girls went to and graduated from high school to a far greater extent than did boys in the early stages of the high school movement and they continued to do so long after the high school movement ended.

The high school movement was, above all, a grassroots movement. It sprung from the people and was not forced upon them by a top-down campaign. In that sense it was a direct extension of the growth of academies and other schools that served youths in the nineteenth century. These private institutions mushroomed in all sections of the nation, wherever parents were willing to pay tuition to continue the education of their children. The publicly funded high school supplanted the academies, which for the most part, folded wherever a public high school opened.

Nor was the high school movement the result of legal compulsion. Our extensive empirical investigation, summarized in Table 6.2, indicates that the expansion of state compulsory schooling and child labor laws can account for only about 5 percent of the large increase in high enrollment rates from 1910 to 1938. The laws were passed or extended after the high school movement had begun and the movement proceeded with such enormous speed that the laws constrained few youths. Only particular parts of child labor and compulsory education laws that covered working youths were effective to some degree. The most effective were "continuation school" laws that compelled working youths, who left school before the compulsory schooling age, to attend school during the work week.

The most important margins at which the expansion took place, before the Great Depression, were the ones that expended public resources. In general, the extensive margin was pushed out. More schools were built, classrooms were created, and teachers were hired. Moreover, we have found that teacher quality was not compromised during the great expansion of high school enrollment. Teacher credentials increased, rather than decreased. The fraction female among high school teachers remained about the same. Yet the salaries of high school teachers did not increase relative to elementary school teachers. The supply of well-educated personnel was extremely elastic.

Not surprisingly the curriculum of high schools, especially in the 1920s, changed. The average student took fewer academic courses and more non-academic courses. But the average student had also changed. The fraction of high school graduates continuing to college decreased, yet the fraction of all Americans who continued to college increased. The expansion of high school was so great that both effects could occur simultaneously.

We will return to the high school in Chapter 9 since our story, thus far, has covered only the period to the mid-twentieth century. We must first discuss the third, and as yet unfinished, transformation to mass college education.

7

Mass Higher Education in the Twentieth Century

In his State of the Union Address delivered on January 11, 1944, President Franklin Roosevelt called for a second bill of rights—an Economic Bill of Rights—that would guarantee economic security to all Americans. The "right to a good education," he said, was among the "economic truths" that have become "self-evident." For the majority of young American men and women in 1944 a "good education" meant one thing—college. Most young Americans had already graduated from high school and the nation was poised for its third transformation in education—that to mass higher education.[1]

The reference to education and to a new bill of rights in Roosevelt's State of the Union Address was prelude to an act that had been planned from almost the start of U.S. involvement in World War II. A good education was central to the Serviceman's Readjustment Act, passed by Congress in May 1945 and known ever since as the GI Bill of Rights.[2] The GI Bill would do much to hasten the transformation to mass higher education, but did not cause it to happen. There is still considerable debate over whether the 1944 GI Bill had a large direct effect on education or simply enabled those who were drafted to achieve the education they would have attained had they not served in the military.[3]

The U.S. higher education system began long before the 1940s. But in 1944, when President Roosevelt proclaimed the "right to a good education," the U.S. higher education system had only recently attained its modern form and was just beginning on the road to greatness and glory. The higher education system that we know today was shaped in the half century that preceded Roosevelt's address, but it became the finest in the world in subsequent decades. In this chapter we explore the twentieth-century evolution of the higher education system—who went to college and when, the size and scope of the institutions, their excellence, and the relative size and role of the public sector. Although a college education became a middle-class entitlement in the mid-twentieth century, by the century's end the third transformation to mass college education was still incomplete.[4] We address why the third transformation is still unfinished.

Going to College

A Century of College-Going Trends

America began to go to college long before it graduated from high school. Because educational systems are hierarchical, the creation of schools at the most basic level and the establishment of those at the very highest were often coincident events. Harvard University, for example, opened its doors in 1638 when the American colonies were in their infancy. Universities were often established even before schools at the secondary level in states that were sparsely settled when they entered the Union. These institutions established their own preparatory departments to train potential undergraduates.

The mass movement to college had to await the transformation to widespread secondary school education. To get a sense of the history of colleges and universities in the United States, we have constructed an extensive time series of the fraction of Americans that went to college and the fraction that completed at least four years of college for those born from 1876 to 1975 (and measured when the individuals were 30 years old) using data from the U.S. federal censuses.

For cohorts born at the start of the twentieth century, the fraction attending any college was about 10 percent and the fraction of males graduating from college was about 5 percent (see Figures 7.1 and 7.2). Although these numbers would appear low, they are high in comparison with data from other nations at the same time. College rates for U.S. males advanced slowly during the next two decades, but soared

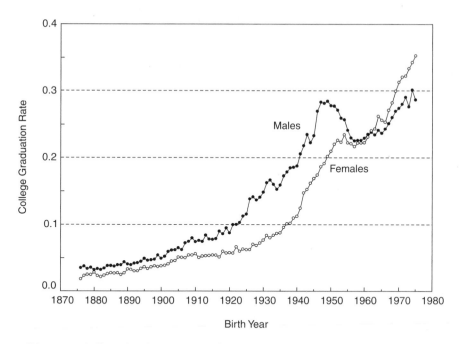

Figure 7.1. College Graduation Rates for Men and Women: Cohorts Born from 1876 to 1975 (by age 30). The figure plots the fraction of each birth cohort, by sex, that had completed at least four years of college by age 30 for the U.S.-born. Since educational attainment data was first collected in the U.S. population censuses in 1940, we infer completed schooling at age 30 for cohorts born prior to 1910 based on their educational attainment at older ages. Because we do not observe all post-1910 birth cohorts at exactly age 30, we use a regression approach to adjust observed college graduation rates for age based on the typical proportional lifecycle evolution of educational attainment of a cohort. The details of age-adjustment method can be found in DeLong, Goldin, and Katz (2003, figure 2-1). College graduates are those with 16 or more completed years of schooling for the 1940 to 1980 samples and those with a bachelor's degree or higher in the 1990 to 2005 samples. The underlying sample includes all U.S.-born residents aged 25 to 64 years. Sources: 1940 to 2000 Census IPUMS; 2005 CPS MORG.

starting with the birth cohorts of the late 1910s to the 1920s. By the 1920 birth cohort the graduation rate had advanced to 10 percent and then doubled to 20 percent for birth cohorts of 1940. The rate for men was almost 30 percent for those born the late 1940s. The increase in the graduation rate was so large for cohorts born in the 1940s and the slowdown so substantial for cohorts born in the 1950s and 1960s that

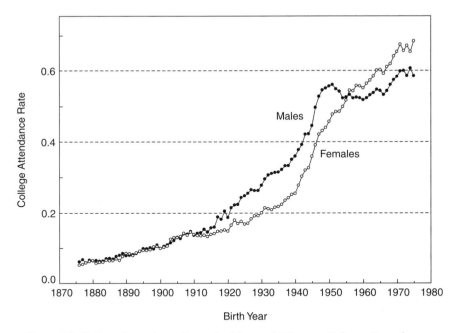

Figure 7.2. College Attendance Rates for Men and Women: Cohorts Born from 1876 to 1975 (by age 30). Sources and Notes: See Figure 7.1. In the 1940 to 1980 samples those who "attended college" had 13 or more years of schooling attended. In the 1990 to 2005 samples those who "attended college" had at least some college.

the graduation rate for males born in 1970 was lower than for males born around 1950.

The reason for the secular increase in college rates is mainly due to the high returns to college, but there are other factors that will be considered later. Some of the change in college rates was due to U.S. involvement in the wars of the mid- to late-twentieth century and the federal response to compensate veterans. Many men in the birth cohorts from the late 1910s to the 1920s fought in World War II, and their education, though initially interrupted by the war, was later funded by the GI Bill. Similarly, many of those born in the 1930s fought in the Korean War and were covered by a similar piece of legislation. The enormously large run up in college graduation and attendance for cohorts born in the 1940s was largely due to college draft deferments. After the Vietnam War, both college attendance and college graduation rates declined for males; the rates increased again only with cohorts born in the 1960s.

Gender Differences

The time trend of college-going for women is similar to that for men but important differences exist. The differences are made clear with reference to the ratio of male-to-female college rates, as shown in Figure 7.3.

In the early part of the century women went to college at rates that were similar to those of men. But, largely because many women attended teacher training schools that were, at the time, often two-year colleges, men's graduation rates from four-year institutions were somewhat higher. Women's college rates advanced at a slower pace than did men's with the cohorts born in the 1910s, and, in consequence, a large college gender gap developed. For cohorts born in the mid-1920s, male college graduation rates were more than double those for females. The ratio dropped as the World War II GI Bill ran its course and dropped again with the Korean War draft. But as veterans returned from Korea with yet another GI Bill to fund colleges, the male advantage increased again.

College rates for women did not increase rapidly enough to reverse the widening trend until the cohorts born in the mid-1930s. Subsequently, college-going and graduation rates for women greatly increased in both absolute and relative terms. Rather than tumbling back with the end of the Vietnam War deferments, as did those for males, the college rates for women flattened out and then soared beginning with cohorts born in the 1960s. The increase for females relative to males was so great that a "new gender gap" in college-going and graduation was created. Rather than lagging behind men in college-going and graduation, women became the majority of college students in 1980. The trend has continued so that at the beginning of the twenty-first century women were 56 percent of all undergraduates. Whereas there was 1 female for every 1.55 males in college in 1960, there was 1 male for every 1.26 females in college four decades later.[5]

Administrative data from colleges and universities are fully consistent with the census data just described (see Figure 7.4 for both series)

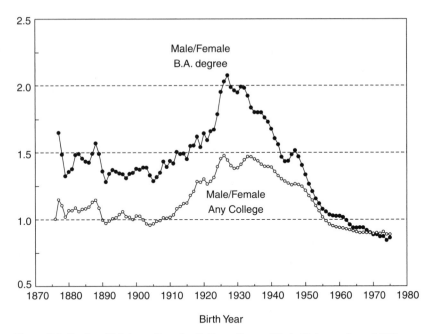

Figure 7.3. Ratio of Male to Female College Rates: Birth Cohorts from 1876 to 1975 (three-year centered moving averages measured at 30 years of age). Sources and Notes: 1940 to 2000 IPUMS; 2005 CPS MORG. College graduates are those with 16 or more completed years of schooling for the 1940 to 1980 samples and those with a bachelor's degree or higher in the 1990 to 2000 samples. Any college includes those with 13 or more years of school attended in the 1940 to 1980 samples and those with some college or more in the 1990 to 2000 samples. The age-adjustment methodology for college graduates is that described in the notes to Figure 7.1. The same age-adjustment approach is used for any college.

and provide a contemporaneous view of gender differences in attendance. The administrative (contemporaneous) data for college attendance reveal relative equality between males and females to the 1930s similar to that from the census. A sharp decline ensued with the drafting of men during World War II. Subsequent to the end of hostilities, around 1946 to 1949, an enormous spike in the ratio of males to females is apparent. The series then began its downward trend, interrupted by yet another war (Korea). The ratio descended rapidly in the 1970s and crossed the line of equality around 1980. Since that date there have been more females than males attending college.

Over the past century the U.S. higher education system has incorporated an enormously large fraction of individuals, and it did so long before other nations. One of the reasons that America could have consid-

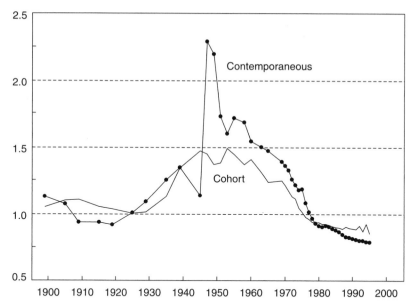

Figure 7.4. Ratio of Males to Females with Any College by Cohort and Year: Census (plus 20 years) and Administrative (Contemporaneous) Data. Sources and Notes: Cohort: see Figure 7.2. Contemporaneous: *Biennials: Opening Fall Enrollments*. Enrollment before 1946 was asked at the end of the year. After 1946 enrollment was asked at the start of the fall term. Graduate, professional, and preparatory students, as well as duplicates, are omitted from the undergraduate totals. In the case of professional students, because some may have been pursuing their first degrees, their omission understates the number of undergraduates. Students attending normal schools were generally enrolled in teacher training, but sometimes not. Up to and including 1930, only the data for the teacher training students were reported. The omission probably understates total enrollment by at most 10 percent and the 1930 number is understated by at most 5 percent. The data to 1955/56 are for "resident college enrollment," meaning individuals registered for a degree. Beginning in 1963 schools also reported non-degree enrollment and separated the enrollment into full-time and part-time. The data given here are for full-time and part-time. Summer session enrollment is not included for any of the groups, and the same is true for enrollment in extension schools and correspondence courses. The cohort and contemporaneous series differ because individuals attend college at older ages and because the contemporaneous series is implicitly weighted by the number of years a group is in school.

erably higher college enrollment is that it had a mass secondary school system by the mid-twentieth century, whereas other nations did not. But there are other reasons as well. These other reasons concern the factors that made up the quintessentially American system of colleges and universities.

A Quintessentially American Higher Education

The Growth of U.S. Higher Education

A high school senior considering going to college at the beginning of the twenty-first century could choose from about 1,400 institutions that awarded a bachelor's degree.[6] Among these colleges and universities 64 percent, or about 900, were privately controlled; about 600 of the 1,400 were liberal arts colleges. In addition, more than 1,500 two-year institutions were available for those who wanted an associate's degree, particular skills, or a second chance to enter a four-year institution.[7]

No nation in the world offers as much choice to potential undergraduates, graduates, researchers, and faculty as does the United States. In England, 102 separate undergraduate institutions existed in the year 2005, equivalent to half the number in the United States on a per capita basis for the relevant age group. In Germany, the 2005 figure was also about 100 institutions or one-third the population-adjusted U.S. figure. The choice offered by American colleges today is truly staggering and, relative to other rich nations, it was even more amazing in the past.

In 1950, England had just 30 institutions of higher education (or one-eighth the U.S. figure, scaled by the relevant population) and Germany had 38 (one-thirteenth the adjusted U.S. figure). In 1900 there were a mere 14 institutions in England (a meager one-seventeenth the U.S. population-adjusted figure) and Germany had 32, or just one-twelfth of that in the United States.[8]

The U.S. higher education system was, almost from the outset, quintessentially American: geographically close to the people, open in various ways, and replete with variety and competition.

Within the private sphere, institutions were founded by large numbers of competing groups. Rather than having one religion found the majority of the institutions, as was often the case for European nations, America contained dozens of religious groups that established competing institutions. In fact, most U.S. private institutions owe their existence to the energies and funds of religious bodies. Among all private four-year institutions of higher education in existence around 1930 almost three-quarters were controlled by a religious group.[9]

Public institutions of higher education were important long before the Morrill Act of 1862, which set up the federal land grant institutions. The College of William and Mary was established as a public institution by Royal Charter in 1693 and was the second colonial American institution of higher education. The University of Georgia, the first state-chartered, publicly supported college, was incorporated in 1785, established in 1801, and graduated its first class in 1804.[10] The University of Ohio was founded in 1804 by the Ohio Company, which purchased a large portion of present-day Ohio from the federal government.[11] Other states that founded state institutions before the Morrill Act include Alabama, Delaware, Indiana, Iowa, Michigan (both the University and Michigan State), South Carolina, Vermont, and Wisconsin. Publicly controlled institutions were 24 percent of those established by 1860. Although the fraction later increased, to 38 percent in 1900 and 40 percent of those existing in 2005, it was remarkably high in the antebellum period.[12]

The American states, therefore, have had an abundance of institutions of higher education in both the public and private sectors. No other nation has had anywhere near the number of private institutions as well as a public sector that provided competition and choice. The breadth and choice and private institutions are reflected in the early histories of many states.

Ohio, for example, had about 40 private colleges in 1930, half of which were founded before 1870. Eight public institutions of higher education also existed in Ohio at that time, of which six opened before 1875 and two—Ohio University and Miami University—were founded before 1825. The higher education system of Ohio, therefore, contained a private sector that was established early in its history and a large, diversified public sector that was primarily set up before 1875. In most states, private and public institutions of higher education coexisted for virtually all of their settlement history in the nineteenth century. The exceptions include some of the more sparsely populated states of the West, which never had a well functioning private sector, and the states of the East, many of which had old and notable private institutions and little in the way of public institutions.

Further evidence of the vast system of public and private institutions can be found in a time series of establishment dates for all four-year institutions of higher education still in existence in the last decade of the twentieth century.[13] Each bar in the graph in Figure 7.5 represents the number of institutions established in the five-year interval ending in the year listed. The darker part of each bar is the number of publicly controlled institutions. The remaining portion is the number of privately controlled institutions established in the five-year interval. From this we see that the peak founding period for American universities and colleges was in the late nineteenth century, from around 1865 to 1895. Because this period is about the same for both public and private institutions, the peak was not just a product of the federal Morrill Land Grant Acts of 1862 and 1890.[14] Almost 60 percent of all four-year institutions that survived to the late twentieth century were founded before 1900, and more than 50 percent of all public institutions existing in the late twentieth century were founded before 1900. In the case of publicly controlled institutions a more recent, but smaller, founding peak is apparent in the 1960s.

Several renowned private institutions of higher education were established in the 1890s, including Stanford, the University of Chicago, and the California Institute of Technology. But few private institutions were founded after the turn of the twentieth century, and those that were have not been as prestigious. Among the 35 private institutions in the top 50 universities in the 2006 rankings by *U.S. News and World Report*, only four began college-level instruction in the twentieth century and just one was founded after 1900. The four twentieth-century institutions are the Carnegie Institute of Technology (later Carnegie Mellon University), established in 1900 with instruction beginning in 1905; Rice Institute (later Rice University), founded in 1891 with college-level instruction beginning in 1912; Yeshiva University in New York City, established in 1886 but with college instruction beginning in 1928; and Brandeis University, founded in 1948.[15] In the top 35 liberal arts colleges (all under private control), as ranked by *U.S. News and World Report*, just three were founded in the twentieth century—Claremont McKenna College (1946), Harvey Mudd College (1955), and Scripps College (1926)—all of which are part of a college system that includes Pomona College, which was founded in 1888.

Something fundamental changed around the turn of the twentieth century making the founding of new and exceptional institutions of higher education, particularly private ones, more difficult. That

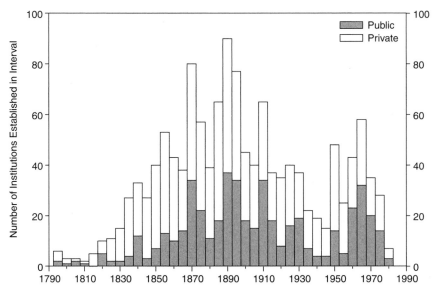

Figure 7.5. Establishment Dates of Institutions of Higher Education, 1790 to 1990. The data refer to institutions in existence in 1992 and founded after 1789, to truncate the thin left tail of the distribution for ease of viewing. The establishment date may be different from the opening date and also from the date at which the institution awarded a bachelor's degree. When an institution was formed out of one or several existing institutions the establishment date of the newly created institution is generally given, which would tend to shift the distribution a bit to the right. Source: Higher Education Publications (1992).

change, we contend, had much to do with barriers to entry stemming from the larger scale and widened scope needed to be competitive. Financial resources became increasingly important and institutional reputation began to matter more. New institutions, therefore, would find it difficult to grow and compete in this environment.

Variety and competition characterized the U.S. system of higher education almost since its origin, but distinction and superiority among the world's great centers of learning would take time. Excellence for American universities finally arrived in the post–World War II era.

Finest in the World

In recent years, the quality of the K-12 educational system in the United States has been seriously questioned, but the U.S. higher educational system has remained the envy of the world. Before World War II the best pupils and the brightest researchers left America to study in Europe. Today, the best and the brightest come to America to study in its colleges, universities, and research institutions. Among the top 20 universities in the world in 2005, just three were not in the United States. Considering the top 50 universities, just 13 were not.[16]

America is host to students from all over the world. In the year 2000, one in 50 undergraduates in a U.S. institution of higher education was a non-U.S. resident. Among all graduate students in the United States, one in eight hailed from another nation, and among those in all science and engineering fields, almost one-quarter were foreign nationals.[17]

American universities were not always among the very finest research institutions and the world did not always flock to study in them. In fact, there was a time when the best American students went abroad to study, especially in the sciences. Before World War II the finest American scientists, as well as scholars in other disciplines, routinely went to European universities and institutes for at least part of their careers.

After the 1950s, the high achievers mainly studied at home. Among U.S. Nobel Prize winners in the fields of chemistry, physics, or medicine who received their Ph.D. before 1936 and were born in the United States, 44 percent did some part of their formal education in Europe. But among Nobel Prize winners with Ph.D.s received after 1935 only 12 percent ever studied formally in Europe. Among those with Ph.D.s received after 1955, thus reflecting more normal, post war conditions in European universities, the fraction studying abroad was still just 13 percent.[18] These data are depicted in Figure 7.6.

Americans make up a greater fraction of the science Nobel Prize winners with Ph.D.s awarded after 1955 than among those with Ph.D.s earned before 1936. Among Ph.D.s received before 1936, the United States accounted for just 18 percent of all the Nobel Prizes in science and medicine; among those with Ph.D.s earned after 1955, 48 percent of the science Nobel Prize winners haled from America. Furthermore, as we just noted, American universities were increasingly responsible for their entire education.

A final feather in the cap of American universities is that a greater fraction of foreign Nobel Prize winners in science and medicine had studied in the United States in the post–World War II era than before. Just 10

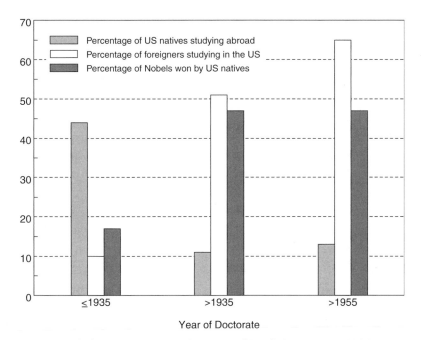

Figure 7.6. Nobel Prize Winners in Science and Medicine, 1901 to 2005: Percentage Studying in the United States, Abroad, and from the United States. Nobel Prize fields used are chemistry, physics, and medicine. Year of highest degree is used for those without a doctorate. U.S. winners are limited to those born in the United States (U.S. natives). The results are similar if those who immigrated as young children are included. Source: For Nobel Prize information and the biographies searched, see http://nobelprize.org/.

percent of the non-U.S. born winners with pre-1936 Ph.D.s studied in the United States at some point in their early careers, whereas 53 percent with post-1935 Ph.D.s and 66 percent with post-1955 Ph.D.s did.

After the 1950s many of the finest American researchers began and finished their studies at home in a higher education system that would rapidly become unrivalled in the world. At the same time that the U.S. higher education system expanded its ability to train an elite corps of scientists, it also extended its reach to the children of middle-income families.

The "Virtues" at the Tertiary Level

How did the U.S. higher education system get to be the best in the world? Many of the virtues of education that made America the twentieth-century leader in secondary education also functioned in the case of higher education. Competition and a laissez faire system, decentralized authority, public funding, gender neutrality, openness, and forgiveness were important in the past for secondary education and continue, in some fashion, to be important for secondary and higher education today.

A laissez-faire system led to the establishment of institutions of higher education that were public and private, secular and sectarian. In most parts of the nation students had enormous choice even if they never ventured outside their state of birth. Because of decentralized control and the absence of federal authority over college, competition existed across the various states, and competition existed even within many of the state systems.

Public support for higher education mattered in many of the ways it had for secondary education. A justification applicable to both cases is that the capital market is imperfect and fewer individuals, than optimal, would attend college if it were not subsidized. As we will discuss below, greater levels of public support did increase the total numbers who attended and graduated from college. Public institutions, furthermore, were more gender neutral. In addition, states that had earlier and better supported public universities had a higher fraction of women among all college and university students.

Openness is a hallmark of American education at all levels and nowhere is it more apparent than at the highest level. In most states entrance to the state university was, at one time, available to any high school graduate. With the onset of national testing, many state universities could become more selective and some states, in consequence, established a hierarchy of universities and colleges. The best example of a hierarchical system is California's, although other states, such as New York, also created a legislated hierarchy among its publicly controlled colleges and universities. Many states have at least two public institutions: a flagship university and a state university, and almost all have community colleges, often the schools of last resort and those that bring college the closest to the people.[19] Earlier in the twentieth century, some universities and colleges—even the University of Chicago—had correspondence courses, similar to Internet courses today, and many offered summer school programs. In all of these ways, college in America was, and still is, unusually open. It is also a highly forgiving or second-chance system, much like that at the lower levels.

When technology (in this case knowledge) advances rapidly, flexible, nonbureaucratic, decentralized institutions that are not beholden to a single funding authority are in a better position to respond. With the explosion of scientific fields in the post–World War II era, U.S. institutions were far better situated to adapt than were the more inflexible institutions in Europe. Monopolistic and bureaucratic universities can become lazy, just like monopolistic and bureaucratic producers.

America's higher education system also benefited from the nation's large population, geographic size, relative homogeneity of income early in its history, and relatively high level of income throughout its history. Just as geographic size was vital to U.S. industrial success and the system of mass production, scale and scope were critical to the initial shaping of the higher educational system and contributed to its eventual greatness.

The Shaping of Higher Education

Size

The 1900 to 1940 period saw higher education take its modern form in America. Among the most important changes was the increase in the sheer size of institutions of higher education, especially in the public sector. Although the number of public and private institutions increased by 1.4 times from 1900 to 1933, the number of students increased almost fivefold. During the next seven decades the number of four-year institutions doubled while the number of students increased by about tenfold.[20] The mean size of an institution of higher education, therefore, increased by about five times. The most important margin at which the system of higher education expanded across the twentieth century was the size of individual institutions. Many public universities, in particular, became behemoths.

In contrast, many private universities around the turn of the twentieth century were larger than the largest of the state institutions. Harvard and Yale, for example, each enrolled more undergraduates around 1900 than did either the University of Michigan or the University of

Minnesota, the two largest public institutions. Smith College had more undergraduates than did the University of Illinois. Public universities, to be sure, were on average larger than were private institutions around 1900 but they were not much larger on an absolute basis. The median private institution had about 130 students whereas the median public institution had about 240, a far cry from differences that would soon develop.[21]

The evolution of scale as it played out in public and private institutions can be summed up easily enough. The ratio of the median number of students in each public institution to that in each private institutions was 1.8 around 1900, but rose to 3.4 in 1923 and then to 4.1 in 1933.[22] A major change in relative magnitudes, therefore, occurred sometime between 1900 and the 1920s. By 1923, public-sector institutions of higher education already included many large, research-oriented universities.[23] Interestingly, more recent data show that the absolute size of institutions has continued to rise, but the ratio of the number of students in the public and private sectors has not changed much. In the early 1990s the median number of students per institution was 1,579 in the private sector and 8,181 in the public sector, for a ratio of 5.2.[24]

Public-sector institutions grew substantially from 1900 to 1933, but they were not yet the giants they would become. Almost half of the largest 25 higher education institutions in 1933, ranked by the number of undergraduates, were privately controlled. Although, as we will soon demonstrate, the relative increase in the public sector in the twentieth century was greatest in the period from 1900 to 1940, the growth of the very largest public institutions would continue well beyond. By the start of the twenty-first century the public sector had so greatly expanded that 24 of the 25 largest B.A.-granting institutions were publicly controlled.[25]

Scope

The formative years of American higher education saw major changes in the scope of institutions, including the emergence of the research university, the demise of independent professional institutions, and the decline of independent schools of theology and denominational institutions in general. For most of the nineteenth century, American institutions of higher education were centers of teaching and learning, not research and innovation. Their mission began to expand in the latter part of the nineteenth century with the founding of the Johns Hopkins University (1876), the first dedicated graduate and research center in the United States, followed by Clark University (1889) and the University of Chicago (1892).[26]

Instruction at European universities took several forms including the classical example of British universities, the scientific training of French grand ecoles, and the graduate and research institutes of Germany. The modern university of the New World was a different creature than its European counterpart for it catered to a broad clientele of students and served the interests of the states. Yet it increasingly strove to be a research center.

The American university became a department store of higher education, with a liberal arts college at the core and graduate departments and various professional schools, including law, medicine, dentistry, pharmacy, theology, and business, at the periphery. But the modern university is far more than a collection of higher education services brought together under one roof. It is a production center in which the research of one part enhances the teaching and research of the other parts. The university form was an organizational innovation enabling the exploitation of technical complementarities among its various components.

The public sector did not have a corner on universities, but for most of the twentieth century it has had a disproportionate share of them. Around 1900, for example, 43 percent of all universities were in the publicly controlled sector even though only 13 percent of both colleges and universities were at the time. The fact that the publicly controlled sector was disproportionately established in the university, research-oriented form gave it a substantial edge over the private sector in the period to 1940, when the overall share of (nonpreparatory) higher education students enrolled in universities, as opposed to colleges, increased from 42 percent around 1900 to 59 percent in 1933.[27]

Certain universities had, in addition, the capacity to bestow reputation on new divisions in untried areas, such as business, and in older disciplines plagued by claims of quackery, as in the case of medicine. Medical schools were increasingly vulnerable in the late nineteenth century with the advent of stricter state licensing, designed in part to replace the "art" of healing with the scientific method. Their numbers thinned further in the wake of a 1910 Carnegie commission report—known as the Flexner Report, after its author Abraham Flexner—which severely criticized many of the 155 medical colleges in the United States and Canada.[28] Thus, the university came to combine the features of a department store, an integrated knowledge-production factory, and a brand name.

As independent professional institutions declined, professional schools associated with universities increased.[29] Around the turn of the twentieth century, 48 percent of students training to be lawyers, dentists, pharmacists, doctors, and veterinarians attended professional schools that were stand-alone entities, independent of any other institution of higher education. At that time, professional schools often did not require their students to have a college degree, and many had not previously attended college at all.[30] By the 1930s, only 19 percent of professional students were attending independent schools.

Thus the informal and apprenticeship programs of the past gave way, in almost all the professions, to scientific, formal, and school-based training. Moreover, the schools in which the training was delivered increasingly became parts of universities rather than existing as separate, independent institutions. The research university had enormous advantages in terms of its productivity and sometimes lent its reputation to occasionally sullied professions.

Changes in the Knowledge Industry

Why did higher educational institutions in the first several decades of the twentieth century increase in scale, scope, public-sector enrollment, and commitment by the states? Our answer for why it became transformed revolves around technological shocks that swept the "knowledge industry" in the late nineteenth and early twentieth centuries.

In the latter part of the nineteenth century, an increasing number of subjects taught in colleges and universities became subdivided and specialized, and the faculty began to define themselves as occupying separate, specialized fields. Changes in each academic subject were brought about by somewhat different factors and at slightly different moments in time. Yet similar factors were at work in the disciplinary profusion.

They include the application of science to industry, the growth of the scientific and experimental methods, and an increased awareness of the social problems of an increasingly industrial and urban society.

In industry after industry in the late nineteenth century a growing dependence on chemistry and physics emerged, most notably in the manufacture of steel, rubber, chemicals, sugar, drugs, nonferrous metals, petroleum, and goods directly involved in the use or production of electricity.[31] Firms that had not previously hired trained chemists and physicists did so at an increasing rate, as did the federal and state governments. The number of chemists employed in the U.S. economy increased by more than sixfold between 1900 and 1940 and by more than threefold as a share of the labor force; the number of engineers increased by more than sevenfold over the same period.[32] Science replaced art in production; the professional replaced the tinkerer as producer.

With greater demand for trained scientists, universities expanded their offerings. With new research findings, the classical scientific disciplines became increasingly fragmented and greater specialization resulted. Greater specialization in biology was driven by changes in empiricism and experimentation earlier stimulated by the appearance of Charles Darwin's *Origin of Species*. Analogous changes appeared in the agricultural sciences, where part of the impetus was the expanding crop variety in the United States as a result of highly specialized farming fueled by the railroad. The social sciences expanded and splintered in the late nineteenth to early twentieth centuries. They were given a mission by the growing social problems of industry, cities, immigration, and economic depression in the 1870s and later in the 1890s. The academic fields were shaped by Darwinian thought and Mendelian genetics, and later by the increased roles of statistics, testing, and empiricism generally.[33]

The increasing specialization in academic disciplines can be illustrated with the founding dates of "learned societies."[34] The first learned society established in the United States, the American Philosophical Society, was founded in 1743. Just five more learned societies came into existence in the hundred years following and an additional six appeared before 1880, making 12. The pace then picked up and 16 such societies came into existence from 1880 to 1899. Another 28 followed in the next 20 years, from 1900 to 1919. Just 10 appeared from

1920 to 1939, although 20 were founded in the 1940 to 1959 period. The final 20-year period in our data set, 1960 to 1979, contains 12 more. It is clear that the greatest period of founding of learned societies was the first several decades of the twentieth century during the time of disciplinary proliferation in the U.S. academy.

The era of the division of labor in higher education had arrived. A reputable college could no longer survive with a mere handful of faculty. Most of the changes served to increase economies of scale in the production of higher education services and thus increase the minimum number of faculty and students required for a college to remain viable. Also important to the story at hand is that those who diffused knowledge increasingly became its creators. Research became the handmaiden of teaching that we believe it is today.[35]

The Role and Impact of State Support

Public Sector's Relative Size and Changing Role

As we already mentioned, the publicly controlled portion of higher education expanded substantially across the twentieth century (see Figure 7.7). From about 1900 to 1940 the fraction of students in four-year publicly controlled institutions increased from 0.22 to about 0.5. Then, from before World War II to 1975, the fraction rose again, from 0.5 to almost 0.7. But since the 1970s the public share of enrollments in four-year institutions has declined somewhat. Overall, then, the full century saw an increase in the fraction of four-year students in the public sector from around 0.22 to 0.65.[36]

The schools included in each of the three lines in Figure 7.7 differ because the functions and types of colleges changed over time. The most important changes concern teacher-training schools and two-year colleges. Teacher-training institutions, which in the period before 1940 were often two-year normal schools or teacher's colleges, became four-year state universities. As the number of two-year teacher's colleges declined, the two-year community colleges increased. Two-year colleges, including community and junior college institutions, were one-quarter of all colleges and universities in 1935, but were just 5 percent in 1917. They continued to increase in relative terms in the late 1960s to 1990 and at the end of the twentieth century the two-year col-

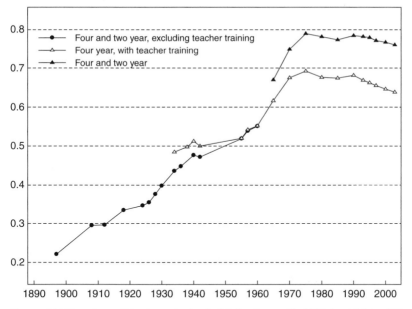

Figure 7.7. Fraction of Students in the Publicly-Controlled Higher Education Sector: 1897 to 2003. Sources and Notes: See Goldin and Katz (1999a, figure 2) updated to 2003.

lege comprised about 40 percent of all institutions of higher education.[37] The inclusion of two-year institutions increases the level of total enrollments in the publicly controlled sector, since about 70 percent of all two-year colleges are now in the public sector. When both two- and four-year colleges are included, the fraction of all students in the public sector reached a twentieth century peak of almost 80 percent by 1980, although it declined slightly in subsequent years.

Even before the historic Morrill Act passed in 1862, two-thirds of the existing 33 states had at least one state-controlled institution of higher education. Four-fifths of the states outside the Northeast did.[38] Early on, many of the state institutions were established to train teachers for the lower grades. But these institutions evolved and eventually provided a larger set of "public goods" for each of the states.[39] State institutions in the nineteenth century were more practically and, often, more scientifically oriented than were their private counterparts, in large measure because of the commitment to provide goods and services of value to citizens and local industrial interests.

Despite the founding of publicly controlled and state-supported institutions, state funding on a per capita or per student basis was measly until the late nineteenth century, when scientific findings became important in agriculture, mining, oil exploration, manufacturing, and construction. In states having a concentration of economic activity by industry or by product, the public sector often invested heavily in training and research in these industries. Wisconsin subsidized work on dairy products, Iowa on corn, Colorado and other western states on mining, North Carolina on tobacco, and Oklahoma and Texas on oil exploration and refining.[40] State institutions of higher education often contained professional training institutes, such as in engineering, and graduate programs in various sciences, including those pertaining to agriculture. With this broad portfolio, the state institutions of higher education attained the status of "university" to a greater extent than did those in the private sector. They contained all the component parts of the university—the liberal arts college, the graduate programs, and the professional schools—and they also had access to research funds from the state at a time when such funds were less available elsewhere.

Among the most striking differences between the curricula of public and private institutions in the formative period was engineering. In 1908, among all public-sector college and university students fully 30 percent were in engineering programs, and 60 percent of all engineering students nationwide were educated in public-sector institutions. Although by 1930 the share of all public-sector students who were in engineering dropped from 30 to 15 percent, the fraction of all engineers enrolled in public institutions rose from 60 to 66 percent. Engineering students were, quite obviously, being produced primarily by the public sector. Private-sector institutions trained engineers in only a handful of states and just three states (Massachusetts, New York, and Pennsylvania) in 1930 enrolled 62 percent of all private-sector engineering students. Because two-thirds of engineers were being trained in the public sector in 1930, the geographic dispersion of engineering students mainly came from enrollments in the public sector. Governments disproportionately hired engineers—almost 25 percent of all engineers in 1940 worked directly for the government.[41] In consequence, the public sector had an interest in their training.

Regional Differences in State Support

U.S. regions differ greatly in their support for higher education and three different models can be discerned. The Northeast can be thought of as an American elite model, the West as an egalitarian model, and the South as a European elite model. The states of the Northeast founded private institutions early in their histories and their public-sector institutions arose relatively late. When public-sector institutions did appear, they were poorly funded in comparison with other regions. The southern states had a disproportionate share of the earliest public institutions. Of the public institutions that were founded before the American Civil War, fully 57 percent were in the South even though the South had only 36 percent of the private institutions.[42] The states of the West, including the Midwest, have a vibrant public sector. Some even have an equally strong private sector. But often in thinly populated western states only a public sector university could survive.

State support for public higher education greatly increased from 1900 to 1940, measured either by the fraction of state spending going to higher education or by the growth of enrollments in public institutions relative to all enrollments or to the college-aged population. For example, total expenditures on state higher-education institutions increased from 5.1 percent to 11.0 percent of state and local government spending from 1902 to 1940.[43] But public funding for higher education, and access to public colleges and universities, varied substantially among states throughout the period. The greatest levels of support were found in the Pacific, Mountain, and West North Central states, and the lowest levels of state support were found in New England and the Middle Atlantic states.

Many of the differences in state support for higher education have persisted—the cross-state correlation between (log) state and local government spending per capita on higher education in 1929 and that in the 1990s was about 0.45, although the correlation decreased somewhat to the early 2000s. It is, therefore, instructive to explore the determinants of state support in the formative years before the expansion of federal support for higher education and just before the effects of the Great Depression on state and local budgets. For these reasons we will examine 1929.[44]

In 1929, state and local government subsidies to higher education averaged $1,230 (in 2005 dollars) per 100 persons across the 48 states, or 6 percent of total state and local government spending. Almost 95 percent of that state and local support for higher education went to publicly controlled institutions; only New York (Cornell) and New Jersey (Rutgers) had any significant level of state support of privately controlled institutions. State and local spending on higher education per 100 inhabitants ranged from a low of $518 in New England to a high of $2,324 in the Mountain states (all 2005 dollars).[45] Enrollments in publicly controlled institutions averaged 3.19 per 1,000 inhabitants: from 0.82 in New England to 6.04 in the Mountain states and 6.09 in the Pacific states. What explains these substantial differences in state support across regions and among individual states?

The public choice decision to provide support for higher education is likely to be affected by the level and distribution of wealth or income in a state, by community stability and homogeneity, and by the importance of industries that capture localized benefits of research at state institutions.

Enrollment in privately controlled institutions in 1900 had a significant depressing effect on state public support for higher education in 1929.[46] In fact, the raw correlation between state spending on higher education in 1929 and the importance of private universities in the state at the start of the twentieth century is –0.69. The magnitude of the effect is also large: the difference between private college enrollments per 1,000 residents in Massachusetts and Iowa in 1900 (3.35–0.99) implies an 84 percent difference in per capita spending on higher education between the two states. State initial conditions around 1900, such as private college enrollments per 1,000 residents, are strong predictors, by themselves, of state support for higher education in 1929.

To sum up, newer states with a high share of well-to-do families and scant presence of private universities in 1900 became the leaders in public higher education by 1930. They remain so today. The tradition of stronger private universities and lower support for publicly controlled universities in the Northeast also continues to the present—the correlation of public college enrollments per capita at the end of the twentieth century with private college enrollments per capita in 1900 is 0.56.[47]

Impact of State Support

For much of the twentieth century a stronger, more generous public sector increased the *overall* rate of college-going and, as a by-product, increased the relative college-going of women. Because college students attend school both in and out of their state of residence, we use data on college attendance by state of residence rather than by state of attendance.[48]

First off, and not surprisingly, there was around 1930 a strong positive relationship at the state level between the (public and private) high school graduation rate and the fraction of state residents (18 to 21 years old) continuing to college (Figure 7.8, upper panel). The states that led in the high school movement of the 1920s also led in the college-going of their residents around the same time. The top 15 states in college-going among young residents in 1931 were in the West (West North Central, Mountain, and Pacific regions), plus Oklahoma. In fact, *all* the West North Central and Pacific states—most of the states that led the nation in the high school movement—were in the top 15 that led in college around 1930. Similarly, all the states having low levels of college-going were either in the South or in the more industrial regions of the East. Both of those areas also lagged during the high school movement. Because we use the college-going rate for young state residents, states that provided generous support for out-of-state individuals would increase the college-going rate for other states, and states that had private institutions that attracted students from other places would do the same.

College-going depends on financial factors, such as the public support given to higher education, and the quality of public and private institutions of higher education, not just on the available pool of college-ready youths. Without subsidization of college through grants or loans, fewer individuals would be able to afford college. As mentioned, many of the states that lagged in public support of higher education, given their high school graduation rates, were in the East. In fact, all six New England states in the upper panel of Figure 7.8 are well below the simple regression line drawn in the diagram. Their college-going rates were substantially lower in 1930 than would have been predicted solely on the basis of their high school graduation rates in 1928. The New England states, it turns out, had among the highest tuition and fee levels in 1930.

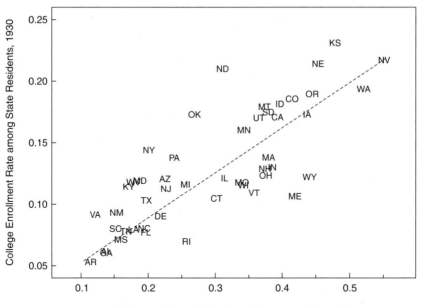

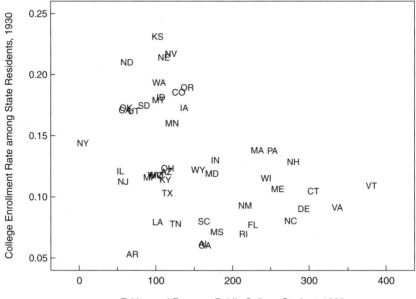

The relationship between public support and college-going among state residents can be better seen in the lower panel of Figure 7.8, which graphs average per student tuition, fees, and residential cost in the public sector against the college-going rate for state residents. The relationship is negative. States that have high tuition and fees have lower college-going in general among state residents, not just lower rates of public college enrollment. Geographic proximity to colleges and universities, particularly in the case of nonresidential institutions, is also of importance. New York State, a clear outlier in the lower panel graph, did not have a state higher education system in 1930 but had heavily subsidized municipal colleges in New York City. The low cost of education, however, could not easily be taken advantage of by residents in the rest of the state.

We have also explored the relationships among the factors just mentioned in a multivariable regression format that includes relevant covariates such as per capita income.[49] The dependent variable is the number of college students by state of residence as a fraction of 18- to 21-year-olds in the state (see Figure 7.8). For 1930, we can estimate the impact of public-sector tuition and fees on the college enrollment rate. Our estimates imply that a decrease in public-sector tuition and fees of one standard deviation in 1930 increased the college enrollment rate of state residents by 1.2 percentage points (or 9 percent of the mean level). The estimation, moreover, holds the fraction of youths graduating from secondary school in the state constant. States that were impecunious with regard to funding higher education might also have been so with regard to secondary education. But even given the state's

---

Figure 7.8. (opposite page) College Enrollment, the High School Graduation Rate, and State Tuition, c. 1930. Sources and Notes: College enrollment rate of state residents: Kelly and Patterson (1934). The numerator is the total number of state residents attending college anywhere in the United States. The denominator is the approximate number of 18- to 21-year-olds in the state. For the sources of the public and private high school rate, see Chapter 6. Per student fees in public institutions, 1930: *Biennial* 1928–30, tables 3b and 6b. The numerator is the total receipts from all student fees (tuition, room and board, and other fees) of publicly controlled institutions. The denominator is total enrollments of publicly controlled institutions.

secondary school graduation rate, having higher college tuition and fees was associated with lower overall college enrollment.

Why certain states had more generous public education systems has much to do with the history of the private sector. America was generally settled from East to West, and the earliest settled states had stronger private institutions of higher education and weaker public ones.

States that entered the Union later had higher (public and private) college enrollment rates of their state residents in 1930 than did those that entered the Union earlier.[50] We have divided the states into four quadrants in Figure 7.9 using the two dividing lines of early states (before 1846) versus late states (after 1845) and high college enrollment rates (>0.15) versus low ones (≤0.15). All of the early states are found in the southwest quadrant (Q IV) and almost all of the later ones are in the northeast quadrant (Q II). That is, all of the early states had "low" college enrollment rates and almost all of the late states had "high" college enrollment rates. Many of the states that entered the Union relatively early—such as Indiana, Maine, Massachusetts, and New Hampshire—had a substantial proportion of their youth prepared to enter higher education, but, in general, their state higher education institutions were few in number, late in coming, and relatively high in tuition and fees.

Almost all of the late entrants to the Union had high levels of college enrollment, including those in the Great Plains and far West. In fact, all of the states in the northeast quadrant (Q II), with the exception of Oklahoma once again, are in the Pacific, Mountain, and West North Central regions of the country. Many of these states were, as we have said, leaders in the high school movement. The differences between the leaders and the laggards in the case of college enrollment are larger than for secondary school. Not only did most of the late-entering states have a large proportion of their young population eligible to attend college, but they also funded higher education so well that they achieved a college-going rate on average about double that of states in the southwest quadrant.

States with later entry dates had higher levels of college enrollment because state funding for higher education was greater and tuition levels were lower. The private sector in the later states was almost nonexistent and thus considerably smaller than in the East among the states that entered earlier. We will never know exactly why eastern

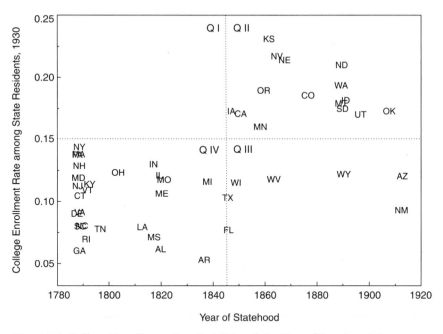

Figure 7.9. College Enrollment Rate in 1930 and the Year of Statehood. Sources and Notes: See Figure 7.8.

states subsidized public institutions to a lesser extent than those in the western part of the country, but the reason probably concerns the lack of perceived need and a bias, possibly in the state legislature, toward the existing private institutions.

Higher levels of public-sector tuition and a smaller public sector did more than just reduce the enrollment rate of high school graduates in any college. A weaker public sector also reduced the relative number of young women in college, among state residents. There is a strong positive relationship between public-sector tuition in the early 1920s and the ratio of males to females in college among state residents and there is a strong negative relationship between the total college enrollment rate and the ratio of males to females in college. In the 1920s, the New England and Middle Atlantic states, with their paucity of public institutions and their tradition of male-only colleges (even though women's colleges were established in the late nineteenth century), had a considerably lower fraction of women among all students. Among the five states with the largest number of males-to-females in college in 1923,

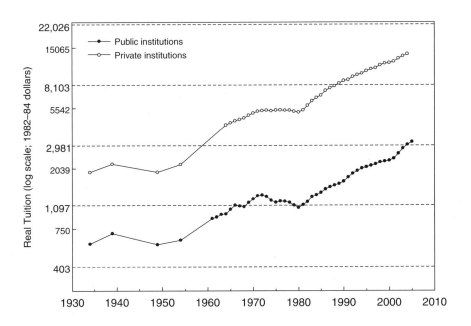

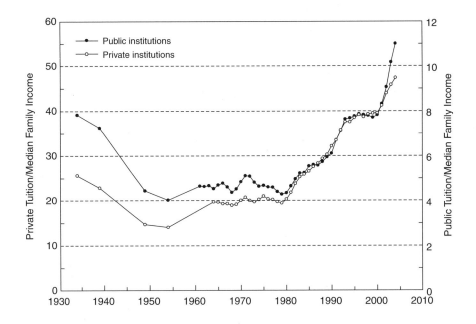

four are in the northeast and all nine are in the top 16 states.[51] New England trailed other regions through the 1930s in this regard, although by the late 1950s the gap began to close.

By the end of the 1930s the U.S. higher education system had universities of grand scale and scope. The public sector provided healthy competition for the private sector. But America's institutions of higher education were not yet the extraordinary research institutions they would soon become. Nor was the federal government yet the player in higher education that it would later become. Whereas the federal government contributed 7 percent of higher education current-fund revenue in 1940 on the eve of World War II, it has contributed between 15 to 20 percent each year since 1950.[52]

Expansion in the Post–World War II Era

Growth of the Public Sector

State institutions of higher education have always been a bargain. Their tuition is far less than that of private institutions since they receive considerably greater subsidies.[53] The tuition of public institutions

---

Figure 7.10. (opposite page) Public and Private College Tuition and as a Percentage of Median Family Income. Tuition includes required fees. In some years, certain states had no tuition but had required student fees. When possible, only the state flagship universities are included in the tuition levels. Flagships and an average of all state tuitions (averaged at the state level) are almost identical from 1964 to 2005. Private institutions include only universities. Sources: Public and private tuition and fees 1934 to 1954: *College Blue Book* (1933), Conrad and Hollis (1955). Public tuition and fees 1961 to 1963: U.S. Office of Education (1961); D'Amico and Bokelman (1963); 1964 to 1971: *Digest of Education Statistics 2005;* 1972 to 2005: Washington Higher Education Coordinating Board (various years). Private tuition and fees 1964 to 2005: *Digest of Education Statistics 2005.* Median family income 1949 to 2005: Nominal median family income before taxes, *Historical Statistics, Millennial Edition,* tables Be67–84 and *Economic Report of the President 2005* translated from real terms into nominal values using the CPI; 1934, 1939: *Historical Statistics, Millennial Edition,* tables Ca20–27 for national income and tables Ae1–28 for the number of families (interpolated for 1934), approximations based on national income per family × 0.58, where the 0.58 figure is based on an average of (median income per family/national income per family) for selected years from 1949 to 1964.

was about one-third that of private institutions in 1934, when our data on public and private tuition first begin, and for most of the years after the 1950s, public institutions were about one-fifth as expensive.[54] Trends in the public and private cost of college-going (in real terms) are fairly similar (see top graph of Figure 7.10), with the exception of the post-2000 period when tuition rose somewhat faster in the public sector. Because the top graph in Figure 7.10 is in log terms, the slope of each of the lines gives the rate of change.

The public sector accounted for about 20 percent of total enrollments around 1900 but accounted for 70 percent of four-year students in 1970, when it reached its twentieth century peak (see Figure 7.7). A key factor driving the relative increase in the public sector was the cost of college. As high school graduation rates rose nationwide and exceeded 50 percent by mid-century, recent high school graduates disproportionately chose the public sector. The reason why is plainly obvious from data on college tuition as a fraction of median family income.

College tuition and related costs have always been a major expenditure for families. Expressed as a fraction of median family income (see the lower graph in Figure 7.10), both public and private colleges were relatively more expensive in 1934 than anytime afterwards until about the mid-1980s. College costs relative to family income came down rapidly in the 1940s and 1950s as incomes soared and college tuitions rose more slowly. Both tuition and family incomes increased at about the same rate from the 1950s to 1980, keeping tuition as a fraction of family income about constant.

During the great expansion in college-going from the 1950s to the 1970s, tuition at a public university was about 4 percent of median family income, whereas that at a private university was around 20 percent. Compared with the same metric before the 1950s and after 1980, college was a real bargain, particularly that in the public sector. Ever since 1980 college has become a far greater financial burden on many families. Tuition at public universities increased to more than 10 percent of median family income by 2005. In the private sector, the average university tuition rose to around 45 percent of median family income in 2005.

Of course, in both the public and the private sectors many college students do not pay "list price." Accounting for fellowships, grants, and other means of support would reduce the levels and slow the growth of costs somewhat, but real tuition rises are a fact. Tuition, furthermore, is just part of the full expense of college and most of the others (e.g., room, board, transportation, books) are fairly comparable for the two types of institutions. Even as tuition levels rose in real terms and as a fraction of median family income, the demand for public and private higher education increased with the returns to education.

Higher education, on a per student basis, is the most expensive form of schooling and, because most of the benefits are privately garnered, state legislatures demand reasons for their funding. Individuals, furthermore, are mobile and a state that invests heavily in higher education could benefit neighboring states. Colleges and universities that provide valued services to the state, such as research on state products, agricultural extension services, and teacher training, have an easier job convincing the state legislature of the need for funds. But as state institutions provided fewer services specific to their state and as more individuals graduated from high schools and were capable of entering college, another way to obtain more funds for higher education had to be found. The answer was to increase accessibility. When the median voter had a reasonable chance of having a child who was a high school graduate and wanted to attend college, state legislatures would more readily pass funding bills to expand higher education. The question was how to accomplish the goal at a reasonable cost to the tax payer.

The solution, many states discovered, was to create extensive hierarchical systems with a select group of universities at the top and a multitude of community colleges at the bottom. The best known of the hierarchical systems is that of California. In the 1950s California began to face the policy problem just described and forged a solution known as the California Master Plan.

The California Master Plan, championed by Clark Kerr in the late 1950s and put into effect in 1960, codified the terms under which individuals could gain entry into California's universities, colleges, and community colleges. Before this plan, fully one-half of all California's college students were attending one of the more highly subsidized universities and colleges in the state. With continuing growth in college enrollments the system would have soon bankrupted the state.[55]

The plan gave the top 12.5 percent of applicants the right to attend one of the more prestigious universities (the pre-plan level was the top

15 percent). The next 20 percent, under the plan, could gain entry to one of the state colleges (the pre-plan level was 35 percent). All remaining students, about two-thirds, could attend a community college and then enter a four-year institution if they merited it. In the years since its adoption the levels set by the plan have been maintained in actuality.[56] In this manner, the Master Plan retained the political base for higher education but cut the cost.

Almost all states have followed California's lead and created a hierarchy of higher education institutions. Community college growth has been one of the outcomes. Two-year institutions enrolled a quarter of all public-sector students in the early 1960s, but almost half in 2005. Few states have attained California's strict ranking of colleges and universities, and the more sparsely settled states have generally retained the primacy of the state's flagship institution. But, by and large, states in the post–World War II era have ensured that tax payers and their children have access to some type of state college.

From "Shaping" to Excellence

In the early twentieth century U.S. universities were not the world's leading research institutions. They became the finest in the land some time after World War II, as we saw earlier using data on Nobel Prize winners in science and medicine. Before the 1930s Nobel Prize winners who were U.S. citizens often studied in Europe but did so far less often after the 1950s. Nobel Prize winners in science and medicine who were not U.S. citizens came infrequently to the U.S. to study prior to the 1930s but ventured in large numbers after the 1950s. We also discussed the fact that U.S. universities are today ranked very highly and that students and faculty from all over the world come to the United States to study and work.

Although we cannot easily compare institutions across nations in other than the more recent period, we can track the excellence of institutions within the United States over time and explore whether public and private institutions differed in their relative excellence at various moments in the twentieth century. Several means exist for such an exploration.

The first method uses the biographical directory *American Men of Science* for 1938 and 1960, two years during which the inclusion criteria did not change. The directories list scientists of note who have been elected to prestigious scientific societies.[57] From 1938 to 1960, a time of expansion of higher education particularly in the public sector, the entire distribution of the listings shows a clear movement toward public universities. Whereas 34 percent of the listings were from public universities in 1938, 41 percent were in 1960. The increase came mainly at the expense of private colleges, not private universities. The top five institutions in each year, by the number of faculty included, show a clear shift toward the public sector, although no winner is obvious in the top 25.[58] Excellence in the sciences, therefore, became more concentrated in the great centers of learning and these shifted toward the public sector.

One possible problem with the method just used, which gives equal weight to each researcher, is that a large department could have a relatively large number of faculty but only a small fraction who are excellent. For that reason we have also used a procedure that ranks departments by excellence. The method might also give disproportionate weight to larger departments, but not as much as the previous one.

We used departmental rankings for the 16 largest Ph.D. fields in the arts and sciences for the years 1928/32, 1969, and 1993.[59] We have produced consistent data for 39 universities from 1928/32 to 1969 and for 74 universities from 1969 to 1993. From the data on field rankings we have constructed university ranks, weighting the fields by their fraction of Ph.D.s. In all three years the private sector did better than the public sector in the top group (top 10, top 20), but movement was slightly toward the public sector.

Across the earlier period, from 1928/32 to 1969, there was a slight shift toward relative greatness in the public sector using the early weights, but there was stability using the later weights. From 1969 to 1993 there was a small shift to the public sector using either the early or later weights. The point is that the fraction of the top group in the public sector hardly changed at all during the first four decades and shifted just slightly toward the public sector during the next quarter century. Across the entire 65 years, there was little shift and the privately controlled institutions retained their strength and vigor.

By and large, the public and private spheres appear to have engaged in very healthy competition across the twentieth century. Public universities grew enormously and that, most likely, is the reason for their increased relative number of entries in *American Men of Science*. But in terms of university excellence, there was little change between the publics and the privates clear across the twentieth century. Some places rose (e.g., Stanford) and some fell (e.g., Columbia), but there was relative stability in general and between the publics and the privates.

Most of the world-renowned U.S. universities are private institutions, according to the list of the leading universities in the world we previously cited. But many are public universities, and something special exists about the states that contain public universities that rank among the finest institutions in the world. A disproportionate number of these states have strong hierarchical structures, similar to that pioneered by California. We judge whether a state has a strong hierarchical structure by the fraction of its enrollment in two-year community colleges. There were 17 state universities in the top 100 universities in the world in 2005 and 12 of them were in states that had an enrollment fraction in two-year colleges above that in the median state.[60] The evidence is suggestive that states with hierarchical structures can allocate more funds to research and to building a great faculty in their flagship institutions than can states with nonhierarchical structures. States with more finely tuned hierarchies can also assemble better peers among undergraduates at the top institutions and have superior graduate programs. At the same time, they can retain the support of tax payers by providing less expensive institutions of higher education for the majority of state residents.

In our discussion of why American universities attained the status of the finest in the world in the second half of the twentieth century, we mentioned several factors. Among the more important are the scale and scope of institutions, the healthy competition that the private sector offered state institutions, and the competition within each of the sectors.

Because the American institutions eclipsed most of the older European ones in the period from the 1930s to the 1950s, there remains the possibility that U.S. institutions of higher education gained at the expense of European institutions when parts of the world were in ruin and scientists and other academics fled to the free world. There is no question that American institutions benefited from European talent and the absence of combat.

But those who have studied the question closely have made a convincing case that the ascent of U.S. institutions of higher education and their sustained greatness were not due to the residual impact of World War II. Rather, they note that many pre–World War II features of U.S. institutions of higher education would have made for global greatness even had European centers of learning not been damaged by the war. These advantages include the scale and scope of American universities, which allowed interactions among large groups of diverse researchers, and the sheer number of researchers in each field, which enabled specialized institutions to arise. U.S. universities spent more on research, particularly in the sciences, than did those in other nations, and the system grew strong because of its diversity, competition, and decentralization.[61]

The Unfinished Transformation: A Summary

Higher education in the United States expanded at extraordinary rates during most of the twentieth century. Whereas 10 percent of all Americans born in 1900 would attend some college, 50 percent of those born in 1950 did. About 4 percent of all Americans born in 1900 would graduate from a four-year college, but 24 percent did among those born in 1950. Although women's college enrollment and graduation was considerably less than men's among cohorts born between about 1910 and 1950, a greater number of females than males have attended and graduated college in more recent cohorts.

Colleges greatly expanded in both size and scope across the twentieth century. Public institutions at the start of the century were not much larger than were private institutions. Universities in both sectors grew enormously in the next several decades, and those in the public sector became giants vastly exceeding the size of most privately-controlled institutions. The growth was both because the core of the institution, the undergraduate college, increased in size and because the university expanded its graduate offerings and added or merged with professional schools of various types. That is to say, it grew in both scale and scope.

But the true hallmarks of the American higher education system relative to its counterparts in other nations have been its decentralized control, the enormous competition between and within the public and private sectors, and its laissez faire orientation. It offers enormous choice and grand variety and gives students a second chance. In all of these ways, it is a quintessentially American institution.

Another hallmark of American higher education has been its world-renowned excellence. But that superiority came only after the 1950s when the best minds chose to remain at home and were joined by Europeans and others who wanted to study in America at the best universities.

By the end of the twentieth century a large group of American colleges and universities were the best in the world. The system had become enormously large with ample choice, variety, and competition for students, faculty, and research funds. America in the latter part of the twentieth century was on the verge of a third great educational transformation. College was becoming the mass institution that high school had become earlier in the century. Why, then, is the third transformation unfinished? If America has a system of higher education that is unparalleled in the world and offers vast choice, why is America losing ground to other nations in putting its masses through college? Chapter 9 continues that discussion. But we must first tie together our findings on inequality, education, and technological change to understand why wage inequality fell during much of the twentieth century but soared in the more recent past.

III

THE RACE

8

The Race between Education and Technology

Two Tales of the Twentieth Century

The "Best Poor Man's Country"

In the late eighteenth century America was deemed the "best poor man's country."[1] Land was plentiful, farming provided ample living standards, and wealth was rather equally distributed. A century later, much had changed. As James Bryce remarked in the late 1880s: "sixty years ago [in the time of de Tocqueville] there were no great fortunes in America, few large fortunes, no poverty. Now there is some poverty, many large fortunes, and a greater number of gigantic fortunes than in any other country of the world" (Bryce 1889, p. 600).[2] Bryce was clearly wrong that poverty did not exist in the 1830s,[3] but he was correct that wealth inequality had increased.

Living standards were considerably higher in 1890 than in 1790, but economic inequality had greatly expanded. Although the full distribution of income is hard to pin down before 1940, wealth had become far less equal by 1870 and considerably unequal by the 1920s, and the very top of the income distribution was relatively richer in 1913 than at almost any time since.[4] Around the turn of the twentieth century, earnings in occupations that required greater levels of schooling were far higher than those that required far less education (see Chapter 2). The economic return to a year of high school or college around 1915 was enormously high and only recently has the college premium approximated its value in 1915. We do not know precisely when in the nineteenth century the premium to schooling increased and whether it was as high even in 1850, but we do know that by 1900 a year of high school or college was an extremely good investment.

The large premium that accrued to those employed in occupations having substantial educational requirements was observed and commented on by close contemporaries in the early twentieth century. The economist Paul Douglas, for one, noted that "during the nineties [1890s], the clerical class constituted something of a *non-competing group*."[5] Douglas's interest in the wage distribution was sparked by a great wage compression that was apparent by the early 1920s. The astonishing change that took place in his own time prompted his comment: "Gradually the former monopolistic advantages are being squeezed out of white-collar work, and eventually there will be no surplus left."[6]

According to Douglas, several factors acted in concert to compress wages beginning in the late 1910s and early 1920s. One was the deskilling of clerical workers through the substitution of office machinery for skill. Another was the reduction in the flow of immigration, which according to Douglas led to an increase in the earnings of the less educated. Finally, the supply of educated and trained workers qualified to assume various white-collar positions greatly increased and thereby depressed their earnings.

Douglas was correct that multiple factors were at work, but the relative increase in the supply of skilled and educated personnel was of far greater importance, we shall soon demonstrate, than were skill reducing factors on the demand side and also more important than the decrease in immigration. The possibility that deskilling led to the large decrease in the relative earnings of the more educated was laid to rest in Chapter 2 when we showed the similarity of wage changes among clerical occupations. Earnings in white-collar occupations that did not undergo much technical change were reduced almost as much as in those that did.

The wage structure began to collapse a short time before 1920 and continued to narrow in various ways until the early 1950s. Earnings of the more educated were reduced relative to the less educated. Those employed in skilled occupations saw their earnings increase less than did those in the lower-skilled jobs. In fact, the wages of every skilled and professional group for which we could uncover consistent time series data declined relative to the wages for lesser skilled workers during the first half of the twentieth century. In Chapter 2 we presented relative wage series for professors of all ranks, engineers, office and clerical workers, and craft positions. There was also a substantial compression in the wage distribution of production workers within each of a large group of manufacturing industries. The returns to a year of schooling, not surprisingly, plummeted from 1915 to the early 1950s.[7] But the returns to schooling were so high prior to the narrowing that even after the decline in the wage premium education remained a very good investment.

Thus inequality and the pecuniary returns to education were both exceptionally high at the beginning of the twentieth century. Yet America remained the "best poor man's country" because it had a considerably higher average income than did other nations, as well as an open educational system and more equality of opportunity than existed in Europe.[8] Certain groups, in particular African Americans living in the U.S. South, remained left out for some time, but even they gained access to improved schooling during the mid-twentieth century and moved into higher paying jobs in the 1960s.

Integrating the Two Tales

By the early 1970s one could say that America "had it all." The U.S. economy had grown at a record pace in the 1960s, when labor productivity expanded at 2.75 percent average annually.[9] The nation's economy was strong. The wage structure had widened only slightly from the late 1940s and the income distribution had remained remarkably stable. Americans were sharing relatively equally in prosperity regardless of their position in the income distribution. Recall from Figures 2.2 and 2.3 of Chapter 2 that the American economy grew rapidly and its people "grew together" from 1947 to 1973.

Each generation of Americans achieved a level of education that greatly exceeded that of the previous one, with typical adults having considerably more years of schooling than their parents. Racial and regional differences in educational resources, educational attainment, and economic outcomes had narrowed substantially since the early

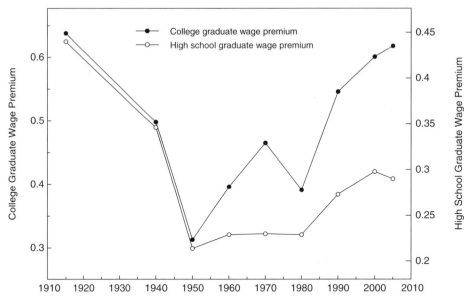

Figure 8.1. College Graduate and High School Graduate Wage Premiums: 1915 to 2005. Sources and Notes: *College Graduate Wage Premium:* The plotted series is based on the log (college/high school) wage differential series in Appendix Table D.1. We use the 1915 Iowa estimate and the 1940 to 1980 Census estimates for the United States. We extend the series to 1990, 2000, and 2005 by adding the changes in the log (college/high school) wage differentials for 1980 to 1990 for the CPS, 1990 to 2000 from the Census, and 2000 to 2005 from the CPS to maintain consistency in the coding of education across pairs of samples used for changes in the college wage premium. *High School Graduate Wage Premium:* The plotted series is based on the log (high school/eighth grade) wage differential series in Appendix Table D.1. We use the 1940 to 1980 Census estimates for the United States. To maintain data consistency, we then extend this series backwards to 1915 using the 1915 to 1940 change for Iowa and forward to 2005 using the 1980 to 1990 change from the CPS, the 1990 to 2000 change from the February 1990 CPS to the 2000 CPS, and the 2000 to 2005 change from the CPS.

twentieth century.[10] Upward mobility with regard to education characterized American society.

But the American economy did not stay the course. Inequality soared from the late-1970s to the early 2000s. Productivity did not continue to advance at the rate it once had, slowing considerably in the mid-1970s and remaining low for about two decades. Although productivity growth eventually resumed its previous rate, rising inequality magnified the impact of the sluggish economy on the vast majority of Americans.

The course of the entire twentieth century, therefore, contains two inequality tales—the first tale is one of narrowing differences whereas the second is one of widening differences. These tales can be clearly observed in the graph of almost century-long series for two key components of wage inequality shown in Figure 8.1: the college graduate wage premium (relative to those who stopped at high school) and the high school graduate wage premium (relative to those who left school at eighth grade), both from 1915 to 2005. Although it would be best to have the complete income and wage distribution for the entire period, these data do not exist for the pre-1940 period. The returns to education, however, can be analyzed in a consistent manner for the period from 1915 to 2005.

The returns to education and other components of wage inequality do not always move in lock step. But from 1940 to 2005 changes in the wage structure were closely correlated with changes in the premium to college (see Figure 2.6, Chapter 2, for the college premium and the 90–10 log wage differential). Furthermore, in recent decades the lion's share of rising wage inequality can be traced to an increase in educational wage differentials.[11] We feel confident that changes in the premium to education are reasonable proxies to those for wage inequality during the 90-year period we explore.

The college wage premium reveals a sharp decline from 1915 to 1950, jaggedness from 1950 to 1980, and a rapid increase after 1980. The premium to a college education came full circle in the twentieth century and by 2005 had returned to its high water mark at the beginning of the high school movement in 1915. The wage premium for high school graduates shows an equally sharp decrease in the pre-1950 era but less of an increase during the rest of the century.

The Race

Why did education returns fall in the first half of the twentieth century but rise at the end of the second half? That is the central question we address in this chapter. We analyze changes in the returns to education using the conceptual framework of a race between education (the supply of skill) and skill-biased technological change (the demand for skill).

We use direct evidence on changes in the stock of skill among the U.S. workforce and infer changes regarding skill-biased technological

change. That is, we do not use direct evidence on skill-biased technological change but deduce it from estimates of changes in the relative
demand for skill. These relative demand change estimates are derived
from relative wages by education, relative supply shifts by education,
and our estimates of the elasticity of substitution between the education groups. In Chapter 3 we documented detailed historical evidence
of the importance of skill-biased technological change in the evolution
of employment opportunities. That evidence includes strong positive
relationships throughout the last century between the utilization of
new and more capital-intensive technologies and the employment of
more highly educated workers. These findings make us confident that
our estimated changes in the relative skill demands are substantially
driven by skill-biased technological change.

The concept of a highly educated worker has changed across the period we analyze. A college graduate or possibly one with a postgraduate degree is considered highly educated today; in 1915, however,
a high school graduate would have been deemed well educated. For
that reason we use two definitions of a more-educated worker in our
analysis, focusing on the college premium for most of the century and
the high school premium for the early part.

In the race between technological change and education, education
ran faster during the first half of the century and technology sprinted
ahead of limping education in the last 30 years. The race produced
economic expansion and also determined which groups received the
fruits of growth.

But which of the two factors caused inequality to decline and then to
rise? Technological change and an increased demand for skilled and
educated workers were common to both periods. There were periods
of more rapid increase and some of slower increase. But, by and large,
the growth rate of the demand for more educated relative to less educated labor was fairly constant over the 1915 to 2005 period.

The major difference across the period was not changes in demand
but in supply. Shifts in the rate of growth in the supply of educated
labor played a critical role in altering inequality trends. Furthermore,
changes in the supply of educated native-born workers have been considerably more important than changes in the stock of immigrants to
the overall supply of skill. That is, changes in home-grown education
supply have been the most important factor in changing the overall

supply of educated Americans. Changes in labor market institutions that have tended to shelter the earnings of low- and middle-wage workers from market forces were key factors during several subperiods, but most of the variation in educational wage differentials can be well explained by a simple supply and demand framework.

We are now ready to offer a fuller analysis of inequality trends in the twentieth century and decompose the change in relative wages by education for the 1915 to 2005 period into its sources. To do so, we construct a framework that contains factors operating on the supply side and the demand side, with changes in wage setting institutions inserted during periods that cannot be fully explained otherwise.

The Supply, Demand, and Institutions (SDI) Framework

To guide the empirical analysis of the factors that altered the returns to education during the past century, we construct a formal supply-demand framework. The framework rests on the central finding in Chapter 3 that skill-biased technical change advanced rapidly throughout the twentieth century and thus that the relative demand for skill increased at a fairly steady rate. Our approach is to determine how much of the evolution in educational wage differentials can be explained by fluctuations in the growth rate of the supply of skills combined with smooth trends in relative demand growth.

A large portion of the evolution of wage differentials, as we will demonstrate, can be explained using the simple framework. But where supply-demand forces alone fall a bit flat, institutional factors can reconcile patterns in the skill premium. In that sense we combine the usual supply and demand framework with institutional rigidities and alterations. The broader framework is most important in understanding wage structure changes during the 1940s and in contrasting changes from the mid- to late 1970s to those of the early 1980s. The wage compression of the 1940s, it appears, went far beyond what can be accounted for by market forces alone and was driven in part by institutional factors of the World War II era, such as the greatly expanded role of unions and the residual impact of the wartime wage-setting policies.

The framework contains two main forces. One is the change in the relative supply of more-educated workers, which has mainly occurred

through changes in the schooling of successive cohorts of labor market
entrants. The second is the change in the relative demand for more-
educated workers, which has been driven largely by skill-biased technological change.

A labor demand framework, in which the aggregate production function depends only on the quantities of skilled and unskilled workers,
guides our analysis. Skilled workers (S) are those with some college
and the unskilled (U) are those without any college. The production
function is assumed to be CES (constant elasticity of substitution) in
skilled and unskilled labor with an aggregate elasticity of substitution
between the two types of labor given by σ_{SU}. Unskilled labor itself is
assumed to be a CES subaggregate that depends on the number of high
school graduates (H) and those without a high school diploma (O), also
called "dropouts," with an elasticity of substitution of σ_{HO}.[12]

The framework is summarized by the following two equations:

$$Q_t = A_t[\lambda_t S_t^\rho + (1 - \lambda_t)U_t^\rho]^{\frac{1}{\rho}} \tag{1}$$

$$U_t = [\theta_t H_t^\eta + (1 - \theta_t)O_t^\eta]^{\frac{1}{\eta}} \tag{2}$$

where eq. (1) is the aggregate production function and eq. (2) is the subaggregate for unskilled labor. In eq. (1) Q is output, A is total factor productivity, S is units of skilled or college labor, and U is units of unskilled
or non-college labor. In eq. (2) H is units of high school graduate labor
and O is units of high school dropout labor. The parameters λ_t and θ_t
give the shares of the different types of labor and are modeled as technology shift parameters.[13] The CES parameters ρ and η are related to the
elasticities of substitution, such that $\sigma_{SU} = 1/(1 - \rho)$ and $\sigma_{HO} = 1/(1 - \eta)$.

Wages for the three skill groups of workers (S, H, O) are derived
using the familiar condition that a competitive equilibrium occurs
when wages equal marginal products. Relative wages for college to
high school workers and for high school graduates to dropouts are
given by:

$$\log\left(\frac{w_{S_t}}{w_{U_t}}\right) = \log\left(\frac{\lambda_t}{1 - \lambda_t}\right) - \frac{1}{\sigma_{SU}}\log\left(\frac{S_t}{U_t}\right) \tag{3}$$

and

$$\log\left(\frac{w_{H_t}}{w_{O_t}}\right) = \log\left(\frac{\theta_t}{1-\theta_t}\right) - \frac{1}{\sigma_{HO}}\log\left(\frac{H_t}{O_t}\right) \tag{4}$$

Thus, relative wages depend on the demand shifters (λ_t and θ_t), the relative supply of the more and less educated groups, and the relevant elasticity of substitution between the two groups (σ_{SU} and σ_{HO}). Equations (3) and (4) are the main estimating equations of the model.

A key assumption in our empirical implementation of the framework is that relative skill supplies are predetermined and thus that, in the short run, labor supply for each skill group is completely inelastic.[14] In addition, the framework assumes that a change in the relative supply of college to non-college labor does not affect the premium to high school graduates relative to high school dropouts. The restriction does not imply that college supplies are unimportant in the determination of unskilled wages, but it does mean that the supply of more educated labor equally affects the wages of the high school graduates and the dropouts.

The framework allows for heterogeneity in worker productivity (efficiency) within each skill aggregate (college, high school graduate, and dropout labor), but it assumes that different workers are perfect substitutes in production within each skill aggregate. The implication is that skill supplies (S, U, H, and O) must be measured in efficiency units (productivity-adjusted hours worked) rather than by hours worked. We measure skill supplies in efficiency units taking into account systematic differences in wages by age, sex, and education and adjusting for changes in the age-sex-education group composition of hours worked within each skill aggregate.[15]

Figure 8.2 is a graphical depiction of how the framework can be used to analyze changes in the wage structure. It is drawn with estimates for the 1960 to 2005 period that we will shortly describe. The SS* lines give the annualized percentage change in the relative supply of educated workers (college educated relative to non-college workers, in this case) for the period noted. The estimation of eq. (3) produces the elasticity of substitution, σ_{SU}, between college and non-college workers and thus the wage elasticity of demand. Given that estimate and the

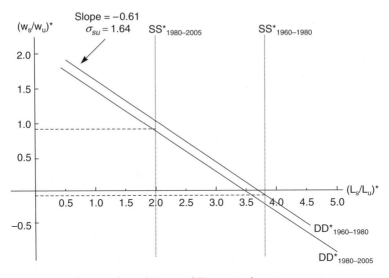

Figure 8.2. The Supply and Demand Framework.

rate of change in the wage premium to college workers, the demand function in rate of change form, DD*, can be identified. The point at which the DD* function crosses the *X* axis gives the change in the relative demand for college workers.

We now take the framework to the data and measure the relative roles of supply forces, demand factors, and institutional changes in affecting the wage premium. We begin with the college wage premium.

Why the Premium to Skill Changed: 1915 to 2005

College Wage Premium

APPLYING THE FRAMEWORK

The facts that any framework of the college wage premium must explain and reconcile are given in Table 8.1 (see also Figure 8.1). They are easily summarized. The college wage premium (col. 1) collapsed from 1915 to 1950 but subsequently increased, especially after 1980. By 2005 the college wage premium was back at its 1915 level. As we noted in describing Figure 8.1, the returns to college have come full circle. The relative supply of college workers (Table 8.1, col. 2) grew

Table 8.1. Changes in the College Wage Premium and the Supply and Demand for College Educated Workers: 1915 to 2005 (100 × Annual Log Changes)

	Relative Wage	Relative Supply	Relative Demand ($\sigma_{SU}=1.4$)	Relative) Demand ($\sigma_{SU}=1.64$)	Relative Demand ($\sigma_{SU}=1.84$)
1915–40	−0.56	3.19	2.41	2.27	2.16
1940–50	−1.86	2.35	−0.25	−0.69	−1.06
1950–60	0.83	2.91	4.08	4.28	4.45
1960–70	0.69	2.55	3.52	3.69	3.83
1970–80	−0.74	4.99	3.95	3.77	3.62
1980–90	1.51	2.53	4.65	5.01	5.32
1990–2000	0.58	2.03	2.84	2.98	3.09
1990–2005	0.50	1.65	2.34	2.46	2.56
1940–60	−0.51	2.63	1.92	1.79	1.69
1960–80	−0.02	3.77	3.74	3.73	3.73
1980–2005	0.90	2.00	3.27	3.48	3.66
1915–2005	−0.02	2.87	2.83	2.83	2.82

Sources: The underlying data are presented in Appendix Table D.1 and are derived from the 1915 Iowa State Census, 1940 to 2000 Census IPUMS, and 1980 to 2005 CPS MORG samples.

Notes: The "relative wage" is the log (college/high school) wage differential, which is the college wage premium. The underlying college wage premium series is plotted in Figure 8.1. The relative supply and demand measures are for college equivalents (college graduates plus half of those with some college) relative to high school equivalents (those with 12 or fewer years of schooling and half of those with some college). The log relative supply measure is given by the log relative wage bill share of college equivalents minus the log relative wage series:

$$\log\left(\frac{S}{U}\right) = \log\left(\frac{w_s S}{w_u U}\right) - \log\left(\frac{w_s}{w_u}\right)$$

where S is efficiency units of employed skilled labor (college equivalents), U is efficiency units of employed unskilled labor (high school equivalents), and w_S and w_U are the (composition-adjusted) wages of skilled and unskilled labor. The log relative wage bill is based on the series for the wage bill share of college equivalents in Appendix Table D.1. The relative demand measure $\log(D_{SU})$ depends on σ_{SU} and follows from equation (3) in the text:

$$\log(D_{su}) = \log\left(\frac{S}{U}\right) + \sigma_{su} \log\left(\frac{w_s}{w_u}\right)$$

To maximize data consistency across samples in the measurement of education, changes from 1980 to 1990 use the CPS, changes from 1990 to 2000 use the census, and changes from 2000 to 2005 use the CPS. The changes for 1915 to 1940 are for Iowa. See Autor, Katz, and Krueger (1998) for details on the methodology for measuring relative skill supply and demand changes.

rapidly for much of the period, although a slowdown of critical importance is apparent more recently, particularly from 1990 to 2005. For
the full period, the growth in relative supply of college workers occurred at a fairly rapid clip—on the order of 2.87 percent per annum.

Because the premium to education at the end of the century was approximately equal to its level at the start, our supply-demand framework implies that the relative demand for skill across the entire century
must have grown at about the same rate as the relative supply of skill.
Even though the race between technology and education was nearly
even over the long haul, the long run conceals crucial short run
changes. What caused the returns to education to decline and then rise
across the past century? We will demonstrate that fluctuations in the
relative supply of college workers together with stable demand growth
go far to explain the higher-frequency movements in the college premium.

We estimate a version of eq. (3) across the 1915 to 2005 period using
data for all the available years: 1915, 1940, 1950, 1960, and annually
from 1963 to 2005.[16] The dependent variable is the wage premium of
those with at least a college degree (16 or more years of schooling) relative to those with exactly a high school degree (12 years of schooling).
The premium is the log of the ratio of the wages for the two education
groups. The relative skill supply measure is that of efficiency units of
college equivalents (those with a college degree plus half of those with
some college) to efficiency units of high school equivalents (those with
12 or fewer years of schooling plus half of those with some college).[17]

A linear time trend allows for secular growth in the relative demand
for college workers. Interactions with specific years enable demand
trend changes. Consistent with our earlier findings concerning the
slowdown in demand growth beginning in the early 1990s (see Chapter
3), we add a term in most specifications to allow the demand trend to
change in 1992.[18] The results are provided in Table 8.2 and graphed in
Figure 8.3.

The most important result from the estimation is that changes in the
relative supply of college workers had a substantial and economically
significant negative impact on the college wage premium across the entire period. Most of the specifications yield similar coefficients for the
relative supply variable (Table 8.2, line 1). That for column 3, our preferred specification, implies that a 10 percent increase in the relative

Table 8.2. Determinants of the College Wage Premium: 1915 to 2005

	(1)	(2)	(3)	(4)	(5)
(College/high school) supply	−0.544 (0.079)	−0.595 (0.093)	−0.610 (0.065)	−0.579 (0.099)	−0.618 (0.079)
(College/high school) supply × post-1949					0.0078 (0.0420)
Time	0.00378 (0.00200)	0.00970 (0.00243)	0.00991 (0.00171)	0.00973 (0.00545)	0.0103 (0.0028)
Time × post-1949	0.0188 (0.0013)				
Time × post-1959		0.0156 (0.0012)	0.0154 (0.0009)		0.0150 (0.0022)
Time × post-1992	−0.00465 (0.00227)	−0.00807 (0.00279)	−0.00739 (0.00196)		−0.00742 (0.00199)
1949 Dummy			−0.137 (0.021)		−0.143 (0.036)
Time^2 × 10				−0.00342 (0.00203)	
Time^3 × 1000				0.105 (0.034)	
Time^4 × 10,000				0.00664 (0.00186)	
Constant	−0.493 (0.168)	−0.645 (0.197)	−0.656 (0.138)	−0.587 (0.210)	−0.674 (0.079)
R^2	0.934	0.917	0.960	0.928	0.960
Number of observations	47	47	47	47	47

Sources and Notes: Each column is an OLS regression of the college wage premium on the indicated variables using a sample covering the years 1914, 1939, 1949, 1959, and 1963 to 2005. Standard errors are given in parentheses below the coefficients. The college wage premium is a fixed weighted average of the estimated college (exactly 16 years of schooling) and post-college (17+ years of schooling) log wage differential relative to high school graduates (those with exactly 12 years of schooling). (College/high school) supply is the log supply of college equivalents to high school equivalents both measured in efficiency units. "Time" is measured as years since 1914. The data for 1963 to 2005 are from the 1964 to 2006 March CPS samples. The college wage premium and relative supplies in efficiency units for 1963 to 2005 use the same data processing steps and sample selection rules as those described in the data appendix to Autor, Katz, and Kearney (2007). The college wage premium for 1963 to 2005 uses the log weekly earnings of full-time, full-year workers. The college wage premium observations for 1914, 1939, 1949, and 1959 append the changes in the college wage premium series from 1915 to 1970 (actually 1914 to 1969) plotted in Figure 8.1 to the 1969 data point from our March CPS series. The log relative supply observations for 1914 to 1959 similarly append changes in the relative supply of college equivalents from 1914 to 1939 for Iowa and for the United States from 1939 to 1949, 1949 to 1959, and 1959 to 1969 from the Census IPUMS samples using the efficiency-units measurement approach of Tables 8.5 and 8.6.

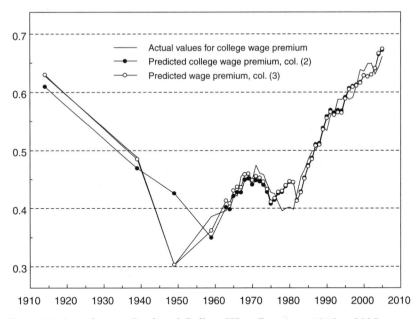

Figure 8.3. Actual versus Predicted College Wage Premium: 1915 to 2005.
Sources and Notes: The actual values for the college wage premium are from the
series used in the regressions in Table 8.2 and documented in the notes to that
table. The two series for the predicted college wage premium are the values of the
college wage premium predicted from the regressions in column 2 and column 3
of Table 8.2, as noted in the figure.

supply of college equivalents reduces the college wage premium by 6.1
percent and translates into an elasticity of substitution between the
skilled and unskilled, σ_{SU}, of 1.64 (= 1/0.61, see eq. (3)). The rapid
growth of the supply of college equivalents from 1915 to 1980 operated to depress the college wage premium despite strong secular
growth in the relative demand for college equivalents. The sharp slowdown in the growth in the supply of college workers since 1980 has
been a driving force behind the rise in the college wage premium.

Overall, simple supply and demand specifications do a remarkable
job of explaining the long-run evolution of the college wage premium.
The predictions from specifications (2) and (3), graphed in Figure 8.3
alongside the actual values for the college wage premium, show that
most of the shorter-run fluctuations can be tracked as well. Two short-
run fluctuations, however, are more complicated. One is the 1940s and
the other is the mid- to late 1970s.

Each of the specifications in columns 1, 2, and 3 uses a different method to account for the 1940s within our general framework. The column 1 specification allows trend demand to differ between the first and second halves of the twentieth century by including an interaction with a post-1949 dummy variable. The trend estimates show slow demand growth for college workers in the first half of the twentieth century, a sharp acceleration after 1949, and a somewhat slower change after 1992. The model over-predicts the decline in the college wage premium from 1915 to 1940 and under-predicts the sharper decline in the 1940s. The specification in column 2 allows the demand trend shift to occur after 1959, rather than 1949.

Figure 8.3 shows that the column 2 specification does a fine job fitting the 1915 to 1940 decline but not the sharp decline in the college premium of the 1940s and the strong rebound of the 1950s. The difficulty in predicting the short-run changes for the 1940s and 1950s probably stems from institutional and cyclical factors. These include the residual effects from World War II wage policies, industrial union strength that increased the bargaining power of the lower-educated, the strong demand for war production workers, and the postwar boom in consumer durables, all of which acted to reduce the relative wage of college workers below the long-run market equilibrium value of 1950.[19]

The decrease in the college wage premium of the 1940s overshot the decrease predicted by changes in the fundamentals, and the increase of the 1950s may have brought the system back into sync. We explore that possibility by including a dummy variable for 1949 to allow temporary institutional factors to impact wage setting in the 1940s (Table 8.2, col. 3). The estimation implies that institutional factors, or temporary demand factors, lowered the college wage premium by 14 log points in 1949. The column 3 estimation in Figure 8.3 fits the data extremely well and provides our preferred specification. The flexible time trend given by the column 4 specification demonstrates the robustness of the coefficient on relative labor supply across the entire period.

Another brief period that is not captured well by the specifications in Table 8.2 is the decline in the college wage premium in the mid- to late 1970s. The period was complicated by the post-1973 productivity slowdown and severe oil price and inflation shocks. Many unions, such as in steel and automobiles, whose members were disproportionately in the non-college group had wage contracts that were fully indexed to inflation and geared to provide real wage increases that tracked expected national productivity growth. Union settlements in the late 1970s were not yet adjusted to slower productivity growth and, in consequence, they led to a relative increase in the wages of the non-college workers. But the deep recession of the early 1980s and changes in employer attitudes toward unions, particularly following Reagan's stand-off with air traffic controllers, led to concession bargaining in the early 1980s and set the stage for the spectacular rebound of the college wage premium. The continued decline of unions and the erosion of the real value of the federal minimum wage in the 1980s may have increased the college wage premium by more than market factors alone would have predicted.[20]

Demand growth for college workers appears to have slowed in the 1990s, as indicated by the negative coefficient on the trend interacted with 1992. Given the rapid spread of information technology in the 1990s and beyond, the finding would appear to be at odds with the skill-biased technological change explanation. But a resolution exists.

Computerization prior to the 1990s largely substituted for non-college clerical and production tasks. More recent advances in information technology have increasingly led to organizational changes that eliminate many lower- and middle-paid college jobs but greatly complement top-end managers and those with strong problem-solving skills. Demand for those who graduated from more selective institutions as well as those with post-B.A. degrees is still soaring and they are doing spectacularly well. But demand for many other college workers is less strong and their earnings have not risen as much relative to non-college workers since 1990.[21] Nevertheless, the college wage premium (even for those with only a B.A.) remains at a historically high level and even "marginal" college graduates earn a very high return to college.[22]

COMPUTING SUPPLY AND DEMAND SHIFTS

The estimated coefficients on college relative supply (that is, σ_{SU}) are used to compute changes in relative demand, as depicted in Figure 8.2. The demand shifts are given in the last three columns of Table 8.1 for three values of σ_{SU}: 1.4 (a consensus estimate from the past literature); 1.64 (our preferred estimate from col. 3 of Table 8.2); and 1.84 (implied by col. 1 of Table 8.2). The results are fairly robust to the choice of parameter values.

On average from 1915 to 2005 supply and demand forces kept pace with each other, as we noted before. Neither education nor technology won the race in the long run. The same was true for the 1960 to 1980 period.[23] But for other periods it was not. Across the earliest periods listed, 1915 to 1940 and 1940 to 1960, supply ran ahead of demand by about 1 percent average annually.[24] For the most recent period, 1980 to 2005, demand outstripped supply. Most important is that for both the early and late subperiods, educational supply changes have been the tail wagging the wage-premium dog. Supply variations were far more important in changing relative wages than were differential demand changes across periods.

That supply factors and not demand factors were the culprits in changing inequality can be seen in Figure 8.2. The relative supply of college workers increased at 3.77 percent per annum from 1960 to 1980 but at just 2 percent per annum from 1980 to 2005. Relative demand, on the other hand, was considerably more stable over the period. Had the relative supply of college workers from 1980 to 2005 expanded at the rate it did from 1960 to 1980, the relative wage of college workers would have *fallen* (the intersection of $DD^*_{1980-2005}$ with $SS^*_{1960-80}$), and not risen at 0.9 percent per annum. Thus, the slowdown in the growth of educational attainment since 1980 is the most important factor in the rising college wage premium of the post-1980 period.

Technology has been racing ahead of education in recent decades because educational growth has been sluggish, not because skill-biased technical change has accelerated. To be sure, relative demand growth for college workers was more rapid in the second half of the twentieth century, particularly in the 1980s, than in the first half, but demand has not been growing rapidly since 1990.[25] We summarized the point in Chapter 3 with the quip: "it's not technology—stupid." We will soon demonstrate that the inequality culprit is also not immigration. Relative supply changes can be affected by changes in the stock of domestically produced workers or by an influx of workers from abroad. The former, it will be shown, was far more important in all periods we examine.

College workers were the most important well-educated group in the second-half of the twentieth century. But in the first-half of the twentieth century college workers were not the only well-educated group and were not the most important quantitatively. A high school diploma was the mark of a well-educated individual in the early part of the twentieth century, just as a college diploma has been from the mid-point onward. We now turn to an understanding of movements in the high school wage premium.

High School Wage Premium

APPLYING THE FRAMEWORK

The high school wage premium collapsed from 1915 to 1950, in an almost identical manner to the college wage premium (see Figure 8.1 and Table 8.3).[26] But the high school wage premium then remained flat from 1950 to 1980 whereas the college wage premium rose, albeit with some jaggedness. The big difference in the two series begins after 1980 when the payoff to college soared and that to high school increased only slightly. By the end of the century the high school wage premium was far lower than it was in 1915.

The primary reason for the collapse of the high school wage premium from 1915 to 1950 was the enormous growth in the relative supply of high school graduates created by the high school movement. From 1915 to 2005 the supply of high school graduates increased at 4.25 percent annually more than the supply of those without a high school diploma (called "dropouts" here), and at 5.54 percent annually more during the high school movement years, 1915 to 1940 (see Table 8.3). The only years of marked slowness in the relative supply of high school graduates are those in the most recent period, 1990 to 2005.

To obtain estimates of the elasticity of substitution between high school graduates and dropouts (σ_{HO}), we estimate a version of eq. (4) for the high school wage premium, similar to the analysis for the college wage premium (see Table 8.4). In the analysis of the college wage premium, the elasticity of substitution (σ_{SU}) was stable throughout the period, varying from 1.6 to 1.8. But, in the case of high school graduates versus dropouts, the elasticity of substitution (σ_{HO}) shifted substantially around 1950.

The shift can be seen by adding an interaction between the relative supply term and a dummy variable for the post-1949 period (Table 8.4, col. 4). In the absence of the interaction the elasticity of substitution is substantial in magnitude (around 5) for the entire period. But

Table 8.3. Changes in the High School Wage Premium and the Supply and Demand for High School Educated Workers: 1915 to 2005 (100×Annual Log Changes)

	Relative Wage	Relative Supply	Relative Demand ($\sigma_{HO}=2$)	Relative Demand ($\sigma_{HO}=3$)	Relative Demand ($\sigma_{HO}=5$)
1915–40	−0.38	5.54	4.79	4.41	3.66
1940–50	−1.32	4.38	1.74	0.42	−2.22
1950–60	0.15	2.72	3.02	3.17	3.47
1960–70	0.01	5.31	5.33	5.34	5.36
1970–80	−0.01	5.65	5.63	5.62	5.60
1980–90	0.44	4.04	4.92	5.36	6.24
1990–2000	0.25	1.87	2.37	2.62	3.12
1990–2005	0.11	1.52	1.75	1.86	2.09
1940–60	−0.59	3.55	2.38	1.79	0.62
1960–80	0.00	5.48	5.48	5.48	5.48
1980–2005	0.24	2.53	3.02	3.26	3.75
1915–2005	−0.17	4.25	3.91	3.75	3.41

Sources: The underlying data are presented in Appendix Table D.1 and are derived from the 1915 Iowa State Census, 1940 to 2000 Census IPUMS, and 1980 to 2005 CPS MORG samples.

Notes: The relative wage is the log wage differential between those with 12 years and 8 years of school, adjusted for demographic factors. This high school wage premium series is plotted in Figure 8.1. The relative supply and demand measures compare exact high school graduates (those with exactly a high school degree or 12 years of completed schooling) to those without a high school diploma (0 to 11 years of schooling). The methodology for constructing the supply and demand measures is the same as described in the notes to Table 8.1 with high school graduates (H) replacing college equivalents (S) and high school dropouts (O) replacing high school equivalents (U). Thus, the log relative supply measure is given by the log relative wage bill share of high school graduates to dropouts minus the log high school wage premium. The log relative demand measure $\log(D_{HO})$ is based on eq. (4) in the text and given by:

$$\log(D_{HO}) = \log\left(\frac{H}{O}\right) + \sigma_{HO} \log\left(\frac{w_H}{w_O}\right)$$

To maximize data consistency across samples in the measurement of education, changes from 1980 to 1990 use the CPS MORG, changes from 1990 to 2000 use the February 1990 CPS and the 2000 CPS MORG, and changes from 2000 to 2005 use the CPS MORG. The changes for 1915 to 1940 are for Iowa.

the interaction reveals that the elasticity of substitution is large only in the post-1949 period and far smaller (around 2) in the previous decades.[27]

The results imply that high school graduates and dropouts are far closer substitutes today than they were prior to the 1950s. Therefore,

Table 8.4. Determinants of the High School Wage Premium: 1915 to 2005

	(1)	(2)	(3)	(4)	(5)
(High school/ dropout) supply	−0.180 (0.059)	−0.193 (0.039)	−0.193 (0.039)	−0.512 (0.071)	−0.352 (0.137)
(High school/dropout) supply×post-1949				0.322 (0.054)	
(High school/dropout) supply×time					0.00496 (0.00218)
Time	−0.00084 (0.00278)	0.00239 (0.00179)	0.00235 (0.00176)	0.0171 (0.0037)	0.0308 (0.0100)
Time×post-1949	0.0132 (0.0011)			−0.0032 (0.0029)	
Time×post-1959		0.0117 (0.0006)	0.0116 (0.0006)		
Time×post-1992	−0.00753 (0.00386)	−0.0109 (0.0026)	−0.0107 (0.0026)	−0.0106 (0.0029)	
1949 Dummy			−0.0278 (0.0192)		
$Time^2 \times 10$					−0.0084 (0.0012)
$Time^3 \times 1000$					0.113 (0.025)
$Time^4 \times 10{,}000$					−0.0055 (0.0015)
Constant	0.088 (0.118)	0.049 (0.078)	0.053 (0.077)	−0.579 (0.142)	−0.282 (0.271)
R^2	0.897	0.953	0.956	0.944	0.971
Number of observations	47	47	47	47	47

Sources and Notes: Each column is an OLS regression of the high school wage premium on the indicated variables using a sample covering the years 1914, 1939, 1949, 1959, and 1963 to 2005. Standard errors are given in parentheses below the coefficients. The high school wage premium is the (composition-adjusted) log wage differential between those with exactly a high school degree (12 completed years of schooling) and those with 8 completed years of schooling. (High school/dropout) supply is the log supply of those with 12 completed years of schooling to those with 0 to 11 years of schooling measured in efficiency units. "Time" is measured as years since 1914. The data for 1963 to 2005 are from the 1964 to 2006 March CPS samples. We use the same data processing steps and sample selection rules as those described in the data appendix to Autor, Katz, and Kearney (2007) in constructing wage series for high school graduates and dropouts and the relative supply measure in efficiency units for 1963 to 2005. The high school wage premium for 1963 to 2005 is for the log weekly earnings of full-time, full-year workers and compares workers with exactly 12 years of schooling to all dropouts. We multiply this high school wage premium series for 1963 to 2005 by 1.44 to make it comparable to a series for the log wage gap between those with 12 and 8 years of schooling. The multiplier of 1.44 is the mean ratio of the log (high school/eighth grade) to the log (high school/dropout) wage differential series in Appendix Table D.1 for 1915 to 1980. The high school wage premium observations for 1914, 1939, 1949, and 1959 append the changes in the high school wage premium series from 1915 to 1970 (actually 1914 to 1969) plotted in Figure 8.1 to the 1969 data point from our March CPS series. The log relative supply observations for 1914 to 1959 similarly append changes in the relative supply of college equivalents from 1914 to 1939 for Iowa and for the United States from 1939 to 1949, 1949 to 1959, and 1959 to 1969 from the Census IPUMS samples using the efficiency-units measurement approach of Tables 8.5 and 8.6.

changes in the relative supply of high school graduates to dropouts today will have smaller effects on the high school wage premium than in the past. High school graduates were once distinctly more skilled than those without a diploma and many positions were reserved for them. Thus the vast increase in high school graduation throughout much of the twentieth century served to reduce the high school wage premium by increasing the relative supply of high school graduates to dropouts.

Earlier in the century firms sought high school graduates as office workers and as blue-collar production workers in many of the high-tech industries of the day. Those hiring employees described certain jobs as requiring a high school diploma or particular high school courses and they viewed high school graduates as vastly superior to those without secondary school training. But today's high school graduates and dropouts are perceived by employers as being close substitutes. The historical facts and our estimates speak to a change in the distinction between a worker with a high school degree and one who is a high school dropout.[28]

There appears to have been some overshooting of the high school premium in the 1940s with a catch-up in the 1950s, as was the case with the college premium. But institutional factors appear far less important than for the college wage premium. The 1949 year dummy, for example, is insignificant in the high school wage premium regression (Table 8.4, column 3).

COMPUTING SUPPLY AND DEMAND SHIFTS

We calculate the relative impact of supply and demand forces in changing the high school wage premium using three values of the elasticity of substitution (2, 3, and 5) that span our estimates (see Table 8.3). Our preferred elasticities are 2 for the pre-1950s and 5 for the post-1950s. The central finding is that the decrease in the high school wage premium from 1915 to 1940 was due mainly to the rapid growth in relative supply.

Relative demand increased greatly from 1915 to 1940, but it grew at a slower pace than supply and the wage premium declined. Relative supply also increased at a rate exceeding demand from 1940 to 1960. The size of the difference will depend on whether one uses the larger elasticity value or the smaller one, since the period spans the shift we observe in the substitution parameter.[29] Also of importance is the moderate increase in the high school wage premium from 1980 to 2005.

Although relative demand growth moderated, the relative supply of high school graduates slowed considerably more.

In the analyses we have done, supply factors were shown to have been more important than demand factors in altering the premium to education in the twentieth century. Changes in the relative supply of educated labor can arise from several sources. We have emphasized changes in the educational attainment of successive cohorts of native-born Americans. But the foreign-born may have been an important contributing force.

Immigration and Demographics

Immigration may have greatly increased the supply of those without a high school diploma in the 1980 to 2005 period, thus reducing the relative supply of high school graduate labor. Immigration may also have reduced the relative supply of college workers, thus serving to increase the premium to college in the post-1980s. Earlier in the twentieth century legislative restrictions that greatly reduced immigration flows could have increased the relative supply of more educated workers. In all cases, immigration forces could have acted in concert with education forces to change the premium to skill. We turn now to a direct estimate of the influence of immigration on skill supplies and the premium to skill from 1915 to 2005.

IMMIGRATION AND THE LABOR FORCE

In the early years of the twentieth century immigrants were a substantial part of labor force growth. By 1915 the foreign-born share of the U.S. labor force (18 to 65 years old) exceeded 21 percent.[30] After the immigration restrictions of the 1920s the foreign-born share of the labor force declined, and by 1970 it was just 5.4 percent.[31] More recently, and especially after the 1965 immigration legislation ended national-origins quotas, the inflow surged again. By 2005 the foreign-born share had risen to 15 percent. The national-origin composition of immigration also shifted in recent decades and the share of immigrants coming from Asia and Latin America (especially Mexico) has increased. In our exploration of the impact of immigration on the skill premium we concentrate on the earlier (1915 to 1940) and later decades (1980 to 2005), when the contribution of immigration to labor force growth was large.

Because immigrants, on average, have been less well educated than U.S. natives, large changes in immigration flows during the twentieth century altered relative skill supplies. Changes in relative skill supplies, in turn, potentially impacted the premium to education. In the first period we consider, 1915 to 1940, the slowdown in immigration would have served to *increase* relative skill supplies. Had immigration continued at its previous rate, there would have been a larger supply of those with less education since the United States was then undergoing its high school movement but Europe, the largest source of immigrants, had not yet done so. In the most recent period, 1980 to 2005, immigration is presumed to have *decreased* relative skill supplies. Immigration today, it is often claimed, is flooding America with workers who compete for jobs with the native-born at the bottom of the education and skill ladder.

How much of the changes in skill supplies have come from immigration and how much from changes in the education of the native-born population? The presumption of many commentators is that immigration greatly increases the premium to skill. But does it?

Our answer is that immigration has had a smaller effect on relative skill supplies than is generally presumed in all periods we examine and that immigration has had only a relatively modest impact on changes in the premium to education. The impact of immigration from 1980 to 2005 was larger than during earlier periods. But our estimates are that immigration was responsible for only 10 percent (about 2.4 log points) of the post-1980s increase in the college to high school wage premium (which was 23 log points). Immigration can explain a considerably larger share (43 percent) of the rise in the high school graduate wage premium, but the domestic education slowdown accounts for more (57 percent).

The reason that immigration is responsible for only a small fraction of the post-1980s increase in the college wage premium concerns the educational distribution of recent immigrants. Many of the foreign-born occupy the very bottom of the education ladder, but some are found at the top with college and graduate degrees. In 2005, 17 percent of the foreign-born population had fewer than nine years of schooling whereas less than 1 percent of native-born Americans did. At the other end of the spectrum, immigrants in 2005 were more likely to have an advanced (post-college) degree and had about the same likelihood of having at least a four-year college degree as did native-born Americans.[32]

Table 8.5. Immigrant Contribution to Labor Supply by Educational Attainment: 1915 to 2005

	Ratio of Immigrants to U.S.-Born Workers						
	High School Dropouts versus High School Graduates			High School Equivalents versus College Equivalents			Immigrant Employment Share
	(1a)	(2a)	(3a)	(1b)	(2b)	(3b)	
Year	Dropouts	Graduates	Immigrant Contribution	High School	College	Immigrant Contribution	
Iowa							
1915	0.223	0.059	0.144	0.198	0.114	0.073	0.156
1940	0.084	0.035	0.046	0.067	0.056	0.010	0.058
U.S.							
1940	0.169	0.075	0.084	0.140	0.088	0.047	0.111
1950	0.124	0.071	0.048	0.103	0.074	0.026	0.086
1960	0.086	0.044	0.039	0.067	0.062	0.005	0.062
1970	0.071	0.040	0.029	0.054	0.063	−0.009	0.054
1980	0.118	0.049	0.065	0.068	0.075	−0.006	0.067
1990	0.291	0.075	0.183	0.106	0.096	0.009	0.093
2005	0.762	0.146	0.430	0.190	0.151	0.033	0.151

Sources: 1915 Iowa State Census, 1940 to 1990 Census IPUMS, and 2005 CPS MORG. The samples include civilian employed workers from 18 to 65 years old.

Notes: The "immigrant contribution" calculation follows the approach of Borjas, Freeman, and Katz (1997) and is derived as follows. The ratio of unskilled (U) to skilled (S) workers can be decomposed as follows:

$$\log\left(\frac{L_{U_t}}{L_{S_t}}\right) = \log\left(\frac{N_{U_t}}{N_{S_t}}\right) + \left[\log\left(1+\frac{M_{U_t}}{N_{U_t}}\right) - \log\left(1+\frac{M_{S_t}}{N_{S_t}}\right)\right],$$

where L_{jt} = supply of workers in skill group j in year t, and N_{jt} (M_{jt}) = supply of U.S.-born (immigrant) workers in skill group j in year t, such that $L_{jt} = N_{jt} + M_{jt}$. The first term of the right side of the equation is the native contribution to the ratio. The second term, in brackets, is the immigrant contribution. We call this term the "immigrant contribution" and it is given in the table in columns 1a and b. The components of the "immigrant contribution" are given in columns 1a and b, $\frac{M_{U_t}}{N_{U_t}}$, and columns 2a and b, $\frac{M_{S_t}}{N_{S_t}}$. The "skilled" groups in the table are high school graduates and college "equivalents"; the "unskilled" groups are dropouts and high school "equivalents," respectively. College equivalents are those with 16 or more years of schooling plus half of those with some college. High school equivalents are those with 12 or fewer years of schooling plus half of those with some college. Worker supplies in columns 1 to 3 are measured in efficiency units: the sum of hours of work weighted by the relative wage of each individual's demographic group in a base year (the average of 1940, 1960, and 2005). We use 60 demographic groups (6 education groups × 5 age groups × 2 sexes). The last column presents the immigrant employment share using raw employment counts not efficiency units.

Immigrants in 1915 expanded the labor supply of dropouts by 22 percent, as compared with 6 percent for those with exactly a high school degree, and by 20 percent for high school equivalents, as compared with 11 percent for college equivalents (Table 8.5, col. 1a). These 1915 data come from our Iowa sample. Figures for the entire United States, if we had them, would probably reveal a somewhat larger immigrant share of employment in each skill group. The differential impact of immigration on labor supply across education groups, however, is likely to have been similar for Iowa and the nation as a whole.[33] In 1940, after immigration restrictions were in place for nearly two decades, the fraction of the foreign-born in each education group had declined substantially.

For much of the post–World War II period, the foreign-born remained a small fraction of the workforce and the distribution of their years of schooling was similar to that of the native-born. In more recent years, however, immigrants have had a much larger impact on skill supplies. In 1990 they increased the number of dropouts by 29 percent, but they increased the number of high school graduates by just 7.5 percent. In 2005 they increased the number of dropouts by an astounding 76 percent and increased the supply of high school graduates by almost 15 percent. The increases in the immigrant share for high school and college equivalents are substantial, but the two are fairly balanced.

IMMIGRATION AND THE EDUCATION GAP

The contribution of the foreign-born to the gap in the supply of more and less educated groups is given in Table 8.5. The "immigrant contribution" gives the fraction of the log difference between the supplies of the unskilled and skilled accounted for by the presence of immigrants.[34] For high school dropouts relative to high school graduates, the fraction is 14.4 percent in 1915, decreases to 2.9 percent in 1970, and then increases for the remainder of the period. In 2005 immigrants expanded the dropout to high school graduate ratio by 43 percent (log points). But the immigrant contribution to the ratio of high school to college equivalents is modest in all years and is greatest for 1915.

We previously noted that there was a large slowdown in the growth of the relative supply of the college educated in the post-1980s. Furthermore, much of the increase in the college wage premium was accounted for by the education slowdown. But how much of the slowdown in skill supplies was due to the increase in immigration?

Table 8.6. Contribution of Immigrants and the U.S. Native-Born to the Growth of Relative Skill Supplies: 1915 to 2005 (100 × Annual Log Changes)

Period	High School Graduates/ High School Dropouts			College Equivalents/ High School Equivalents		
	Total	Immigrant	Native-Born	Total	Immigrant	Native-Born
1915–40	4.80	0.39	4.41	2.82	0.25	2.57
1940–60	3.49	0.22	3.26	2.96	0.21	2.75
1960–80	5.61	−0.13	5.74	3.89	0.06	3.83
1980–2005	2.49	−1.46	3.95	2.27	−0.16	2.43

Sources: See Table 8.5.

Notes: Each cell in the table is the annualized percentage change, from the beginning to the end of the period, of relative skill supplies measured in efficiency units. The "total" column gives the overall growth in relative skill supply. The immigrant and native-born columns decompose the overall relative skill supply growth into the immigrant and native contributions defined in the notes to Table 8.5. The immigrant column can be computed from the data in Table 8.5 columns 3a and b, "immigrant contribution," which is the immigrant contribution to the relative skill supply. For example, from 1980 to 2005 the "immigrant contribution" for high school dropouts versus high school graduates went from 0.065 to 0.430 (Table 8.5, col. 3a). If there had been no foreign-born in 1980, the log ratio of high school graduates to dropouts would have increased by 6.5 log points and in 2005 it would have increased by 43 log points. Thus, the annualized contribution of immigrants to changes in log(H/O) from 1980 to 2005 is given by [(0.065 − 0.430) × 100/25] = −1.46. See the notes to Table 8.5 for the definitions of college and high-school equivalents and efficiency units. It should be noted that the relative supply numbers given here differ slightly from those in Table 8.1 for (college/high school) equivalents in efficiency units and Table 8.3 for (high school graduates/dropout) equivalents in efficiency units. To compute the impact of immigration we used a somewhat different method of computing efficiency units. In Table 8.6 we employ a set of fixed weights (see Table 8.5), but in Tables 8.1 and 8.3 we use different weights for each year.

The answer is that just 14 percent of the college supply slowdown was due to the increase in the foreign-born. The relative supply of the college educated expanded at 3.89 percent per year from 1960 to 1980 but decreased by 1.62 percent per year to an average of 2.27 percent annually from 1980 to 2005 (see Table 8.6). Of that decrease, 1.40 percentage points (= 3.83 − 2.43) or 86 percent of the total (= 1.4/1.62) was due to the slowdown in the relative supply of the college educated among native-born Americans, and therefore 14 percent was due to immigration.

But how much of the increase in the college wage premium was due to immigration? Immigration decreased the relative supply of college equivalents by 3.9 log points from 1980 to 2005 (col. 3b of Table 8.5). Using our preferred estimate of σ_{SU} (= 1.64), the change in relative supply implies an increase in the college wage premium of 2.4 log points or only 10 percent of the overall increase, a fact we noted earlier.

Thus, the slowdown in the growth of relative college supply from the native-born was *nine* times more important than was immigration in explaining the rise of the college wage premium from 1980 to 2005.[35]

Not surprisingly, the impact of immigration on the supply of high school graduates relative to dropouts is larger than for the college group. Immigrants were a substantial fraction of all dropouts in 2005, although they were far less important before 1980. But even in the case of the less-educated groups, the impact of immigration on relative skill supply was of less quantitative significance than was the slowdown in high school graduation among the native-born population.[36]

The relative supply of high school graduates increased by a staggering 5.61 percent per year from 1960 to 1980 but dropped to a sluggish 2.49 percent per year from 1980 to 2005, for a decrease of 3.12 percent per year. Of that rather large decline, 1.79 percentage points (= 5.74 − 3.95) or 57 percent of the total (= 1.79/3.12) was due to the slowdown in the relative supply of U.S. high school graduates. The increase in the foreign-born concentrated in the low-end of the education distribution contributed the remaining 43 percent of the change.

What about the impact of the curtailment of immigration in the earliest of the periods examined, 1915 to 1940, on the relative supply of educated labor? The sharp reduction in immigration starting in the mid-1910s increased the relative supply of educated workers. But the increased schooling of the native-born was by far the more important factor in the rapid relative growth of skill supplies and thus in the decrease in the skill premium. Of the 4.8 percent annual growth in the relative supply of high school graduates to dropouts from 1915 to 1940, 4.41 percent was from the increased educational attainment of the native-born and just 0.39 percent was from the decline in immigration (see Table 8.6). Therefore, the curtailment of immigration accounted for less than 10 percent of the expansion of the relative supply of high school graduates to dropouts during the period. Similarly, less than 9 percent of the increase in the ratio of college to high school equivalents from 1915 to 1940 was due to immigration restrictions.

The main conclusion of this section is that immigration had only a minor impact on the growth in the relative supply of the college educated and a moderate effect on the supply of high school graduate workers relative to dropouts during the 1980 to 2005 period. The slowdown in the growth of educated Americans, domestically produced, was of far greater consequence. As a result, immigration played only a modest role in the surge in the college skill premium in the post-1980s. Similar conclusions were drawn for the earliest of the periods considered, 1915 to 1940, when immigration was sharply curtailed.

Cohort Change

Now that we have shown that changes in relative skill supplies were determined primarily by domestic educational forces, we are led to a question concerning demographics. How much of the variation in the growth of relative skill supplies of the U.S.-born (shown in Table 8.6) was driven by changes in the growth of educational attainment of successive birth cohorts and how much to changes in the size of entering cohorts arising from baby booms and busts? We can answer the question by decomposing the growth of relative skill supplies of the U.S.-born into educational attainment growth across cohorts and changes in cohort size.[37]

We find that changes in the growth rate of educational attainment *across* successive cohorts of the U.S.-born were far more important than were changes in cohort size in altering the growth rate of home-grown relative skill supplies. A few examples will make the point clear.

Consider first the rapid growth rate of the relative supply of college equivalents of 3.83 percent per year from 1960 to 1980 for the native-born. Of the total, 3.51 percent was due to educational upgrading across cohorts and 0.32 percent came from the increasing size of younger and more educated cohorts who entered the labor force with the baby boomers in the 1960s and 1970s. That is, fully 92 percent of the total was due to the educational advancement of successive cohorts. Consider next the slower growth in domestic college supply of 2.43 percent per year from 1980 to 2005. Of the total, 2.54 percent arose from cohort educational upgrading and −0.11 percent from smaller entering cohort sizes.

Of the total decline in the growth rate of the domestic college supply of 1.4 percent per year (3.83 − 2.43) from 1960–80 to 1980–2005, almost 70 percent (0.97 percent per year) was due to the slowdown in the growth of educational attainment across successive birth cohorts. In fact, the deceleration in the growth rate of educational attainment of the U.S.-born explains a 0.59 percent per year increase in the college wage premium (assuming $\sigma_{SU} = 1.64$) out of the actual increase of 0.90 percent per year from 1980 to 2005.

Non-competing Groups: 1890 to 1930

The Premium to Skill and the Relative Supply of Educated Workers

We had previously selected 1915 as the starting date to analyze changes in the premium to education because we were able to compute reasonably comparable estimates of relative skill supplies and skill returns from 1915 to 2005. But an earlier period is of sufficient importance in the history of relative skill supplies that we will make do with a somewhat different measure of skill returns. The period includes the years from 1890 to 1915, termed by Paul Douglas as the era of non-competing groups, as well as the years from 1915 to 1930 when non-competing groups began to fade.

The measure of skill returns that we use is one that we introduced in Chapter 2—the ratio of the wage in an occupation that required some secondary school or higher to the wage in an occupation that did not. We can more finely track the movement of occupational wage ratios prior to 1930 than the returns to education. We showed in Chapter 2 that the premium to various types of office and professional work declined starting around 1914 and continuing to the early 1920s. Although the ratio for some of the series increased a bit at the end of the 1920s, the wage premium for white-collar work never returned to the levels that existed before 1914. What factors were responsible for the substantial premiums to skill and education in the period of non-competing groups and for the sharp and persistent erosion of the premiums after 1914?

We must first provide estimates of the change in wage ratios by skill and supplies of educated workers. To make comparisons over time, we divide 1890 to 1930 into two periods of equal length: 1890 to 1910 and 1910 to 1930. We aggregate the various skill premium series presented in Chapter 2 using employment weights.[38] The wage premium for white-collar work computed in this fashion was fairly steady during the first two-decade period but decreased by 25.7 log points (or about 23 percent) during the second two-decade period. That is, from 1910 to 1930 the skill premium fell by 1.28 percent per year on average.

The stock of high school graduates prior to 1940 must also be constructed. Our preferred approach is to use the administrative data presented in Chapter 6 on the annual flow of new high school graduates at the national level. In constructing the stocks of high school graduates in

Table 8.7. High School Graduates as a Share of the Labor Force (≥ 14 years old)

	Administrative Records	Census
	(1)	(2)
Year		
1890	0.040	0.063
1900	0.044	0.080
1910	0.054	0.102
1920	0.079	0.150
1930	0.123	0.212
Change in high school graduate share		
1890 to 1910	0.014	0.039
1910 to 1930	0.069	0.110
Change in log relative supply		
1890 to 1910	0.315	0.523
1910 to 1930	0.899	0.857
Annualized log relative supply change × 100		
1890 to 1910	1.57	2.62
1910 to 1930	4.49	4.28

Sources: The estimates in column 1 are from Goldin and Katz (1995, table 8). The estimates in column 2 use the 1915 Iowa State Census and the 1880 to 1940 Census IPUMS.

Notes: The relative supply measure is the ratio of high school graduates to those with less than 12 years of schooling. The column 1 estimates use the administrative data on flows of new high school graduates from Figure 6.1 (Chapter 6) to build up stocks of high school graduates following the methodology described in the notes to table 8 of Goldin and Katz (1995).

The column 2 estimates use individual-level data on all labor force participants (those reporting a gainful occupation) aged 14 years or older in each Census IPUMS from 1880 to 1930. We impute the probability that a labor force participant in the 1880 to 1930 Census IPUMS is a high school graduate based on high school graduate shares by birth cohort and sex in the 1915 Iowa State Census (for pre-1890 birth cohorts) and the 1940 Census IPUMS (for 1890 to 1916 birth cohorts). The Iowa estimates for pre-1890 birth cohorts are multiplied by 0.8, the mean ratio of the high graduate share for the overall U.S. to Iowa residents for 1870 to 1890 birth cohorts in the 1940 IPUMS. We assume that the labor force participation rate from 1880 to 1930 was the same for male high school graduates and less-educated males. We assume that the labor force participation rate of adult female high school graduates (those 21 years and older) was 1.4 times the rate of less-educated adult females for 1880 to 1930. These assumptions are based on the labor force participation rates by education, sex, and cohort in the 1915 Iowa sample and 1940 IPUMS. We adjust downward the high school graduation rates of those 14 to 19 years old to reflect the lower labor force participation rates of those continuing in school. The 1890 estimate of the high school graduate labor force share is the average of the 1880 and 1900 estimates since there is no 1890 Census IPUMS sample.

each year from 1890 to 1930 using the administrative data, we assume that the high school graduate share of the workforce was 4 percent in 1890 and add the flows of new high school graduates each year to the existing stock in the workforce.[39] We adjust our measure of the stock of high school graduates in the workforce in each year to account for differences in labor force participation rates between high school graduates and other adults. Based on tabulations from the 1915 Iowa State Census and the 1940 IPUMS for the relevant cohorts, we take the labor force participation rate for male high school graduates to have been the same as the overall male participation rate, and that it was 40 percent higher for female high school graduates than for females who had not completed high school.[40]

The implied estimates from the administrative data of the high school graduate share of the U.S. labor force are presented in column 1 of Table 8.7. The stock of high school graduates in the United States increased slowly to 1910, when they were 5.4 percent of the U.S. labor force. But after 1910 the stock increased far more rapidly, not a surprise given the high school movement. From 1890 to 1910 the change in the relative supply of high school graduates to those with less than a high school degree in the labor force was 31.5 log points and from 1910 to 1930 it was 89.9 log points, almost three times as large. These data translate into a 1.57 percent average annual increase in the relative supply of high school graduates from 1890 to 1910 and 4.49 percent per year from 1910 to 1930.[41]

The census and administrative estimates imply similar growth rates in the relative supply of high school graduates from 1910 to 1930, but the census estimates of relative supply growth are considerably faster for 1890 to 1910. Both approaches imply a sharp acceleration in the growth of the relative supply of high school graduates after 1910. We place more confidence in the administrative estimates for the period prior to 1910 and we will use them in the analysis to follow.[42]

Explaining the Skill Premium Decline: Education,
Immigration, and Demand

Douglas suggested several possible factors that could account for the decrease in the skill premium beginning in the late-1910s: a relative increase in educated workers; a decrease in immigration (thus fewer less-educated workers); and a decrease in the relative demand for skill due to the deskilling of various office positions. We assess each of these explanations using the aggregate measure of the change in the skill premium, changes in the stock of educated workers including immigrants, and our estimate of the elasticity of substitution between skilled and unskilled workers, σ_{SU} (which is the wage elasticity of demand for skill).[43]

Because there was no change in the premium to skill from 1890 to 1910, relative supply and demand must have increased at the same rate. The relative supply of high school graduates increased by 1.6 percent annually (31.5 log points) during those decades (using the administrative data estimates in col. 1 of Table 8.7) and thus demand must have increased at the same rate. But during the next decades, from 1910 to 1930, relative supply grew at an astounding 4.5 percent annually (by 89.9 log points) and the premium to skill decreased by 1.3 percent annually (25.7 log points). Furthermore, given our preferred estimate of $\sigma_{SU}=1.64$, relative demand grew at 2.4 percent annually (47.8 log points) from 1910 to 1930. Our estimates imply that the relative demand for high school graduates grew at a rate that was 0.8 percent more per year from 1910 to 1930 compared with 1890 to 1910.[44]

Thus the large decrease in the wage premium to educated workers was caused by the enormous increase in the supply of educated workers. At the same time, relative demand, rather than slowing, had actually accelerated. But the increase in high school graduates to dropouts could have been caused by immigration restrictions as well as by the high school movement. What was the role of immigration restriction, as opposed to schooling advances, in this early period?

The foreign-born were almost 22 percent of the U.S. workforce between 1890 and 1910. With the passage of immigration restrictions in the 1920s, and the substantial cessation of international labor mobility during World War I, the foreign-born became a smaller fraction of the labor force. By 1930 they were about 16 percent of the labor force. The decrease in immigration would have served to increase the fraction of the labor force with high school education since immigrants were less well-educated than the native-born workforce. But what was the actual impact? The actual impact of the large change in immigration was much smaller than one might have expected.

We simulate the impact of immigration on the supply of high school graduates from 1910 to 1930 by analyzing what would have happened if the immigrant share remained constant at 22 percent from 1910 to

1930 rather than declining to 16 percent. We use data from our 1915 Iowa sample showing that immigrants had, on average, one-third the high school graduation rate of the U.S.-born. Thus, the high school graduation expansion of the native-born was more than ten times as important as was immigration in explaining the growth of the high school graduate share of the workforce from 1910 to 1930. Using our administrative data the immigrant decline can explain only a 0.5 percentage point increase in the growth of the high school graduate share of the workforce from 1910 to 1930 as compared with a 5.9 percentage point increase from the rising educational attainment of the U.S.-born.[45]

The increase in the education of native-born workers was so great after 1910 that even had the foreign-born remained at their 1910 level from 1910 to 1930, the relative supply of educated workers would have increased by 85.2 log points as compared with its actual increase of 89.9 log points. Thus, schooling gains among the U.S.-born were more than *eleven* times more important than immigration in explaining the faster skill supply growth after 1910 and were consequently the major reason for the collapse in the white-collar wage premium from 1910 to 1930.[46]

Recapitulation: Who Won the Race?

Technological change can create winners and losers. Such distributional problems are more likely when technological change is skill biased; that is, when new technologies increase the relative demand for more educated, skilled, and advantaged workers.

A nation's economy will expand as technology advances, but the earnings of some may advance considerably more than the earnings of others. If workers have flexible skills and if the educational infrastructure develops sufficiently, then the supply of skills will expand as their demand increases. Growth and the premium to skill will be balanced and the race between technology and education will not be won by either side and prosperity will be widely shared. External factors can also alter the demand and supply of skills. The immigration of workers who are disproportionately at the bottom of the skill distribution could greatly impact the earnings of those who are their closest substitutes. Changes in international trade patterns and off-shoring opportunities can also alter skill demands.

We began this chapter with a summary of the returns to skill and education first developed in Chapter 2. The premium to education and skill was extremely high in the late nineteenth century but decreased at several junctures until the 1940s. By the 1960s America was growing rapidly and the fruits of economic growth were being shared fairly equally across the income scale. But the story quickly and abruptly changed in the late 1970s and early 1980s when rapidly rising inequality took hold and productivity growth was sluggish at best. The twentieth century contains two inequality tales. This chapter has been a search for an explanation to them.

The estimates of relative skill supplies provided in Chapter 1 have been used in the quest to uncover why the relative premium to skill changed. We did so by estimating the elasticity of substitution between various groups of workers by skill or education. We then used these estimates to compute the degree to which relative labor demand and supply shifted.

The supply and demand framework we employed does an extremely good job in explaining changes in the premium to skill. There were times when we appealed to institutional changes and rigidities but, by and large, the framework allows us to tell a consistent and coherent story that reconciles the two inequality tales of the twentieth century. We now summarize the major findings of that analysis beginning with the college wage premium.

The college wage premium ended up in 2005 at about the same place it had been in 1915. Thus over the very long run the relative supply for skilled workers grew at the same rate as did demand. But that does not help us understand the two tales. Only a detailed analysis of the subperiods will. From 1915 to 1980 education raced far ahead of technology and that served to reduce skill premiums and to lessen the economic power of what Paul Douglas termed non-competing groups. From 1915 to 1940 supply outstripped demand by 1.41 times (3.19 percent average annually versus 2.27); from 1940 to 1960 it did so by 1.47 times (2.63 percent average annually versus 1.79). In both periods supply increased by about 1 percent per year more than demand. But a big reversal occurred around 1980. Had the relative supply of college workers increased from 1980 to 2005 at the same rate that it had from 1960 to 1980, the college premium, rather than rising, would have fallen. Late in the twentieth century, education lost the race to technology.

Similarly for the high school graduate premium, we found that from 1915 to 1940 supply raced ahead of demand, again by about 1 percent per year (5.54 percent average annually versus 4.79 with $\sigma_{HO}=2$) and considerably more from 1940 to 1960 (3.55 percent average annually versus 1.79 with $\sigma_{HO}=3$). The rapid increase in high school graduates caused the high school graduate premium to plummet in the pre-1950 period.

We questioned whether some of the supply changes we measured were really due to changes in immigration rather than to changes in domestically supplied schooling. The issue is most important for the earliest of the periods we studied, when immigration was high and then became restricted, and also for the most recent period, when immigration surged again. We noted that during the critical period 1980 to 2005, when the college premium increased by an astonishing 25 percent, immigration could account for only 10 percent of the surge or just 2.4 percent. Most of the increase was due, instead, to the slowdown in college-going among the native-born population. In fact, educational changes among the native-born population were nine times more important than was immigration in explaining the rise in the college wage premium.

Immigration was more important for the relative decline in supply at the bottom end of the skill distribution. But even in that case, educational slowdowns among the U.S.-born were more important quantitatively.

Earlier in the century, the high school movement was considerably more important than immigration restrictions to the reduction in the skill premium. Had the fraction foreign-born in the labor force remained at its high early twentieth-century level and the high school movement had occurred, as it did, the relative supply of educated workers would have grown at 95 percent of its actual rate (85.2 versus 89.9 log points) from 1910 to 1930.

We noted that the wage structure and the returns to skill have exhibited important discontinuities. Most of the narrowing in wage differentials, for example, took place in the 1910s and the 1940s, periods close to or coinciding with the two world wars. They were times of increased demand for the lower skilled, great innovation, and union activity. Although the discontinuities in the wage structure suggest structural change, the fact that the wage structure remained in place though the institutions changed suggests the importance of fundamental changes in both education and technology.

Our central conclusion is that when it comes to changes in the wage structure and returns to skill, supply changes have been critical, and changes in the educational attainment of the native-born have driven the supply side. This fact was true in the early years of the twentieth century when the high school movement made Americans educated workers and in the post–World War II decades when high school graduates became college graduates. But the same is also true today—the slowdown in education at various levels is robbing Americans of the ability to grow strong together. We now address what it takes to win the race for shared prosperity.

$\backsim$ *9*

How America Once Led
and Can Win the Race
for Tomorrow

American Leadership in the Human Capital Century

Once a Leader

Not long ago the United States led the world in education and had done so for quite some time. In the nineteenth century the United States pioneered free and accessible elementary education for most of its citizens. In the early to mid-twentieth century it extended its lead with the high school movement, when other nations had just discovered mass elementary school education. In the immediate post–World War II era, higher education became a middle-class entitlement in America. A further capstone to the U.S. lead in education in the immediate postwar years was that its universities became the finest in the world. By the 1950s the United States had achieved preeminence in education at all levels and its triumphant lead would remain undisturbed for several decades.

But sometime in the early 1970s indicators of educational attainment in the United States began to change. Secondary school graduation rates reached a plateau; college graduation rates slid backwards; educational attainment by cohort reached a standstill. After the mid-1980s educational attainment for young Americans did begin to rise again, largely driven by a surge in college-going, especially for young women. But this has not been enough to brighten the overall picture.

College completion and high school graduation rates have been sluggish and overall years of schooling have risen more slowly than in the past. Is something the matter?

Possibly. But maybe not. An upper bound exists for a graduation rate; it cannot exceed one. Perhaps there is also a ceiling on the number of years one can endure attending classes. Has the United States reached some natural limit to educational advance? It has not. We have arrived at that conclusion by two routes.

First, we can look at educational advance in comparable nations. These comparisons demonstrate that the long-standing U.S. lead in education has disappeared. The United States is no longer the first in the world in high school and college graduation rates and lags considerably in K–12 quality indicators.

Next, we can examine whether high returns exist from obtaining more education. Maybe more education is not such a good thing. Perhaps we have simply slid down the marginal benefit function for education and that it is not economical to educate the next individual. We will demonstrate, to the contrary, that education is still a very good investment. In fact, the marginal individual today who does not graduate high school, who does not continue to college, and who does not complete college, is leaving large amounts of money lying on the street. The difficult question is why there are large bills being left on the sidewalk and how we can get the youth of America to pick them up by getting more education.

The slowdown in the growth of educational attainment, as we showed in Chapter 8, is the single most important factor increasing educational wage differentials since 1980 and is a major contributor to increased family income inequality. If technology continues to race ahead (and history suggests it will) and educational attainment does not begin to increase rapidly, we are likely to see continued increases in inequality. For many reasons, then, the United States *must* find a way to increase the stock of educated Americans.

International Comparisons

Secondary School and College Completion Rates

U.S. high school graduation rates suddenly stopped increasing after the early 1970s. At the same time, secondary schooling was fast becoming mass education in other parts of the world. Taken together, by the early

2000s the United States, once the leader in secondary school education, had an upper secondary school graduation rate that put it in the bottom third of the 26 OECD nations.[1] Just seven nations had graduation rates that were lower than the U.S. rate; 18 had graduation rates that were higher. The average upper secondary school graduation or diploma rate among European Union nations was 83 percent in 2004 as compared with 75 percent for the United States.

Educational attainment in the United States can be placed in a somewhat better light using a measure of high school completion at various ages rather than by measuring the contemporaneous high school graduation rate. The difference is mainly due to high school equivalency certificates. Some youths who do not graduate on time from secondary school later obtain a General Education Development equivalency degree (GED), or attend community colleges even though they do not have a standard high school diploma.

Using the metric of high school completion by age, the United States was seventh out of the 20 richest OECD nations, graphed in Figure 9.1, in terms of the fraction of individuals 25 to 34 years old in 2004 that had completed upper secondary school.[2] Note that for individuals 55 to 64 years old in 2004, the United States was at the very top of the heap. These older individuals were of high school graduation age around the 1960s when educational attainment in the United States was far ahead of that in Europe, Asia, and Latin America.

The growth of secondary school completion across the globe was rapid from the 1960s to the early 2000s, as revealed in Figure 9.1. Many nations that were significantly behind the United States at that time, such as Finland, Ireland, Japan, and Sweden, narrowed or closed the gap entirely by the early 2000s. While high school graduation rates had seriously stagnated in America they took off in Europe and in other parts of the world.

College-going rates among 20- to 24-year-olds, to be certain, have increased substantially in the United States—from 44 percent in 1980 to 61 percent in 2003 largely in response to the post-1980 rise in the college wage premium.[3] But college *completion* rates have not kept pace and the United States has fallen to the middle-of-the-pack among OECD nations in four-year college completion rates for recent cohorts.[4] For young people in 2004, four-year college graduation rates from administrative data show the United States at about the OECD

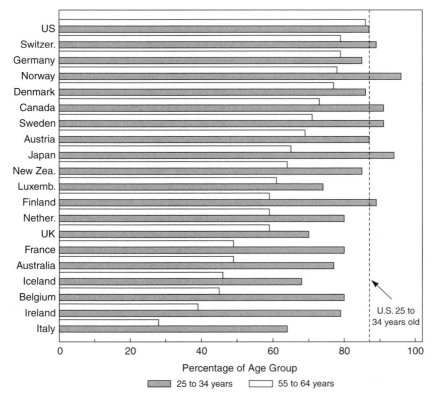

Figure 9.1. Population Attaining (Upper) Secondary School for 20 Nations, by Age: 2004. Data are for the population that attained at least upper secondary education in 2004. We have selected the 20 nations with the greatest per capita GDP in the year 2000 from the 30 OECD nations listed with educational data. The countries are arrayed by the secondary school rate for 55- to 64-year-olds. Sources: Secondary school data, OECD (2006, table A1.2a); country per capita GDP data, Penn World Tables http://pwt.econ.upenn.edu/php_site/pwt61_form.php

average, placing it behind 12 nations.[5] For the slightly older group of 25- to 34-year-olds, the United States trailed four nations, Israel, Korea, Netherlands, and Norway.[6] But the United States had the highest college graduate share of those aged 55 to 64 years, again reflecting its historical lead in the move toward mass higher education.

Clearly, the United States no longer leads the world in the education of young adults. The OECD's summary measure of educational attainment (mean years of schooling) for individuals 25 to 34 years olds in 2004 finds that America was 11th out of 30 countries for males and in 10th place for females.[7] The slowdowns in the growth of the high school graduation rate and the growth of the college completion rate for post-1950 U.S. birth cohorts have been the main reasons for the evaporating U.S. lead in educational attainment.

Quality of Education

As many nations caught up to the United States in secondary school completion some of the rougher edges of American education were exposed. As the *quantity* of education became more equal across nations, the *quality* of U.S. K–12 education became subject to greater scrutiny. Scores on internationally comparable standardized achievement tests revealed that American education paled in comparison.

At first, the results showing that the United States lagged in important areas such as in science and mathematics were discredited and attributed to sample selection bias. Those who attended secondary schools in many other countries, it was maintained, were highly selected whereas those who attended secondary school in America were not. It was no wonder, some claimed, that other nations did better since they weeded out the dolts. The criticism was probably well founded for some of the earlier comparisons. Yet even as secondary schooling became universal in other nations and the international exams were better monitored to maintain comparability in the choice of students in the sample, the gap in test scores persisted.

The results of the "gold standard" of international testing—the Third International Math and Science Study (TIMSS) administered in 1995—clearly showed that the United States was far behind other nations in twelfth grade math and science. Of the 20 nations included in the TIMSS, 14 nations had general mathematics scores significantly above those for the United States. Although AP calculus students in the United States scored very well relative to the advanced mathematics students in almost all other nations, the average U.S. senior received a failing score relative to those in other nations.[8] More recent results from the 2003 Program for International Assessment (PISA) also showed U.S. 15-year-olds to be substantially below the OECD average in mathematics literacy, problem solving, and scientific literacy.[9]

Implications for Cross-Country Inequality Trends

The relative slowdown in the growth of U.S. education has an important implication for comparative inequality trends. Recent technological advances have been widely shared among rich nations, suggesting that they have experienced similarly rapid skill-biased technological change as in the United States. Because education continued to advance in Europe precisely when it lagged in the United States, wage inequality should have increased far less in Europe than in the United States. That is precisely what happened. Even though wage inequality has increased in most OECD nations since 1980, the increase was greater in the United States.[10]

Rapid increases in the relative supply of college workers led to falling educational wage differentials throughout the OECD in the 1970s. But the wage structures of rich nations then diverged. Countries with substantial slowdowns in the growth of skill supplies—the United States and United Kingdom—had large increases in educational wage differentials and overall wage inequality after 1980. In contrast, nations with educational supplies that continued to grow rapidly, such as France and Germany, had almost no increase in educational (or occupational) wage differentials and more modest increases in overall wage inequality.[11]

Institutional factors, to be sure, have played a role in the different inequality experiences among rich nations.[12] Market forces toward increased inequality after 1980 were reinforced in the United States and the United Kingdom starting under the administrations of President Reagan and Prime Minister Thatcher by the decline of unions and the erosion of other labor market institutions that once protected low- and middle-income workers. However, the greater growth of wage inequality in the United States has been substantially driven by the slowdown in skill-supply growth combined with flexible wage-setting institutions and a less generous social safety net.

Unfinished Transformations

Back to High School

As discussed in detail in Chapter 6, in the years between the onset of the Depression and the start of World War II the median youth in most regions of the nation became a high school graduate. Leader and laggard states had existed in the early twentieth century. Although the manufacturing North caught up by 1940, the southern states, for black and white youth alike, had relatively low rates of high school enrollment and graduation until the 1950s. Finally, the gap between the South and the rest of the nation began to narrow. By 1970 the national public and private high school graduation rate was 77 percent and that for the South was about 70 percent. Although differences across regions still existed, they were a fraction of what they were earlier in the century.

Starting around 1970 a disturbing trend became apparent in the high school graduation rate. The fraction of young Americans graduating from public and private high schools began to backslide. Although the backsliding appears to have now ended, the high school graduation rate did not increase noticeably from 1970 to 2004. The conventionally measured high school graduation rate (the ratio of public and private secondary school graduates to the number of 17-year-olds), depicted in Figure 9.2, reveals a steady increase to 1970 and then a sudden change.[13] The high school graduation rate, measured in this fashion, actually decreased at times, even as the fraction of youths continuing to college increased.

We term this measure of the high school graduation rate "conventional" because the procedure counts diplomas received from typical bricks-and-mortar high schools, both public and private. But a nonconventional rate can also be constructed that includes high school equivalency certificates. Certification is through the GED exam, which was introduced during World War II and has been greatly expanded since.

The GED is administered at the state level and consists of a battery of five examinations that individuals can repeat if one or more are failed. The examination was proposed before World War II and gained support when some GI's returned to civilian life worldly wise but without a high school diploma. GED certification was first offered to civilians in 1952.[14] In 1961 the GED accounted for more than 4 percent of all diplomas or GED certificates in the reporting states. By 1971, when GED data are first available at the national level, GED certificates were fully 7 percent of the total and in 1995 they accounted for 16 percent.[15] GED certification declined significantly after 2000 and as a percentage of the total has returned to its late 1970s level of around 10 percent.[16]

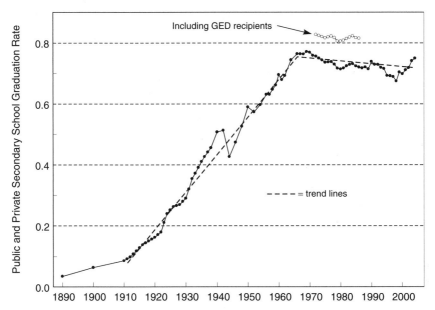

Figure 9.2. Public and Private Secondary School Graduation Rates: 1890 to 2004. GED recipients include individuals receiving a GED certificate. They have been allocated to the closest year in which they would have been approximately 18 years old. Therefore, an individual receiving a GED at age 25 in 1990 would be included in the data for 1983. Only GED certificates awarded up to age 34 are included; in most years more than 85 percent of GED certificates are awarded to individuals less than 35 years old. The secondary school graduation rate is the number of public and private high school diplomas awarded in a year divided by the number of 17-year-olds in the United States. Sources: 1890 to 1970, see Chapter 6 and Appendix B. 1971 to 2004, *Digest of Education Statistics, 2004* for public and private secondary school graduates and the number of 17-year-olds; American Council of Education (various years) for GED certificates and age distribution of those certified.

Some youths who drop out of high school later obtain a GED certificate, often when they are considerably older than 18 years. More than 35 percent of GED certificates are earned by individuals older than 24 years of age, although just 15 percent are older than 34 years.[17] By using the ages of those who received the GED certificate we can estimate the number of GED recipients who had been 18 years old from 1972 to 1986 and add them to their peers who earned a high school diploma on time. We use only the group who received a GED certificate before 35 years old.[18]

From 1972 to 1986, the years we can consider given the availability of the data, the GED increased the overall high school diploma and certification rate by between 7.2 and 9.5 percentage points, if a GED certificate is treated equal to a regular high school diploma. The aggregate high school graduation rate from 1972 to 1986 is the upper line in Figure 9.2, when GED certificates are attributed to their recipients in the year they were 18 years old. The aggregate high school diploma and certification rate increases from about 74 to 82 percent, but the flat trend of the original high school graduation rate is unaffected.

It is not obvious, however, that the number of GED certificates awarded should be added one-for-one to the number of diplomas. An extensive literature shows that those who earn a GED certificate do less well in the labor market than those who earn a conventional high school degree, conditional on various observables including ultimate educational attainment and cognitive test scores. But GED recipients tend to do better than those who drop out of high school and do not get a GED. The most probable reason that GED certificate holders perform poorly on various outcomes relative to those with high school diplomas, but no further education, is that GED recipients often lack various noncognitive skills such as punctuality, responsibility, and the ability to concentrate. They do, however, score well on skills tested by the GED, such as mathematics and reading comprehension.[19]

The main conclusion we draw from this is that adding GED certificate holders to those who received a high school diploma raises the aggregate graduation rate and is the major reason why the cross-country data in Figure 9.1 are higher for the United States than are data from contemporaneous measures of graduation that include only diplomas received from conventional high schools. But the fact still stands that the rate of secondary school completion has been rather flat in the United States for more than three decades and that many other nations now have rates that are higher.

The high school graduation rates in comparable nations are not only higher than the conventional U.S. high school graduation rate, they are also higher than the nonconventional rate that includes both regular diplomas and the GED. The growth of second chance or GED systems could, potentially, be responsible for some part of the slowdown by providing an alternative for impatient and troubled youths;

however, many who opt for the alternative never take advantage of it and they often become dropouts forever.

The Role of Immigration

What, then, can account for the sluggishness of the U.S. high school completion rate in recent decades? The large number of foreign-born who have low levels of education is one reason that has been offered. The fraction foreign-born in the American population increased from below 5 percent in 1970 to 10 percent by 2000. In 2005 the foreign-born share of the population was about 12 percent, including estimates of the illegal population, or very close to its level in the heyday of immigration before the open door was closed in the 1920s.

In Chapters 1 and 8 we addressed the impact that the foreign-born have had on the educational attainment of the U.S. workforce. Their impact on the high school graduation rate concerns two factors. The first is that the denominator of the conventional high school graduation rate includes some individuals who entered the United States as teenagers but who never attended high school in the United States. The second is that the children of less-educated foreign-born parents are likely to attain a lower level of education themselves than the children of U.S.-born parents. This change in composition serves to lower the high school graduation rate, as well as other education indicators.

The rapid growth, since 1970, in the fraction of youth who entered the United States during their teenage years can affect the interpretation of the high school graduation rate series, such as that given in Figure 9.2. Under that scenario, a recently arrived 17-year-old immigrant youth will be included in the denominator of the high school graduation rate. But some of those who arrived as teenagers will never attend U.S. schools and their educational outcomes should probably *not* be attributed to the U.S. school system.

During the last few decades many more youths have arrived in America as teenagers. From 1970 to 2005 the share of all 17- and 18-year-olds who were recent immigrants (that is, had immigrated within the past five years) grew from 0.8 percent to 4.6 percent.[20] However, even though the group greatly increased in size, its impact on the graduation rate has not been large, and that holds even if an upper-bound correction is employed. To produce the upper-bound correction we remove all recent immigrant youths from the denominator of the conventional high school graduation rate. The correction is an upper bound because we do not remove from the numerator immigrant teens who actually did graduate from a U.S. high school.[21]

The correction increases the graduation rate from 77.5 percent in 1970 to 78.3 percent in 2004. The uncorrected figures had shown a modest decline, from 76.6 percent to 74.9 percent, across the same years. Thus, even with the upper-bound adjustment for recent immigrant youth, the share of U.S. youth earning standard high school degrees has been fairly flat since 1970.

Hispanics have been among the largest immigrant group in the post-1970s period and they have mainly come from Mexico. The Mexican-born population increased from 0.4 percent of the total U.S. population in 1970 to 3.3 percent in 2000. Their growth was even more significant within the U.S. foreign-born population, increasing from 8 percent of the U.S. foreign-born population in 1970 to about 30 percent in 2000. Hispanic immigrants and native-born Hispanics have lower high school graduation rates than do non-Hispanics.[22]

Can the increase in the Hispanic share of the population explain the cresting of the high school graduation rate around 1970 and its stagnation in subsequent decades? A rising population share of a group with lower rates of high school graduation mechanically (or compositionally) serves to lower the overall graduation rate.[23] To calculate the compositional impact of the increased share of Hispanics on the high school graduation rate, we use U.S. federal population census data, which contain high school completion rates by race, ethnicity, and country of birth.[24]

From 1970 to 2000 the overall high school graduation rate (including GED recipients) for U.S. residents 20 to 22 years old increased by just 1.4 percentage points, from 79.5 percent in 1970 to 80.9 percent in 2000.[25] The graduation rate of non-Hispanics aged 20 to 22 years increased by 5.1 percentage points, from 80.7 percent in 1970 to 85.8 percent in 2000; for Hispanics the graduation rate increased by 2.2 percentage points, from 55.8 percent in 1970 to 58.0 percent in 2000. That is, the high school graduation rates of *both* non-Hispanic and Hispanic youths increased by more than the overall rate.

The Hispanic share of those aged 20 to 22 years skyrocketed by 12.7 percentage points between 1970 and 2000, from 4.7 to 17.4 percent, while the average difference between the Hispanic rate of high school graduation and that for non-Hispanics was 26.3 percentage points. The compositional effect of the increased share of Hispanics from 1970 to 2000 reduced the growth in the high school graduation rate by 3.4 percentage points (0.127×26.3). In other words, the aggregate (nonconventional) high school graduation rate would have been 3.4 percentage points greater in 2000 (84.3 percent rather than 80.9 percent) had the fraction Hispanic remained at its 1970 level.[26] Therefore, even if the composition of the population had remained constant, the high school graduation rate would have increased only modestly in recent decades and its level would still have been lower than that achieved in some other high-income countries.

The educational attainment of Hispanics is currently far below that for non-Hispanics, but the U.S. historical record gives reason to be optimistic. Previous groups of immigrants and their children also lagged behind the native-born with respect to education, but substantial intergenerational progress occurred resulting in eventual educational convergence.[27] Far less clear are the magnitudes and skill mix of future U.S. immigrant waves, and that will depend on U.S. immigration policy and the economic and social conditions in key source countries such as Mexico.

The American Dream as an Unfinished Transformation

As we saw in Chapter 7, the share of young adults graduating from a four-year college soared in America in the 1950s and the 1960s, but college completion rates slowed considerably in the 1970s and even reversed for young men until the mid-1980s. The slowdown and reversal were so extreme that college graduation rates for young men born in the mid-1970s are no higher than for those born in the late 1940s. College graduation rates for young women also slowed for cohorts born in the 1950s but picked up again with those born in the mid-1960s.

The slowdowns in both high school and college completion rates have meant slower growth in educational attainment since 1980. Mean years of schooling among U.S.-born 30-year-olds had once increased by about one year per decade—by 2.4 years from 1930 to 1955 and by 2.3 years from 1955 to 1980. But from 1980 to 2005 the educational attainment of 30-year-olds increased by just 0.8 years.[28] Educational attainment grew rapidly beginning with cohorts at the start of the high school movement (those born around 1900) through the early baby boom cohorts (those born around 1950). Each generation was far better schooled than their parents, but that is no longer the case, particularly for males.

Interestingly, college graduation rates today are far below what high school graduation rates were in 1940. In the 1950s America believed it was on the path to universal college and the third educational transformation. But the transformation stalled. The question is why.

We previously asked whether we have reached some natural limit to educational attainment. Our first answer was that evidence from comparable nations suggests that we have not. The second way to answer the question is to measure the returns to education.

Standard measures of rates of return to education, particularly to college completion and to graduate and professional training, are exceptionally high today. They have increased substantially since 1980 and are currently at historically high levels. Our estimates from Chapters 2 and 8 imply about a 13 to 14 percent rate of return to a year of college in 2005. The true economic rate of return would remain high even after adjusting for the direct resource costs of providing a college education. Thus, investments in schooling would appear to make enormous economic sense. What is preventing America from crossing the finishing line?

One possibility is that some young people might *not* actually benefit from going to college. The rate of return we have estimated may not be applicable to some young people who do not currently attend or complete college. The average wage gap between college and high school workers may, therefore, overstate the returns to those on the margin of going to college. But that possibility appears not to be the case.

Recent estimates of the rate of return to a year of schooling have used "natural experiments" from policies that have increased access to college, changed college tuition subsidies or merit aid, and altered compulsory schooling laws. These carefully executed studies using plausibly exogenous variation in educational attainment find high rates of return to further schooling.[29] Because these returns would accrue to the marginal youth affected by such policy interventions, often an individual of modest means, they reinforce our conclusion that returns could be extremely high for many individuals currently not finishing college or even not finishing high school.

In addition, similar quasi-experimental empirical methodologies find that the social returns to education extend far beyond the labor market returns. Increased schooling substantially improves health, reduces criminal behavior, and increases political participation and these effects appear to be causal.[30] More education is beneficial not only for the current generation. Increased educational attainment of parents is also of enormous value for the health and educational outcomes of their children.[31]

The slowdown in educational attainment is worrisome. What is keeping young people from choosing to further their education?

In order to answer that question, it is instructive to return to the virtues that enabled U.S. schools to spread and its children to attend school. The same characteristics that were virtuous in the past may be stifling change in the present. Perhaps it is time to review the characteristics that once served us well and ask whether shifting circumstances demand a new set of virtues.

Virtues of the Past, Virtues of the Present

Recall from Chapter 4 that there have been six outstanding virtues of U.S. education in America's history: (1) decentralization with many fiscally independent districts, (2) public provision, (3) public funding, (4) separation of church and state, (5) gender neutrality, and (6) an open and forgiving educational system.

Decentralization and Public Provision

The United States has one of the most decentralized educational systems in the world at all levels. The central governments of most European nations have exercised far more control, especially concerning the funding of schools and teachers. In some nations, most famously France, even the curriculum is uniform and is set by the national government. Although the United States is far larger than any one of the nations of Europe, American states smaller than many European countries also have highly decentralized educational systems with regard to the collection of revenue, expenditures, curriculum, and standards.

During the high school movement in America individual school districts as small as a township could decide to fund a secondary school even if the majority of districts in the state did not decide to fund their own schools. As we noted in Chapter 6, the decentralized authority to build and staff schools meant that local governments could undertake projects that the state government would not have. To take an extreme example, assume that the median voter in a particular school district wants to fund a new school in the district but the median voter in the state does not want to fund additional schools. If the electorate has sufficiently varying tastes across the state, a decentralized educational authority could produce a higher level of education than a centralized authority.

We also mentioned in Chapter 6 that there are times when the centralized authority can hasten change. To take a counter example, assume that the median voter in the state wants to increase the number of schools but the majority of voters in some districts disagree. A centralized authority can hasten change. The adoption of "free tuition" laws by many states in the 1910s and 1920s was a way of accelerating change when many districts had not yet built a high school. These laws mandated that districts without high schools were fiscally responsible to pay tuition to other districts when their children attended schools outside their district. Before the law, many parents paid tuition directly to other districts.[32]

Similarly, college and university education in the United States today, as in the past, is considerably more decentralized than in other nations. There is no federal university in the United States.[33] Most of the decisions regarding public universities and public colleges are made at the state level. In some states, moreover, local communities make decisions about community colleges. Many states have various public universities and colleges that compete with each other for students and faculty. Another aspect of decentralization is that a vibrant private higher education sector arose early in many states.

The decentralization of fiscal authority gave localities, often very small communities, the right to determine the amount that they paid in taxes, what was spent on schools, where schools were built, which texts were used, and which teachers and principals were hired. It was a system that generally worked well.[34]

Decentralization was a virtue at all levels of schooling for much of America's educational history and served to expand schooling through several transformations. But the independent financing of school districts has always meant that some have been richer than others and some considerably poorer. One of the more recent complaints lodged against the decentralized system of American education is that it reinforces the inequities that plague our society and economy.

Rich people live together and poor people live together, and their school districts often do not overlap. The inequality of expenditures per pupil was made the subject of several contentious court cases beginning in the early 1970s with *Serrano v. Priest* (1971), and a series of related *Serrano* cases in California. The court decisions and state legislation that followed in most U.S. states aimed to have state governments redistribute revenue from rich to poor districts and, in turn, raise per pupil educational expenditures in poor districts without lowering that in the richer districts. That goal was reached in some states, but certain school finance equalization systems appear to have been poorly designed and public educational expenditures per pupil may actually have been lowered as a consequence.[35]

Local control, we contend, was important for the early expansion of schools. If local control was such a virtue in the early period, did it produce less inequality in per pupil expenditures than today? The answer is a resounding "no," backed up by the available data from the early period, which vastly understate inequality in expenditure. Thus, the actual inequality in per pupil expenditure was far greater in the past than today.[36]

The data we have are of two kinds. One consists of per pupil expenditures by urban school district for K–8 and 9–12 pupils separately for the 1920s to the 1930s (see Appendix C). The other data are per pupil expenditures by state for all K–12 pupils from the early 1900s to the 1950s.[37] We use these two data sets in tandem to make comparisons with school district data from the 1970s to the 1990s.

Our urban data set reveals a level of variation across districts comparable to that more recently. But because the urban data from the 1920s and 1930s exclude small cities as well as all of the smaller and poorer rural school districts, many of which had public secondary schools, we will seriously understate the variation across districts. This is where the all-inclusive state data from the 1900s to the 1950s come in handy. We know from the state-level data that there was considerable convergence in expenditures per pupil by state in the first half of the twentieth century. Thus the variation in expenditures per pupil across *all* school districts—even just for those having secondary schools—must have been considerably greater in the 1920s and 1930s than it is today.[38]

We observed expenditures for elementary and high school students separately by school district for cities with more than 20,000 people in the 1920s and 1930s. By that time, each urban area, even the largest, was a single separate school district. Various measures of inequality across school districts can be used and we give three in Table 9.1 that are found in the more recent literature on education-finance reform—the 90/10 and 95/5 ratios and the coefficient of variation. We aggregate the data for 1923 and 1927, as well as those for 1933 and 1937, because of the possibility of short-term fluctuations in expenditures. In a balanced panel of cities for all years the inequality measures are similar for secondary schools and elementary schools and expand during the Great Depression.[39] A balanced panel is necessary because southern cities drop out of the sample in the 1930s due to nonreporting. The 1920s data show that the measures of inequality are greater when the South is included and more cities are reporting. For all of the reasons given, actual inequality of expenditures by pupil was much greater than we can report.

Even though the inequality measures for the 1920s and 1930s given in Table 9.1 are vast understatements of inequality across all secondary school districts and certainly across all school districts, the measures are comparable to those for the early 1970s to early 1990s. For example, the coefficient of variation (in percent) is between 25 and 30 in the earlier data and is also between 25 and 30 in the more recent data.[40] For the three census regions that contain many of these cities, we find that the variation within region is also substantial. Therefore, were we able to include the smaller and poorer districts as well as the South in our sample, inequality in expenditures by pupil would surely have been considerably *greater* in the 1920s and the 1930s than it has been in recent decades.

For a long time in U.S. educational history the expansion of educational opportunity took precedence over disparities in educational expenditures per pupil. But as educational opportunity diffused, the nation became more concerned with unequal spending per pupil even though inequalities always existed and had once been considerably greater. Decentralization once served the nation well in expanding educational opportunity. But a multitude of small, fiscally independent districts has always entailed large inequities.

Table 9.1. Inequality in Expenditures per Pupil among City School Districts: 1920s and 1930s

| | Grades K through 8 | | | | Grades 9 through 12 | | | |
Sample	90/10	95/5	Coeff. of Var.	Number of Cities	90/10	95/5	Coeff. of Var.	Number of Cities
All U.S. Cities								
1920s	2.02	2.45	25.3	272	2.00	2.41	25.0	267
1930s	1.91	2.44	28.9	244	2.00	2.34	30.0	223
Balanced Panel								
1920s	1.73	2.16	22.1	240	1.72	2.09	21.9	219
1930s	1.93	2.49	28.9	240	2.01	2.39	27.9	219
New England								
1920s	1.51	1.67	17.1	47	1.63	1.73	19.4	46
1930s	1.63	1.74	18.6	47	1.67	1.93	22.8	45
Middle Atlantic								
1920s	1.76	2.02	22.8	61	1.70	2.00	23.1	61
1930s	2.16	2.19	26.0	61	1.93	2.30	26.1	60
E. North Central								
1920s	1.59	1.88	17.8	66	1.62	1.85	17.3	64
1930s	1.94	2.21	21.9	69	1.58	1.69	17.5	66

Sources: City Level Secondary School Data Set; see Appendix C.

Notes: All cities with populations over 20,000 are included. The data set includes four years of data in the 1920s and 1930s. The data for 1923 and 1927 were averaged, as were the data for 1933 and 1937. Total expenditures are divided by the average number of pupils in daily attendance. Pupils in junior high schools were allocated to the elementary and high school grades. The balanced panel of cities includes only those that were present in all four years: 1923, 1927, 1933, and 1937. Most southern cities did not report to the U.S. Office of Education in the 1930s and were not included in the *Biennial* data for those years. Coefficient of variation gives the standard deviation as a percentage of the mean.

Separation of Church and State

The nineteenth-century common school movement in the United States was a crusade to ensure that the nation's children would have a "common" educational experience. That commonality of experience would be guaranteed not only because schools would be free of tuition but also because the control of schools would be secular. The crusade, therefore, was also one for nonsectarian (though not necessarily god-less) schools.

As we discussed in Chapter 4, the common school crusade succeeded in ridding the nation of the rate bills. It also succeeded in the passage, eventually by every state in the nation, of laws and constitutional amendments that forbade state funding of religious schools. The separation of church and state in the funding of schools was a virtue that helped guarantee a common experience for America's children.

But the virtue is now being questioned. In many of America's poorest neighborhoods the public school system does not appear to be adequately fulfilling its responsibility, and there are few or no private schools to which parents can turn. Even if there were private schools, poor families could hardly afford the tuition. Certain municipalities and some private donors have responded in recent years by making available limited vouchers for poorer families to use toward private school tuition. The motivation for such vouchers is not only to give youths in failing schools the opportunity to attend a better school but also to provide competition so that the failing schools might improve.

A potential problem with such plans is that private schools, except for Catholic and other denominational schools, often do not exist in many poor areas.[41] Therefore, several municipalities not only granted vouchers but also allowed parents to use the vouchers to pay for tuition at church-controlled schools. In Cleveland, that expenditure was challenged on the grounds that it violated the separation of church and state in the provision of education. But the U.S. Supreme Court in *Zelman v. Simmons-Harris* (July 27, 2002, 536 U.S. 639) upheld the voucher system because the state was not providing the funds to the church-run schools. Rather, the state provided funds to the parents who, in turn, gave them to the church-run schools.

There are, currently, about a half-dozen states with voucher systems similar to that in Cleveland and a substantial voucher pilot program is taking place in the District of Columbia. Some states have had to abandon their systems because of court challenges. But more systems may surface in the years to come.[42] Although the 2002 U.S. Supreme Court decision would appear to guarantee the legality of voucher systems that allow parents to use church-controlled schools, not all are being upheld at the state level. In a case decided after the U.S. Supreme Court's ruling on the Cleveland system, the Florida Supreme Court did not uphold the use of vouchers that would go to church-run schools (*Bush v. Holmes*, January 5, 2006, 919 So.2nd 392). But the reason offered by the court was not that such expenditures violated state law forbidding the funding of church-run schools; rather, the court struck down the program on the grounds that it diverted funds from the existing system of free public education and thereby violated the state's guarantee of quality public education for all its children. A Milwaukee voucher system, on the other hand, was upheld by the Wisconsin Supreme Court in 1998 and the U.S. Supreme Court has refused to hear a legal challenge.

Our point is that the separation of church and state in the provision of a "common" education is a virtue that is currently being challenged. But a trade-off may exist between the provision of a common education and the provision of an *adequate* education, especially for children in poor neighborhoods with failing public schools.

An Open and Forgiving System

A key virtue of the U.S. educational system has been its open and forgiving quality. Many European systems in the early part of the twentieth century tested youths at 10 or 11 years old to decide whether or not they could advance further with their education. Americans did not test students for such consequential tracking at early ages.[43] As we noted before, European visitors in the early twentieth century commented that the U.S. educational system wasted resources by educating the masses. But Americans viewed their educational system as egalitarian and essential to providing equality of opportunity.[44]

The U.S. system was, and continues to be, a forgiving second chance system at all levels. Youths who do poorly in elementary school might do well in secondary school. Those who fail to graduate high school can obtain a GED even years after they dropped out. Some attend a community college without high school certification. And, in many states, those who do poorly in high school can still continue with some form of higher education by taking remedial courses at colleges and universities, and aim even higher should they succeed.

The openness of the system is related to the absence of strict standards, which has historically marked U.S. education. In opposition to many European systems that have often involved national testing, the U.S. system has left virtually every aspect of education up to the states and most states have only recently imposed graduation standards on students and used state tests to award high school diplomas.

In fact, states have historically made few demands on high schools with regard to graduation standards. Because most state universities, early in their histories, were required to accept the graduates of state certified high schools, states had a distinct stake in graduation standards, apart from the instruction of secondary school pupils. In 1925,

according to a compilation of state high school graduation subject requirements, all states required a minimum number of courses or course
credits for graduation.[45] The majority of states required high school
graduates to have taken various standard fields. Most states (45 of the
48) required three or more years of English; 41 required at least one
year of history; 27 required a course in algebra, geometry, or both; 10
required at least one course in biology, chemistry, or physics; and 20
required some science course. But few states mandated many specific
courses. Most states did not have demanding and lengthy requirements
and fully 17 states required fewer than three basic subjects, excluding
general mathematics, general science, and physical education. Massachusetts, a leader in education, required only American history and
physical education. Rhode Island had no state requirements and Connecticut's high schools worked out their own programs. States with
well-funded university systems generally had stiffer graduation requirements but California, a state with an extensive public higher education system, had among the least demanding standards.

The general point is that, until recently, states left much concerning
graduation standards up to localities, often townships and municipalities.
Even with regulations concerning what courses had to be taken, there
were few exit examinations set by the state. The New York State Regents
has administered examinations ever since 1865 both to set standards for
the curriculum and to award a Regents' diploma.[46] But the New York
State system, the oldest in the nation, is an exception to the rule.

By the mid-1990s state course requirements for high school graduation had expanded far beyond their levels in 1925, but other aspects
of the requirements for graduation did not change. Several states still
left almost all requirements to local boards. More importantly, few
states had standard examinations for graduation that tested students
on high school material; instead, most had examinations of "minimum
competency."

In the decade that followed there has been an increase in what is
known as high stakes examinations. These exams require that, to graduate, students show mastery of high school material, but often no
higher than tenth grade. Students take the test early in their high
school years and, if they do not pass, they can take it over again in subsequent years. Whereas 13 states had adopted these tests by 1996, 22
states had them in 2006. Early analyses of the new state requirements

of exit exams for high school graduation suggest they have little impact on the overall high school graduation rate, although they increase dropout rates for blacks and for those in urban and high-poverty districts and reduce dropout rates in low-poverty and suburban districts.[47]

The virtues of openness and forgiveness served Americans well when educational attainment was low. As educational attainment increased and the quantity of education has expanded, the tide has turned. Forgiveness and an absence of strict standards might further years of schooling but they do little to increase the quality of education. Furthermore, a second-chance system can lead some to delay finishing their education. The GED, for example, may have led to a decrease in conventional high school completion rates rather than an increase.[48]

Leaders and Laggards of the Past and Present

Significant convergence in high school graduation rates occurred among states during the high school movement and just after. But, as shown in Figure 9.3, there has been remarkable persistence in the leading and lagging states from the end of the high school movement until today, and the persistence is not entirely due to the fact that the states of the South continue to lag. The persistence of educational excellence is demonstrated by graphing the high school graduation rate by state in 1938 against an index of educational performance by state in the 1990s, where the index incorporates high school graduation rates and various achievement test scores. The raw correlation of the two variables by state for 1938 and the 1990s is 0.72.

The persistence of high education rates in many states for more than a half century speaks to the importance of many of the virtues of the past and cautions against altering these characteristics. That said, several extreme outliers are obvious, including California and Nevada. Both California and Nevada have had rapid population growth and large influxes of Hispanics that have strained K–12 educational resources. But, by and large, there is remarkable persistence in their state educational outcomes.

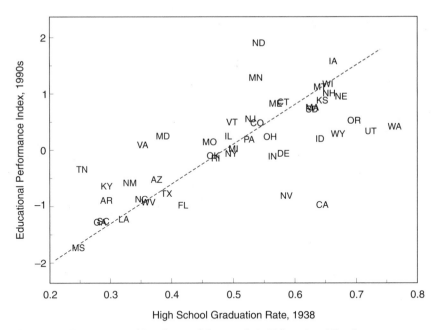

Figure 9.3. Persistence of Leaders and Laggards in Educational Performance: 1938 and 1990s. The dashed line is the regression line

Educational Performance Index, 1990s $= -2.02 + 4.09 \cdot$ High School

$\qquad\qquad\qquad\qquad\quad$ (0.388) (0.588)

Graduation Rate, 1938, $R^2 = 0.513$; standard errors in parentheses. Sources: The state (public plus private) high school graduation rates for 1938 are documented in Appendix B. The educational performance index for the 1990s is from Braatz and Putnam (1997). The index averages three components: (1) a combination of seven National Assessment of Educational Progress scores for 1990, 1992, and 1996; (2) the average Scholastic Aptitude Test score in 1993, adjusted for participation-rate differences among states; and (3) a measure of the high school dropout rate for 1990 to 1995.

How America Can Win the Race for Tomorrow

Causes and Some Solutions

Around the turn of the twentieth century less than one in ten American youths was a high school graduate. Graduates were often an elite group socially and a noncompeting group economically. The returns to secondary school education were substantial and had been so for some time. The transformative effects of education were apparent to parents in the late nineteenth century and they sent their children to academies, if they had the means. As the demand for educated workers grew in the early twentieth century, the beneficial effects of secondary schooling led to grassroots efforts to establish universally accessible and free public high schools. A secondary school diploma became the norm by mid-century, with the important exception of African Americans who often lacked access to high schools.

As the supply of educated workers rose, the premium paid to an additional year of education decreased. The economic returns to a year of education reached their lowest levels in the last century from the 1950s to the 1970s. After the 1970s the education premium climbed.

The economic returns to high school today are substantial and the economic benefits to college and post-college training are at historically high levels. But the educational attainment of American youth at the turn of the twenty-first century is not rising as rapidly as it did a hundred years ago in response to strong economic incentives.

Given the poor economic prospects of high school dropouts, one wonders why so many American youth leave school before obtaining a high school diploma and, given the enormous economic returns to obtaining a college degree, why the share of young adults completing college is not rising more rapidly. High school completion is not yet universal and college completion rates are still lower than high school graduation rates were in the mid-twentieth century.

Two factors appear to be holding back the educational attainment of many American youth.[49] The first is the lack of college readiness of youth who drop out of high school and of the substantial numbers who obtain a high school diploma but remain academically unprepared for college.[50] The second is the financial access to higher education for those who are college ready.

One view of the underlying source of the lack of college preparedness focuses on the role of school K–12 *resources*. According to this view, the growth of resources in many states has not kept pace with the increased disadvantages of many students and greater career opportunities for potential female teachers.[51] Studies show that smaller class size, higher teacher salaries, and summer learning opportunities improve student outcomes.[52] The results of the large-scale, random-assignment Tennessee STAR experiment strongly suggest that smaller class size in the early grades improves academic performance particularly for poor and minority children.[53]

But critics of the insufficient school resources hypothesis counter that increased average per pupil spending and reduced average class size in recent decades have neither increased test scores nor enhanced educational attainment.[54] Their view is that K–12 education suffers from a *productivity* crisis related to several factors. As state financing relative to local financing rose, with school finance equalization plans starting in the 1970s, the link between accountability and funding was reduced. In addition, incentives for the selection and retention of the most talented teachers are often nonexistent, and bureaucratic inflexibilities are legion in many large school districts and are occasionally imposed by teacher unions. The solutions to the productivity crisis would be to enhance accountability through testing and standards, devise alternative approaches to teacher selection and retention, and increase parental choice.[55]

The K–12 system is less than perfect for many students, but it is important to recognize that schools are essentially failing particular students. Those left behind by the system are mainly minority children in inner-city schools who become the youths who are not college ready.[56] One of the original virtues of American education—the reliance on small and medium-sized school districts—may be to blame. The system once operated to provide healthy competition through residential location choices. But the system may not work well for many poor, inner-city residents who cannot easily relocate to new jurisdictions. Expanded schooling options that do not require residential mobility (such as public school choice, charter schools, and vouchers) could improve the situation for low-income families, although the existing evidence on the effectiveness of such policies is mixed.[57]

The high economic returns to doing well in school and the pathways through which schooling can lead to labor market success may not be salient (or even known) to children from disadvantaged backgrounds. These youths lack adult role models who have succeeded in the labor market, and their peers are frequently hostile, to say the least, to students who achieve academically.[58] Policies that provide more immediate financial incentives for doing well in school hold the promise of breaking down the barriers facing disadvantaged children who want to learn and excel in school.[59]

But even policies that target school-age children may come too late for kids from troubled families and inadequate early learning environments. A potential source of the slow growth in U.S. educational attainment is the large increase since 1970 in the share of American children living in poor families and single-parent households.[60] It may be difficult for schools to overcome the lack of school readiness without earlier interventions.

Parenting programs and early childhood health and education interventions, such as Head Start (the federal government's largest preschool program), are likely to be complementary with later human capital investments.[61] The bulk of existing research indicates large returns from investments in high-quality early childhood education programs targeted at low-income families.[62]

Even when we are able to prepare youth for college, a second hurdle remains. Some who are prepared to attend college may lack the family resources to afford a college education and may find it difficult to access financial aid or borrow sufficient funds. Facilitating access to college could boost academic expectations and increase incentives for youth to become college ready by taking more difficult courses. Public and private college tuition has increased rapidly since 1980, even relative to typical family incomes (Figure 7.10 of Chapter 7), and financial aid has not kept pace for families with moderate income.[63]

College costs have a substantial impact on the college attendance and completion rates of youths from families with below median income.[64] College attendance increased substantially since 1980, in response to substantial college returns, but differences in college attendance rates by parental income, race, and ethnicity are large even among students with similar academic grades and achievement test scores.[65] The combination of the high cost of college, credit market constraints, and student debt aversion leaves many youth from poorer and middle-income families behind in the pursuit of a college education.[66]

Demographics plays a key role in changes in college access. The large baby boom cohorts who reached college age in the 1970s and early 1980s put significant strains on government resources available for higher education. The demographic bulge in college-age individuals reduced available public subsidies per student and created pressures for higher tuition in public universities and colleges. Reduced resources per student in higher education are strongly associated with reductions in the number of students attending and completing college.[67] Even with the smaller cohort sizes since the mid-1980s, in many states higher education increasingly competes with rising demands for more public spending on medical care, criminal justice, and other programs.

Rising college costs to students and their families have also increased the fraction of college students who are employed while attending college. Among all 18- to 22-year-old college enrollees, the employed rose from 38 percent in 1970 to 52 percent in 2003, while the average hours worked of those employed increased from 21 to 24 hours per week.[68] The increase in hours of paid work by U.S. college students is indicative of greater financial constraints and reduced time available for studying.[69]

To make matters worse, the financial aid system can be harder to crack than Fort Knox, creating further barriers to college attendance for disadvantaged youth.[70] More generous college financial aid for low-income youth and a more transparent financial aid system have the potential to expand college-going and completion.

Policies not directly related to education may also boost the college-readiness of disadvantaged kids. The substantial growth in the geographic concentration of poverty in inner cities and the sharp rise in residential segregation by family income since 1970 may have served to depress human capital investments in children from low-income families.[71] Policies to promote residential mobility, such as a greater availability of housing vouchers, might improve educational outcomes for children from low-income families.[72] Mentoring programs that offer social and emotional support to poor kids, as well as financial assistance for post-secondary training, can play a modest role in improving the likelihood of college-going for a low-income youth.[73] Second chance job training programs, such as the Job Corps, for disadvantaged youth who have dropped out of high school have proved promising as have Career Academies and other programs to better connect high schools to employers and labor market realities.[74]

In sum, three main types of policies are needed to increase the growth rate of U.S. educational attainment and the relative supply of college workers. The first policy is to create greater access to quality pre-school education for children from disadvantaged families. The second is to rekindle some of the virtues of American education and improve the operation of K–12 schooling so that more kids graduate from high school and are ready for college. The third is to make financial aid sufficiently generous and transparent so that those who are college ready can complete a four-year college degree or gain marketable skills at a community college.

Many credible evaluations show positive (although, at times, modest) impacts on educational attainment and later economic and social outcomes from individual policy interventions operating on each of these margins. These policies, furthermore, complement each other. When more kids are ready to learn upon school entry, improving K–12 schooling is easier. When more youth from modest-income backgrounds are college ready, there will be a bigger "bang" from each financial aid "buck." Thus, the economic gains from a coordinated set of policies working on all three margins (early childhood education, K–12 schooling, and college financial aid) may be greater than evaluations of a single program in isolation would imply. In addition, the short-run fiscal burdens of increased spending on education are likely to be more than offset in the long run with increased tax revenue from a more productive workforce and lower public spending to combat social problems.[75]

Educational investments can have a range of beneficial effects for those who go further in school. They can also benefit the entire nation through an increase in economic growth and a slowdown, or reversal, of inequality trends. But the impact of greater educational investments on inequality can take a long time, especially if the investments start with young children. Furthermore, education-based policies might not have much effect on the share of national income accruing to the very top of the income distribution (the top 1 percent).[76] The nation, therefore, may want to complement greater educational investments with policies that have a more immediate impact on the distribution of the benefits of economic growth.

The progressivity of the U.S. tax system has greatly diminished since the early 1980s.[77] A modest increase in tax rates at the very top end of the income distribution can provide revenue to fund payroll tax relief for lower-wage workers, a more generous earned income tax credit, and greater health care access. This approach could provide an immediate move toward greater economic equity.[78] The erosion of labor market institutions (such as the minimum wage and unions) has exacerbated the market forces that have driven the recent rise of U.S. wage inequality.[79] Strong institutional interventions in wage setting may prove costly in terms of employment opportunities when market forces are shifting strongly in the opposite direction. But some enhancement of institutions to buffer the earnings of workers with modest incomes could work well when accompanied by policies that expand education and the overall supply of skills.[80]

Why (and How) We Must Prepare for Further Skill-Biased Technological Change

Throughout the volume we have emphasized the existence of an on-going and relentless race between technology and education. Economic growth and inequality are the outcomes of the contest. As technological change races forward, demands for skills—some new and some old—are altered. If the workforce can rapidly make the adjustment, then economic growth is enhanced without greatly exacerbating inequality of economic outcomes. If, on the other hand, the skills that are currently demanded are produced slowly and if the workforce is less flexible in its skill set, then growth is slowed *and* inequality widens. Those who can make the adjustments as well as those who gain the new skills are rewarded. Others are left behind.

It is, therefore, imperative to know what new skills will be demanded in the future. We emphasize that such a prediction is fraught with difficulties. We saw that over the course of the twentieth century new technologies rewarded general skills, such as those concerning math, science, knowledge of grammar, and ability to read and interpret blueprints. In a similar fashion, the skills that are in the greatest demand today are the analytical ones. But globalization has produced a new challenge.

Today skills, no matter how complex, that can be exported through outsourcing or offshoring are vulnerable. Even some highly skilled jobs that can be outsourced, such as reading radiographs, may be in danger of having stable or declining demand. Skills for which a computer program can substitute are also in danger. But skills for non-routine employments and jobs with in-person skills are less susceptible. The general point is that having desired skills for which there are only imperfect (domestic or international) substitutes provides the greatest security.

Thus, we see great demand today for the highly analytical individual who can think abstractly and who understands such disciplines as finance, nanotechnology, and cellular biology in a deep, not routine, manner. We have also seen an increased demand for those who provide skilled in-person services, such as nurses and other medical specialists. College is no longer the automatic ticket to success. Rather, degrees in particular fields and advanced training in certain areas are now exceedingly important. Interpersonal skills, possibly garnered from being in diverse college peer groups and interacting with educated people, also matter a lot. The general point today is somewhat different from the past. No longer does having a high school or a college degree make you indispensable, especially if your skills can be imported or emulated by a computer program.

America is at a crossroads with education flagging both relatively and absolutely. The virtues of education once served us well. America educated its masses, grew economically, and reduced inequality. In this chapter we have questioned whether the virtues of the past continue to serve us. A central point of the volume is that we should not lose track of history. We must shed our collective amnesia. America was once the world's education leader. The rest of the world imported its institutions and its egalitarian ideals spread widely. That alone is a great achievement and one that calls for an encore.

Appendix A:
The 1915 Iowa State Census Sample

The 1915 State Census of Iowa is a unique document. It is the first census in the United States to include information on education and income prior to the U.S. Federal Census of 1940. It contains considerable detail on other aspects of individuals and households, some of which were never asked in a U.S. Census. The 1915 Iowa census of is a complete sample of the residents of the state. The returns were written by census takers (assessors) on index cards, one for each individual. These cards were kept in the Iowa State archives in Des Moines and were microfilmed in 1986 by the Genealogical Society of Salt Lake City.

The census cards (see facsimile on page 356) were sorted by county, although large cities (those having more than 25,000 residents) were grouped separately. Within each county or large city, records were alphabetized by last name and within last name by first name. Our project sampled the records for three of the largest Iowa cities (Davenport, Des Moines, and Dubuque) and for ten counties that did not contain a large city. The counties were chosen by grouping the ninety-nine counties in Iowa into four equal units by the mean educational levels of their adult population and then randomly taking three from each of the four groups. None of the counties contained a large city. The ten resulting counties were determined by the quality of the microfilm. These rural counties span the geography of the state: Clay and Lyon in the northwest, Mitchell in the north central, Johnson and Buchanan in the east

Card No	Name .... Age ....
Sex: Male Female	County .... PO..................................
Color	Town or Township .... Ward ....
Marital Status	Occupation .... Months in 1914 Unemployed ....
Months Schl 1914	Total earnings for 1914 from occupation ....
Public .... High........	Extent of Education: Common......... yrs Grammar school........yrs
Private......College.....	High school .... yrs College........yrs
Read	Birthplace .... Do you own your home or farm? yes no
Write	Incumbrance on farm, home $......... Value of farm, home $ ....
Blind ...... Deaf	Milit Service: Civil War........ Mexican......... Spanish........ Infantry......
Insane ...... Idiot	Cavalry..... Artillery..... Navy..... State ..... Regiment..... Company....
If Foreign Born	Church Affiliation ....
Naturalized?......	Father's Birthplace .... Mother's Birthplace ....
Yrs IA...... Yrs US	Remarks .... Signed................ Assessor

central, Marshall in the central, Wayne in the south central, Adair and Montgomery in the southwest, and Carroll in the west central.

The tabulations use sampling weights to reflect the differing sampling rates in the urban and rural samples. The weighted tabulations are intended to be representative of the entire state of Iowa (except for individuals in the rural areas of counties containing large cities). Our urban sample contains 26,768 observations or 5.5 percent of Iowa's population in large cities, and the rural sample contains 33,305 observations or 1.8 percent of the population in counties without large cities.

All variables on the census cards were recorded. These include (in their order on the card): card number, sex, color, marital status, months of schooling in 1914 by type of school (public elementary, private elementary, high school, college), whether individual could read and/or write, whether handicapped (blind, insane, deaf, idiot), if foreign born whether naturalized, years in Iowa and years in the United States, full name, age, address (county, post office, town or township, ward), occupation, months unemployed and total earnings from occupation for 1914, extent of education (years in common, grammar, high school, college), birthplace, whether person owned home or farm, encumbrance on and value of the same, military service, church affiliation, father's and mother's birthplaces, the assessor's name, and remarks.

Appendix B:
Construction of State-Level
Secondary School Data

In this appendix we describe the procedures used to construct the state- (and regional) level public and private secondary school enrollment and graduation numbers for 1910 to 1970. The graduation and enrollment rates by census region are given in Tables B.1 and B.2 and the fraction in private schools is given in Table B.3.

General Comments

Data on the number of individuals enrolled in and graduating from secondary schools were collected by the U.S. Commissioner of Education, who requested such data from each secondary school on record with the U.S. Office [Bureau] of Education.[1] Grades nine through twelve were included in the secondary school group.[2] Such data were collected as early as 1870 and were published in the annual *Report of the Commissioner of Education*, also known as and termed here, the *Annuals*, and, after 1917, in the *Biennial Surveys of Education*, also known as and termed here, the *Biennials*. We term these "the school survey data" to distinguish them from the data that the U.S. Office of Education later received from the states. Each state independently collected similar data, although coverage varied by state and over time.

The secondary school data presented here begin with 1910. Before that date the proportion of secondary schools responding to the Office

Table B.1. Graduation Rates by Nine Census Regions: 1910 to 1970

Year	New England	Middle Atlantic	South Atlantic		East S. Central		West S. Central		East N. Central	West N. Central	Mountain	Pacific
			All	White	All	White	All	White				
1910	0.155	0.080	0.041	—	0.038	—	0.043	—	0.125	0.106	0.088	0.117
1911	0.165	0.088	0.046	—	0.042	—	0.051	—	0.130	0.111	0.102	0.132
1913	0.187	0.102	0.059	—	0.056	—	0.062	—	0.153	0.136	0.130	0.171
1914	0.201	0.110	0.065	—	0.062	—	0.068	—	0.162	0.146	0.129	0.192
1916	0.228	0.129	0.075	—	0.074	—	0.079	—	0.183	0.174	0.164	0.229
1918	0.246	0.142	0.085	—	0.084	—	0.110	—	0.215	0.199	0.185	0.248
1920	0.253	0.139	0.086	—	0.086	—	0.120	—	0.209	0.205	0.198	0.273
1922	0.272	0.165	0.105	—	0.103	—	0.130	—	0.250	0.246	0.239	0.309
1924	0.309	0.197	0.135	—	0.133	—	0.166	—	0.298	0.304	0.293	0.359
1926	0.337	0.217	0.156	—	0.139	—	0.184	—	0.311	0.387	0.312	0.394
1928	0.359	0.223	0.171	—	0.162	—	0.198	—	0.334	0.394	0.353	0.432
1930	0.394	0.254	0.192	[0.24]	0.170	[0.21]	0.225	[0.25]	0.360	0.410	0.380	0.447
1932	0.478	0.316	0.225	[0.29]	0.184	[0.23]	0.276	[0.31]	0.430	0.476	0.454	0.551
1934	0.512	0.389	0.250	[0.30]	0.210	[0.25]	0.295	[0.33]	0.494	0.504	0.481	0.567
1936	0.586	0.474	0.307	[0.38]	0.245	[0.30]	0.343	[0.38]	0.530	0.545	0.532	0.625
1938	0.599	0.516	0.353	[0.43]	0.260	[0.31]	0.381	[0.42]	0.543	0.578	0.557	0.678
1940	0.602	0.543	0.377	[0.44]	0.303	[0.36]	0.427	[0.47]	0.572	0.615	0.576	0.711
1942	0.616	0.579	0.409	[0.48]	0.309	[0.36]	0.463	[0.52]	0.618	0.624	0.589	0.666
1944	0.454	0.514	0.355	[0.41]	0.260	[0.30]	0.376	[0.42]	0.541	0.543	0.476	0.532
1946	0.563	0.553	0.310	[0.35]	0.292	[0.34]	0.383	[0.42]	0.579	0.582	0.523	0.597
1948	0.625	0.622	0.369	[0.41]	0.347	[0.40]	0.450	[0.50]	0.637	0.636	0.580	0.653
1950	0.665	0.631	0.408	[0.44]	0.388	[0.44]	0.450	[0.50]	0.638	0.640	0.573	0.638
1952	0.591	0.567	0.387	[0.42]	0.379	[0.42]	0.443	[0.48]	0.584	0.623	0.539	0.557
1954	0.610	0.561	0.433	[0.46]	0.423	[0.45]	0.471	[0.50]	0.604	0.635	0.566	0.583
1956	0.610	0.566	0.462	—	0.455	—	0.515	—	0.625	0.668	0.588	0.599
1958	0.622	0.592	0.475	—	0.478	—	0.531	—	0.623	0.680	0.595	0.624
1962	0.682	0.652	0.542	—	0.523	—	0.579	—	0.656	0.702	0.651	0.716
1968	0.795	0.774	0.677	—	0.668	—	0.680	—	0.767	0.829	0.766	0.772
1970	0.818	0.778	0.687	—	0.684	—	0.685	—	0.788	0.845	0.770	0.774

Table B.2. Enrollment Rates by Nine Census Regions: 1910 to 1970

Year	New England	Middle Atlantic	South Atlantic All	South Atlantic White	East S. Central All	East S. Central White	West S. Central All	West S. Central White	East N. Central	West N. Central	Mountain	Pacific
1910	0.282	0.185	0.119	—	0.109	—	0.137	—	0.237	0.220	0.230	0.288
1911	0.292	0.201	0.130	—	0.116	—	0.158	—	0.244	0.226	0.260	0.313
1913	0.324	0.221	0.151	—	0.137	—	0.177	—	0.268	0.265	0.308	0.376
1914	0.351	0.237	0.158	—	0.143	—	0.190	—	0.281	0.281	0.319	0.400
1916	0.393	0.277	0.176	—	0.166	—	0.205	—	0.321	0.318	0.356	0.499
1918	0.408	0.300	0.207	—	0.177	—	0.266	—	0.360	0.357	0.402	0.518
1920	0.431	0.305	0.228	—	0.194	—	0.309	—	0.401	0.414	0.452	0.604
1922	0.465	0.369	0.259	—	0.213	—	0.316	—	0.468	0.457	0.497	0.669
1924	0.484	0.400	0.290	—	0.252	—	0.355	—	0.496	0.492	0.517	0.680
1926	0.513	0.438	0.325	—	0.265	—	0.391	—	0.516	0.551	0.555	0.735
1928	0.546	0.449	0.345	[0.48]	0.295	—	0.416	—	0.555	0.582	0.594	0.739
1930	0.598	0.498	0.379	[0.55]	0.310	[0.39]	0.455	[0.51]	0.601	0.608	0.628	0.768
1932	0.719	0.610	0.432	[0.57]	0.344	[0.44]	0.495	[0.56]	0.683	0.662	0.681	0.821
1934	0.758	0.715	0.469	[0.61]	0.378	[0.46]	0.506	[0.56]	0.725	0.695	0.706	0.840
1936	0.795	0.764	0.504	[0.66]	0.390	[0.48]	0.559	[0.63]	0.738	0.717	0.731	0.858
1938	0.798	0.803	0.537	[0.66]	0.410	[0.49]	0.599	[0.67]	0.752	0.748	0.745	0.908
1940	0.786	0.836	0.577	—	0.445	[0.52]	0.636	[0.70]	0.812	0.776	0.770	0.921
1942	0.785	0.859	0.600	—	0.457	—	0.647	—	0.784	0.787	0.754	0.895
1944	0.730	0.792	0.523	[0.60]	0.421	—	0.573	—	0.734	0.715	0.663	0.791
1946	0.752	0.830	0.549	[0.59]	0.459	[0.52]	0.562	[0.62]	0.780	0.738	0.691	0.852
1948	0.790	0.851	0.557	[0.64]	0.505	[0.56]	0.584	[0.63]	0.764	0.771	0.722	0.854
1950	0.868	0.862	0.619	[0.62]	0.559	[0.62]	0.618	[0.66]	0.789	0.802	0.756	0.880
1952	0.790	0.772	0.600	[0.68]	0.560	[0.60]	0.629	[0.66]	0.777	0.776	0.733	0.786
1954	0.801	0.791	0.652	—	0.616	[0.66]	0.686	[0.72]	0.784	0.808	0.792	0.805
1956	0.836	0.807	0.691	—	0.670	—	0.733	—	0.817	0.851	0.825	0.823
1958	0.876	0.865	0.749	—	0.726	—	0.776	—	0.880	0.890	0.867	0.856
1962	0.896	0.898	0.876	—	0.787	—	0.830	—	0.888	0.935	0.929	0.909
1968	—	—	—	—	—	—	—	—	—	—	—	—
1970	0.927	0.905	0.848	—	0.845	—	0.875	—	0.900	0.940	0.917	0.911

Table B.3. Fraction of Secondary School Graduates in Private Schools, by Region: 1910 to 1970[a]

Year	New England	Middle Atlantic	South Atlantic	East S. Central	West S. Central	East N. Central	West N. Central	Mountain	Pacific
1910	0.194	0.216	0.308	0.349	0.189	0.113	0.167	0.196	0.154
1911	0.202	0.217	0.276	0.360	0.182	0.115	0.165	0.201	0.153
1913	0.192	0.205	0.237	0.284	0.168	0.110	0.149	0.145	0.135
1914	0.187	0.194	0.237	0.279	0.145	0.111	0.145	0.161	0.121
1916	0.186	0.177	0.202	0.209	0.114	0.105	0.121	0.136	0.0900
1918	0.173	0.176	0.182	0.189	0.0953	0.0918	0.115	0.148	0.0875
1920	0.188	0.168	0.189	0.182	0.0768	0.0938	0.111	0.124	0.0792
1922	0.179	0.154	0.151	0.171	0.0747	0.0842	0.107	0.108	0.0816
1924	0.185	0.144	0.141	0.154	0.0637	0.0862	0.0949	0.0808	0.0733
1926	0.214	0.139	0.129	0.152	0.0769	0.101	0.0731	0.0867	0.0854
1928	0.169	0.143	0.0978	0.116	0.0586	0.107	0.0832	0.0490	0.0840
1930	0.164	0.139	0.0930	0.0979	0.0489	0.113	0.0810	0.0496	0.0765
1932	0.193	0.129	0.0956	0.101	0.0467	0.0891	0.0813	0.0441	0.0590
1934	0.160	0.110	0.0677	0.0822	0.0405	0.0664	0.0671	0.0391	0.0541
1936	0.156	0.095	0.0601	0.0761	0.0408	0.0660	0.0673	0.0320	0.0499
1938	0.163	0.097	0.0550	0.0670	0.0377	0.0726	0.0643	0.0344	0.0477
1940	0.142	0.101	0.0530	0.0536	0.0343	0.0728	0.0573	0.0349	0.0445
1942	0.149	0.099	0.0470	0.0571	0.0326	0.0759	0.0608	0.0349	0.0473
1944	0.216	0.117	0.0520	0.0733	0.0413	0.0966	0.0748	0.0443	0.0590
1946	0.185	0.113	0.0572	0.0698	0.0418	0.0994	0.0744	0.0413	0.0524
1948	0.177	0.105	0.0460	0.0628	0.0365	0.0985	0.0723	0.0381	0.0477
1950	0.175	0.107	0.0398	0.0598	0.0376	0.107	0.0760	0.0394	0.0486
1952	0.197	0.119	0.0419	0.0612	0.0382	0.117	0.0780	0.0419	0.0558
1954	0.248	0.162	0.0494	0.0733	0.0482	0.149	0.100	0.0523	0.0744
1956	0.238	0.168	0.0504	0.0748	0.0481	0.149	0.109	0.0601	0.0703
1958	0.255	0.222	0.0723	0.0747	0.0672	0.171	0.124	0.0705	0.0871
1962	0.233	0.202	0.0633	0.0683	0.0616	0.162	0.120	0.0643	0.0759
1968	0.213	0.175	0.0557	0.0629	0.0550	0.139	0.109	0.0553	0.0710
1970	0.204	0.171	0.0538	0.0616	0.0529	0.131	0.105	0.0515	0.0673

a. Includes preparatory departments of public and private colleges and universities.

of Education's request for data was low (see Table B.4, col. 5) and evidence for many states is difficult to obtain. Further, before 1910, a large fraction of secondary students were in private schools or academies and in the preparatory departments of colleges and universities, all of which were incompletely counted.

The intent of both the states and the U.S. Office of Education was to survey all public schools and as many private schools as could be found. The Office of Education received lists of public and private schools from the states and it checked and augmented these lists in various ways. Not all states collected information on private schools for each of the years, and the U.S. Office of Education recognized that the private school data were probably the most deficient.

Undercounts present the greatest potential problem with the school survey data collected by the U.S. Office of Education. Prior to 1920 the U.S. Office of Education relied solely on its school survey data. Enrollment data in various state reports were greater than those in the Office's school survey data, but there is scant commentary in the annual reports of the Office regarding the possibility of an undercount. There was mention that the number of schools responding was less than the total, generally somewhere around 85 percent in the 1910 to 1920 period. But comment was also made that most schools that did not respond were small and, by implication, that the percentage undercount of students was far less than the percentage undercount of schools.

In 1920 the U.S. Office of Education attempted to bring their data into line with those reported by the states by requesting information from the states and publishing it in a separate section of the *Biennials* called "Statistics of State School Systems." Thus the *Biennials* from 1920 to 1938 contain two sets of numbers for both the public and private schools. One is from the school surveys and has considerable detail on students, teachers, and schools. The other is from the states and, although it lacks detail, it has been treated as more accurate by the Office of Education. Oddly enough, there is no discussion in the *Biennials* about the two series and their differences. Thus from 1920 to 1938 the *Biennials* contain two sets of state enrollment estimates, often with large differences. In 1932–34 the U.S. Office of Education began to augment the school survey information with data from the records of the various state departments of education. The enrollment data from the two surveys are nearly identical from that point onward.

Table B.4. Public High School Enrollments from State and Federal Reports: 1890 to 1934

Year	(1) Bureau of Education State Survey	(2) Bureau of Education School Survey	(3) (1)/(2)	(4) 1/(3)×100	(5) % of Schools Reporting[a]
1890	—	202,963	—	—	60.8
1895	—	350,099	—	—	70.3
1896	—	—	—	—	75.2
1900	—	519,251	—	—	77.4
1905	—	679,702	—	—	—
1906	—	—	—	—	84.0
1910	—	915,061	—	—	85.2
1911	1,156,995	984,677	1.175	85.1	—
1912	1,200,798	1,105,360	1.086	92.1	84.6
1913	1,333,356	1,134,771	1.175	85.1	—
1914	1,432,095	1,126,456	1.271	78.7	84.0
1915	1,564,556	1,328,984[b]	1.177	85.0	—
1916	1,710,872	1,456,061	1.175	85.1	84.5
1917	1,821,974	—	—	—	—
1918	1,933,821	1,645,171	1.176	85.0	87.2
1919	2,057,519	—	—	—	—
1920	2,181,216[c]	1,849,169	1.180	84.8	—
1922	2,725,579[d]	2,220,306	1.228	81.5	—
1924	3,176,074[d]	2,529,889	1.255	79.7	—
1926	3,541,254[d]	3,047,690	1.162	86.1	85.9
1928	3,911,279[d]	3,335,690	1.173	85.3	—
1930	4,399,422[d]	4,129,517	1.065	93.9	92.9
1934	—	—	—	—	95.6

Notes and Sources: The year given is the end year in the *Biennials*, e.g., *Biennials* 1915–16 is 1916. This appears to be the procedure used by the Office of Education.

Col. 1, 1890–1920: *Biennials* 1918–20, chapter 1, p. 46, table 1; estimated by U.S. Office of Education from data provided by the states.

1922–1930: *Biennials* (various years).

Col. 2, Data reported by schools to the Office of Education: *Annuals* (various years) and *Biennials* (various years). Only grades 9 to 12 are included; postgraduate and special students are subtracted.

Col. 5, 1890–1918: *Biennials* 1916–18, Bulletin no. 19, "Statistics of Public High Schools, 1917–18," by H. R. Bonner, pp. 12–13, table 1; 1920–1934: *Biennials* (various years).

a. In 1911 the Office of Education stopped tabulating schools with fewer than 10 pupils (the cutoff was 5 pupils before 1911).

b. There was no report for 1915.

c. The total, which appears in *Biennials* 1929–30, also includes vocational and normal schools.

d. The totals from the series with grade reported have been used. The totals without grade reported appear to include students attending continuation and certain evening schools.

To summarize, the data on enrollments and graduation before 1920 were obtained by the Office of Education through their school surveys. From 1920 to 1938 the Office obtained data both from schools and from the states. The data obtained from the states contain only enrollments, not graduates, although enrollments were given by grades. Therefore, from 1920 to 1938 the state data can be used to revise those from the school surveys, but there are no easily obtainable state-level data for the period before 1920.

Graduation rates have attracted the most interest, in large measure because they are considered the single most important statistic regarding secondary school performance. Graduation rates have also received attention because official data, such as in *Historical Statistics* series H 598–601, contain graduation rates back to 1870.[3] The *Historical Statistics* data from 1870 to 1930 are substantially different from those in the original reports from the U.S. Office of Education for those years. The source in *Historical Statistics* for 1870 to 1930 is "table 15" of the "Statistical Summary of Education, 1929–30," an obscure document that provides no information on the exact method for the adjustment.[4] To find the underlying sources for "table 15" one has to go back to earlier documents of the Office of Education and even then, the exact adjustments made in 1929–30 are elusive.

The earliest note regarding an adjustment to the original data appears in the *Biennials* 1918–20 (U.S. Office of Education 1918–20, table 2). The note states that the table is "largely estimated" and that the "enrollment in 1912, 1918, and 1920 reported to the Office of Education from the departments of the several states" forms the basis for the adjustment. "Enrollment for other years [is] computed from enrollment reported to the bureau . . . multiplied by the ratio (1.175) which the high school enrollment reported for 1918 by the departments of education of the States bears to the enrollment for that year reported by the high schools [to the U.S. Office of Education]." These enrollment numbers, however, were not largely incorporated in the subsequent document to which we just referred ("Statistical Summary of Education, 1929–30," chapter 1 of *Biennials* 1929–30, table 3). The only revision adopted was that for 1920.

Thus the adjustment to the enrollment data appears to have come from a belief by those working at the Office of Education that their undercount was about 85 percent. That is, they believed that the state enrollment numbers were 1.175 times those collected by the Office of Education. As Table B.4, column 3 indicates, there were years when the ratio of the state to the school survey numbers was above that figure and there were times when it was below. It should also be noted that the ratio is much closer to one (it was 1.065 in 1930) as the number of schools reporting (col. 5) increased. The percentage of schools reporting and the ratio of state-to-school survey numbers, columns 4 and 5 in Table B.4, are very similar. But if the reporting of smaller schools was disproportionately greater, then the actual undercount would have been less than that given in column 3.

It appears that the Office of Education took the adjustment factor 1.175 from the enrollment numbers and applied it as well to the graduation numbers for both public and private secondary schools. Although there is no mention that this was the procedure used, one can virtually duplicate the national graduation numbers given by the Office of Education using that procedure. However, there is no reason to assume that the undercount of graduates would have been the same as the undercount of enrollments.

We have, thus far, commented solely on the possibility of undercounts. There are also problems with missing data, particularly as regards the number of graduates in certain years. These adjustments are detailed below and are based on straightforward extrapolation procedures, some using independent evidence from Catholic schools.

Adjusting the State Education Data

Public Schools: 1910 to 1922

The adjusted data were published by the U.S. Office of Education at the national level for the 1910 to 1920 period. Interest in the present work centers on the data at the state or regional level. There are various ways of adjusting the state numbers. One could use the ratio of the enrollment reported by the state to that reported by the schools for each of the states in a suitable year and apply it to the 1910 to 1920 period. The state data are conveniently listed for the 1920s in the various *Biennials*. The earliest year for secondary school enrollments by states is 1920, but both it and that for 1922 produce inconsistent ratios for various states, possibly due to the impact of World War I on enrollments. Because of that, we have used the 1924 state data, constructed a ratio to the numbers from the schools in the *Biennials*, and applied the undercount (there were some overcounts) by state to all data for the 1910 to 1922 period.

It is likely that this procedure overstates the enrollment in and graduation from public high schools. The implied undercount of students is almost identical to the percentage of schools not reporting (see Table B.4, compare cols. 4 and 5), yet the schools that did not report were smaller than average. In the *Biennials* 1924–26, the Commissioner of Education noted that the schools not reporting (of which there were 3,064 or 14.1 percent of the total for that year) were "small schools" (U.S. Office of Education 1924–26, p. 1037). Thus the adjustment to the number of students should be *less* than the undercount of schools. The question thus arises of whether the data reported by the schools to the states were in excess of those reported by the same schools to the U.S. Office of Education.

The schools had little incentive to overstate their enrollments and graduation numbers to the federal government, but they may have had an incentive to do so at the state level. One cannot assess the possibility by comparing the responses of identical schools because there are no known surviving records. Because the size distribution of the nonreporting schools is also unknown, we cannot use an adjustment that weights the schools by their student populations. The data suggest that the undercount of students is probably less than the undercount of schools.

The results that we obtain from the various procedures, outlined in more detail below, virtually duplicate at the national level the data given by the Office of Education in their eventual revision of the numbers on graduates. We repeat that the Office of Education gave very little information on their adjustments and thus we are not following their formulae. By using the state data to correct the data reported by the schools, we have followed the lead of contemporaries who worked for the U.S. Office of Education. It is most likely that they knew the answers to many of the questions raised here. Why no one left a record of the answers is another question.

If our corrected data err, they probably are on the high side, particularly the graduation numbers. Because much of our work demonstrates a large increase over time in secondary schools, a bias that increases rates early in the period would be preferred to one that lowers the

Table B.5. Private High School Enrollments and Percentage of Schools Reporting

	(1)	(2)	(3)	(4)
Year	Office of Education State Survey	Office of Education School Survey	(2)/(1)	% of Schools
1920	213,920	184,153	86.1	—
1922	225,873	186,641	82.6	—
1924	254,119	216,522	85.2	—
1926	295,625	248,076	83.9	—
1928	341,158	280,449	82.2	—
1930	—	—	—	84.7

Notes and Sources: Neither the state data nor those from the school surveys contain numbers on preparatory students in colleges and universities.

Cols. 1 and 2, 1920–1928: *Biennials* (various years), "Statistics of State School Systems."

Col. 4, 1930: *Biennials* 1928–30, chapter 1, p. 1. The figure of 84.7 percent is estimated using the number of forms sent out plus the number of schools in existence in the previous year but not listed in 1929–30.

rates. Particularly when we compare the data with those from the 1940 Census, we would much prefer that any bias in the contemporaneous data create an upper bound to the actual estimates.

Private Schools: 1910 to 1922

In Table B.5 we compare the private school data in the school and state surveys of the U.S. Office of Education. The undercount, shown in column 3, is somewhat larger than that for public schools. Private schools were apparently harder to track than were those in the public sector. As in the data for public schools, the U.S. Office of Education published state data on private schools beginning with 1920. The adjustment uses the data by state for 1924 for 1910 to 1922, similar to the correction for the public school numbers.

Public and Private Schools: 1924 to 1940

The adjustments for the 1920s are based entirely on the data reported by the states in the *Biennials* for 1924 through 1938. After 1930 the Commissioner of Education used the state data to adjust their own when schools were missing. Therefore, after 1930, the difference between the state reports and those from the schools is very small.

Public school graduation data are missing for various years in the 1930s and have been estimated from the data for twelfth-grade students using the relationship in previous years between enrollments in the last year of high school and graduates in that year. The adjustments are summarized below.

Private Schools: 1940 to 1970

Graduation and enrollment data for private schools are given in the *Biennials*, but the graduation data terminate in 1934. Data on private school enrollments exist for 1936 to 1940, 1946 to 1950, 1958, 1962, 1966, and 1970, but there are no data on students by grade in private schools. Data for the number of graduates exists for most years in the 1960s. The private school enrollment data are used to estimate the number of graduates. We have computed the number of private school graduates in 1952, 1954, and 1956 based on Catholic school data from the *Summary of Catholic Education* (National Catholic Welfare Conference various years).

Preparatory Departments of Colleges and Universities: 1910 to 1936

Another undercount concerns students in the preparatory departments of colleges and universities. The Office of Education included preparatory students in the college category because schools were surveyed by type. The college and university data, therefore, were not included in any of the enrollment and graduation data in the *Biennials*.

In the nineteenth and early twentieth centuries, when the public high school system was in its infancy, many colleges and universities trained secondary school students. These preparatory departments were founded to ensure that college students had the appropriate training. Many preparatory students were in denominational schools, which were included in the college survey because they had graduate programs. These were often schools with hundreds of secondary school students and only a few graduate student priests. Many of these institutions were in the Midwest and it may be that local boosterism favored calling them colleges rather than high schools. Other high schools in this group (such as Hunter High School in New York City)

Table B.6. Preparatory Student Enrollments in Colleges and Universities, 1900–1936

Year	(1) Colleges and Universities	(2) Normal Schools	(3) [(1)/Total Private Enrollments] × 100
1900	56,285	—	—
1903	53,794	13,995	—
1905	64,085	15,324	—
1907	76,370	12,831	—
1909	70,834	11,037	—
1910	66,042	12,890	31
1914/15	67,440	13,504	27
1920	59,309	22,058	22
1922	67,649	—	24
1925	[58,703]	[12,470]	18
1928	50,588	—	13
1930	47,309	11,978	10
1934	23,188	—	—
1936	27,680	—	—

Notes and Sources:
Columns 1 and 2:
1900: *Biennials* 1934–36, "Statistics of Universities and Colleges"
1903 to 1909: *Annuals* (various years).
1910, 1915, 1920, 1930: *Biennials* 1928–30 (p. 5, table 3), and virtually identical to those from the original sources. Note that for 1910 the figure 60,392 is given in the original report, but that figure does not include women's colleges.
1925: extrapolated on the basis of data for other years
1922 to 1936: *Biennials* (various years), "Statistics of Universities and Colleges" and "Higher Institutions"

had once been part of a system in which the high schools were governed by the state higher education bureaucracy.

The national figure for preparatory students in college and universities is given in the *Biennials*, but is not graded and has no graduation data. Enrollment data by state also exist for various years. We have used those for 1910, 1922, and 1928 in making the adjustments. For each of these benchmark years the proportion of the national total for preparatory students is allotted to each state. The aggregate number is then assigned to each state according to the closest benchmark year.

The number of preparatory students, as a fraction of all private high school students, was quite large until the 1920s when the high school movement took off. In 1910 preparatory students in colleges and universities were 31 percent of all private high school students and in 1920 they were 22 percent (see Table B.6). There are, then, important adjustments to be made.

We have not included preparatory students in normal schools because the figures appear inconsistent from year to year, and we have not been able to find estimates of the grade distribution and graduation rates from these schools. Omitting the normal schools will decrease the enrollment and graduation rates of girls far more so than of boys. The understatement is likely to be quite small even before 1920 (see Table B.6, col. 2).

To obtain the number of graduates from the preparatory departments one must know the percentage of the total enrollment that the graduates formed. Such figures were not collected by the U.S. Office of Education, but we have located them for one state—New York. Graduates in these data formed about 16 percent of the total enrollment in a given year. The 16 percent figure is used in Table B.7, column 4, to estimate the number of graduates from preparatory departments. For the period 1910 to 1930, the preparatory departments provided a substantial fraction of all private school enrollments and graduates, although the fraction declines over time (see Table B.6, col. 3). Public high schools increased significantly during the period, making preparatory students a far smaller fraction of the total. And, with the expansion of public high schools, many colleges and universities no longer had reason to have their own preparatory departments to train youths for college.

Enrollment and Graduation by Race: 1930 to 1954

Segregated schools existed in the 17 states of the U.S. South until 1954, and the *Biennials* list the numbers of black students in high school. The number of white students was obtained by subtraction from the totals. The black student data exist from 1916 but with incomplete coverage until 1930. Enrollment data by grade exist from 1930, but the number of graduates begins in 1940. The number of graduates is estimated from the data on students by grade prior to 1940.

Summary

Secondary school enrollment and graduation data collected by the U.S. Office of Education from the various schools in the period prior to 1920 require adjustment because about 15 percent of public and private schools did not return the surveys. Further, the preparatory departments of colleges and universities were never included in the Office's surveys on secondary schooling. Complicating the matter is that the states performed their own surveys, and the enrollment and graduation numbers submitted by the states were often higher than those from the U.S. Office of Education school surveys, even allowing for the school surveys that were not returned. Sometime in the 1920s, the Office began to accept the state data and adopted a method for making revisions to the national data. That procedure was never fully described, although the method was hinted at in various reports. The method that we have devised results in estimates that are very much in line with those of the Office.

As Table B.7, column 1, shows, the percentage difference between our revisions of the number of graduates (col. 6) and those revised by the U.S. Office of Education (and adopted by *Historical Statistics,* col. 1) is very small for the 1910 to 1930 period (see col. 7). The difference between the two series is at most 5 percent, and the average across all years is a mere 0.4 percent. Because high schools were growing rapidly during the period, a difference of 5 percent means that the series are off by just one year in the number of graduates or enrollments. After 1930 the Office of Education began to fill in missing data with information from the states and the adjustments become less important.

The adjustments made by the Office of Education prior to 1930 were very poorly documented in the Office's reports. So meager was the documentation that extremely able personnel at the current Department of Education made fundamental errors in interpreting them. An otherwise informative publication, *120 Years of American Education: A Statistical Portrait* (U.S. Department of Education 1993) gives a largely inaccurate historical series on graduation rates. The revised data (such as that found in *Historical Statistics*) are used for the number of public and private graduates, but in *120 years* the graduates from private high schools are computed as a residual, by subtracting from the total graduates the number graduating from public high schools, as given in the original *Annuals* and *Biennials.* The procedure results in an extremely large, and erroneous, figure for the private graduation numbers for the period prior to 1930. As we already stated, the public graduation numbers were largely revised by the U.S. Office of Education

Table B.7. Public and Private Secondary School Graduation Data

	(1)	(2)	(3)	(4)	(5)	(6)	(7)
Year	Historical Statistics (000)	Office of Education Public Schools	Office of Education Private Schools	Preparatory Departments	(2)+(3)+(4)/ (1)×100	Public and Private Revisions	(6)/(1)× 100
1910	156	111,636	14,409	10,567	87.6	154,804	99.2
1911	168	—	—	—	—	167,609	99.8
1912	181	—	—	—	—		
1913	200	—	—	—	—	203,435	101.7
1914	219	160,606	20,303	—	87.5	219,514	100.2
1915	240	—	—	10,790	—		
1916	259	—	—	—	—	257,921	99.6
1917	272	—	—	—	—		
1918	285	—	—	—	—	297,993	104.6
1919	298	—	—	—	—		
1920	311	230,902	24,166	9,489	85.1	305,530	98.2
1921	334	—	—	—	—		
1922	357	—	—	—	—	372,445	104.3
1923	426	—	—	—	—		
1924	494	—	—	—	—	470,306	95.2
1925	528	396,003	38,547	9,392	84.1	544,712	97.1
1926	561	—	—	—	—		
1927	579	—	—	—	—		
1928	597	—	—	—	—	610,229	102.2
1929	632	—	—	—	—		
1930	667	591,719	51,447	7,569	97.6	681,420	102.2

Notes and Sources:

Col. (1) *Historical Statistics*, series H 598–601.

Cols. (2), (3) *Biennials* (various years).

Col. (4) Preparatory departments: secondary school students in the preparatory departments of colleges and universities. The *Biennials* collected such data as part of the college and university survey. Although the aggregate numbers are not disaggregated by year and graduation, similar data for New York State have been used to estimate the fraction graduating. We have used 16 percent of the total figure for the fraction graduating in that year. The data for preparatory and normal (public and private) schools are given in Table B.6.

on the basis of state survey data. Therefore, the total graduation data in *Historical Statistics* already reflect revisions to the published data of the U.S. Office of Education. Thus, if the adjustments in this appendix appear beyond comprehension for those unfamiliar with education data, they have been equally unintelligible for those in the agency that originally produced the data.

The Adjustments in Detail

There are three main types of adjustments to the data: those to the public secondary school data, those to the private secondary school data, and those to the preparatory school data from colleges and universities. Most of the adjustments render the school survey data consistent with the data from the states. Others produce numbers that were never collected for certain years.

The school year is taken to be the end year given in the documents and of the school term. For example, the enrollment number for 1924 is from the *Biennials* 1922–24, which reports data for the school year 1923–24. The number of graduates is often given in the surveys for both the current and preceding year, but only the current year is used here. Secondary students are those in grades 9 through 12.

1910 to 1922: Public and Private Secondary Schools

The data from the school survey are used in each year as a base for enrollments by grade and sex and the number of graduates by sex (all by state, for public and private schools separately). The adjustments for the 1920s make the total enrollment numbers consistent with those from the state-level data. But the Office of Education did not conduct state surveys in the 1910 to 1920 period. The ratio of the state numbers in the 1924 report (by sex) is used to adjust the public and private school data from 1910 to 1924 (each year of secondary school and the graduation numbers). In most cases there is an undercount, although in a few there is an overcount. The largest undercount is in the South, although some non-South states (e.g., California and New York) also have large undercounts. We chose not to use 1922 to make the adjustment, even though the Office conducted a survey in that year, because there are large differences between this survey and that for 1924 in many states.

For the public secondary schools, the state survey contains data on total enrollments and for each of the four grades (9 through 12), although not by sex. The state survey data do not contain graduation numbers. The school survey graduation data were revised using the state and school data for 12th grade.

Private school state data exist only for enrollments. Enrollments for each grade and for the number of graduates are all adjusted using the ratio of total enrollments in the state survey to those in the school survey. The Utah private school numbers are not adjusted and are left at the levels reported by the schools. Adjustments make them unreasonably large. The data for the Mountain region were overly inflated using the procedure outlined and all its numbers are divided by 1.12 to bring them in line with the data for 1926.

1924 to 1958: Public and Private Secondary Schools

The state data in the *Biennials* are used to adjust the public and private school data in each year. The procedure is similar to that for 1910 to 1922, but the contemporaneous year is used for the adjustment.

Public and private school graduate numbers are missing for various years. The public school data were often extrapolated on the basis of information on 12th grade enrollment and the fraction, by region, that advanced from 12th grade to graduation. The private school data were generally obtained by constructing a ratio (by sex) of graduates to enrollments for each region during a prior survey year and multiplying it by the enrollment data for the missing year (by sex and region). The missing years and data are:

- 1932 public and private school graduates
- 1934 private school graduates
- 1936 public and private school graduates
- 1938 private school graduates
- 1940 public school graduates by sex
- 1952 public school graduates by sex

Private school enrollment and graduation data are added for 1952, 1954, and 1956 using data on Catholic schools, as mentioned above. The ratio of total private school enrollments and graduates to those in Catholic schools were obtained from the nearest year with data on all private schools and Catholic schools. Private school data for the three years were obtained by multiplying the ratios by the Catholic school numbers. The aggregate number enrolled was found to be nearly equal to the numbers in *120 years* (U.S. Department of Education 1993, tables 9, 19), but those for the number of graduates were found to be too small. A factor of proportionality for the nation as a whole was used to inflate each state's number of private school graduates so that the national figures were equal to the accepted national totals.

1910 to 1936: Preparatory Students

The data on preparatory schools is added to the private secondary school enrollment and graduation numbers. Total figures for preparatory school enrollments exist for 1910, 1911, 1913, 1914, 1918, 1920, 1922, 1924, 1926, 1928, 1930, 1932, 1934, and 1936. The distribution of enrollment by state was obtained for 1910, 1922, and 1928. The nearest year is used to distribute the totals by state. Data from New York State indicate that graduates were about 16 percent of enrollments, and that fraction is used to estimate the number of graduates for the 1910 to 1936 period.

Appendix C:
Construction of City-Level Secondary School Data Set

The city-level secondary school data set was constructed from tabular information in U.S. Office of Education, *Annual Report of the Commissioner of Education Reports for 1915*, "City School Systems," and U.S. Office [Bureau] of Education, *Biennial Survey of Education for 1923* [1922–24], *1927* [1926–28], *1933* [1932–34], and *1937* [1936–38]. Our convention throughout this volume is to call the reports issued annually to 1917 the *Annuals* and all subsequent ones the *Biennials*, which are the terms used by the Office of Education.

The data in the *Annuals* and *Biennials* are listed by school district arranged by groups of cities ranging from those with populations between 10,000 and 30,000 to those with populations greater than 100,000. Data exist for smaller cities, but these were more difficult to collect and we have not done so. Almost all the cities in the data set had single or unified city school districts. In the few cases where they were not, we have combined the two districts in the data set. Information regarding public schools only is covered in the reports.

The information given in the reports is quite detailed. For each type of public school (kindergarten, elementary, junior high, and high) data exist on the number of schools, supervisors and principals, teachers, enrollments, average daily attendance, and average school term (in days). Also given are the expenses for the salaries of supervisors and principals, teachers, other expenses of instruction (e.g., textbooks), expenses for the operation and maintenance of plant, capital outlay, and various fixed charges such as insurance and rent. The number of years covered by each school was given. In most cases a junior high school covered grades 7 through 9 and the high school in the same district would cover grades 10 through 12. But some high schools included grades 9 to 12, especially when a junior high did not exist in the district and there are other configurations. Our procedures readjust the data to include information only on grades 9 to 12. That is, the numbers of teachers, principals, students, and so forth was allocated to include only grades 9 to 12.

The number of cities in the total sample, that is those having a population that exceeded 10,000 persons in 1910, is 289. The distribution by region is given below.

Region	Number of Cities	Percentage of Cities
New England	48	16.6
Middle Atlantic	64	22.2
South Atlantic	31	10.7
East S. Central	11	3.8
West S. Central	17	5.9
East N. Central	70	24.2
West N. Central	22	7.6
Mountain	8	2.8
Pacific	18	6.2
Total	289	100.0

In some cases cities merged, and when they did they were added together in the data for all years prior to the merger. Similarly, in a few cases cities split in two and they were also merged in all years. In some cases these anomalous cities had to be dropped from the sample. In 1937 and to a lesser extent in 1933, many cities in the South are missing data in the *Biennials*. The Office of Education did not give a reason for the missing information.

Enrollment and average daily attendance numbers are transformed into rates using the approximate population of youths 14 to 17 years old in the city. In some cases, generally for small cities, the computed enrollment and attendance rates exceed 1. The most plausible reason is that youths from rural areas outside the city boundaries attended city schools.

The city-level secondary school data set has been merged with data from several other sources to provide city-level information on variables such as population, fraction foreign-born, taxable wealth per capita, fraction Catholic, and economic activity. Information on taxable wealth is available for 242 cities with a population greater than 30,000 in 1920.

Appendix D:
Construction of Wage Bill Shares and Educational Wage Differentials, 1915 to 2005

Table D.1. Wage Bill Shares and Educational Wage Differentials: 1915 to 2005

	Wage Bill Shares (percent)			Educational Wage Differentials		
	High School Dropouts	High School Graduates	College Equivalents	College/ High School	High School/ Eighth Grade	High School/ Dropout
Iowa						
1915	80.9	9.1	7.4	0.638	0.370	0.243
1940	58.1	23.9	13.4	0.498	0.276	0.185
United States						
1940 Census	58.3	20.6	16.7	0.498	0.346	0.242
1950 Census	52.1	25.0	17.4	0.313	0.214	0.149
1960 Census	42.4	27.1	23.4	0.396	0.229	0.159
1970 Census	29.7	32.3	29.7	0.465	0.230	0.167
1980 Census	17.0	32.5	39.3	0.391	0.229	0.179
1980 CPS	15.4	34.2	39.5	0.356	0.223	0.170
1990 Feb. CPS	7.8	29.8	50.0	0.540	0.349	0.243
1990 CPS	8.6	29.9	49.4	0.508	0.267	0.207
1990 Census	8.0	26.8	51.0	0.549	0.284	0.213
2000 CPS	5.4	25.5	56.1	0.579	0.374	0.285
2000 Census	5.4	22.7	57.4	0.607	0.309	0.255
2005 CPS	5.0	24.4	57.6	0.596	0.366	0.286

Sources: 1915 Iowa State Census; 1940 to 2000 U.S. Census IPUMS; 1980, 1990, 2000, and 2005 CPS MORG samples; and February 1990 CPS.

Notes:

Wage Bill Shares: Wage bill shares, defined as the share of total labor earnings paid to each education group, are calculated for samples that include all individuals 18 to 65 years old employed in the civilian

Table D.1. (continued)

workforce at the survey reference date. Since employment at the survey reference date is not available in the 1915 Iowa State Census, we include all individuals with occupational earnings in 1914 in our calculations of wage bill shares for Iowa in 1915. The earnings of wage and salary workers and the self-employed are included in calculating wage bill shares in all years and samples. In those samples for which the earnings for the self-employed are not available (the 1940 Census IPUMS, the CPS MORG samples, and the February 1990 CPS), we impute the hourly earnings of the self-employed using the average earnings of wage and salary workers in the same industry-education-year cell following the approach of Autor, Katz, and Krueger (1998). High school dropouts are those with 0 to 11 years of completed schooling. High school graduates are those with exactly 12 years of completed school and no college. College equivalents include all of those with at least a four-year college degree (16 or more years of completed schooling) plus one-half of those with some college.

Educational Wage Differentials: The log college/high school wage differential is a weighted average of the estimated college (exactly 16 years of completed schooling or bachelor's degree) and post-college (17+ years of schooling or a post-baccalaureate degree) wage premium relative to high school graduates (those with exactly 12 years of completed schooling or a high school diploma) for the year given. The weights are the employment shares of college and post-college workers in 1980.

The log (high school/eighth grade) wage differential is the estimated wage premium for those with exactly a high school degree (12 years of completed schooling) and those with exactly 8 years of completed schooling. Changes in education coding in the census and CPS lead us to include workers with 5 to 8 years of completed schooling in the eighth grade category for the 1990 and 2000 Census, February 1990 CPS, and the 2000 and 2005 CPS MORG samples.

The log (high school/dropout) wage differential is a weighted average of the estimated wage premium for those with exactly a high school degree (12 years of completed schooling) relative to four groups of "dropouts," those with exactly 8, 9, 10, and 11 years of completed schooling. The weights are the employment shares in 1980 of dropouts with 8, 9, 10, and 11 years of completed schooling.

Educational wage differentials for the United States for 1940 to 2005 are estimated in each sample using a standard cross-section regression of log hourly earnings on dummies for single years of schooling (or degree attainment) categories (some schooling categories contain multiple years starting in 1990), a quartic in experience, three region dummies, a part-time dummy, a female dummy, a nonwhite dummy, and interaction terms between the female dummy and quartic in potential experience and the nonwhite dummy. The educational wage differentials are directly taken from the coefficients on the dummy variables for schooling categories. The regression samples include civilian employees from 18 to 65 years old. The regression specification and the specific data processing steps follow the approach of Autor, Katz, and Krueger (1998, table 1).

Estimates of educational wage differentials for Iowa from 1915 to 1940 required a different treatment based on our concerns with the meaning of college education for older cohorts in the 1915 Iowa State Census, and difficulties in measuring the returns to education for women in the early twentieth century given the potential importance of unpaid family work. These issues are discussed in detail in Goldin and Katz (2000).

We use our preferred estimates of the returns to a year of college for young men (18 to 34 years old) in 1914 and 1939 from Chapter 2, Table 2.7, to estimate the change in the log college high school wage differential from 1915 to 1940. The return to a year of schooling for young men decreased by 0.033, from 0.148 in 1915 to 0.115 in 1940, which implies a decline in the log (college/high school) wage differential of 0.140 from 1915 to 1940 after proportionally scaling up the 1940 return to a year of college for young men by a factor of 4.307 to equal the 1940 national (college/high school) wage differential of 0.498 for all workers aged 18 to 65 years.

The log (high school/eighth grade) and (high school/dropout) wage differentials for Iowa in 1915 and 1940 are estimated from samples of non-farm, full-year male civilian workers aged 18 to 65 years in the 1915 Iowa State Census and from those residing in Iowa in the 1940 Census IPUMS. These measures of

Table D.1. (continued)

the high school wage premium are taken from cross-section regressions of log annual earnings on dummy variables for single year of schooling categories, a quartic in potential experience, and dummy variables for nonwhites and for foreign-born status. Hours and weeks of work are not available in the 1915 Iowa State Census but information on months of unemployment in 1914 is available. Full-year workers for 1915 are those with earnings in 1914 but no unemployment in 1914. Full-year workers in 1940 are those who worked at least 50 weeks in 1939.

Notes

Introduction

1. *The American Invaders* was written by Fred A. McKenzie (c. 1901, pp. 137–138).

2. That education was gender neutral in various ways might come as a surprise. But enrollment in common and elementary schools was at parity for girls and boys until about age 14 in 1850 and until age 15 by 1880. Girls attended academies and public high schools to a greater degree than did boys in the late nineteenth century. The early period will be discussed in Chapter 4. Girls did not continue to college at the same rates as boys in this early period, but few went to college in the overall population.

3. We have taken the metaphor from Tinbergen (1974; 1975 chapter 6).

1. The Human Capital Century

1. Chapter 6 addresses the education of blacks in secondary schools.

2. See Chapter 4 for a discussion of the virtues of American education that had origins in the nineteenth century.

3. Judd (1928, p. 9). Charles Hubbard Judd was an educational psychologist who taught at various universities and was head of the department of education at the University of Chicago from 1909 to 1938.

4. Referring to England, Scotland, and France, Kandel notes that these nations claimed to "equalize conditions at least for pupils of ability, and . . . recruit talent, irrespective of class considerations, by a generous provision of scholarships and maintenance grants" (1934, p. 21).

5. Kandel, referring to the early twentieth century noted: "In both France and England it was clear that the opportunity for education beyond the level of elementary schools depended upon the accident of residence as well as on family

circumstances . . . this condition may be contrasted with the American attitude described by the authors of *Middletown* [1929]: 'If education is sometimes taken for granted by the business class, it is no exaggeration to say that it evokes the fervor of a religion, a means of salvation, among a large section of the population' " (1955, p. 91).

6. "The purpose of secondary and higher education in France is the preparation of an intellectual elite 'to whom shall be entrusted the direction of the intellectual interests and the social and political destinies of the nation.' . . . not more than one-tenth of the children of France continue their education beyond the elementary stage," according to Swift (1933, vol. 1, p. 82) in his multivolume and exhaustive work on European education in the 1930s.

7. The "convergence club" refers to a group of nations whose per capita incomes or labor productivity converged on the leader nation during some period of time. The notion of convergence is implicit in many growth theory models. The absence of convergence is often due to political, religious, and other cultural factors. For an early statement of convergence, see Baumol, Blackman, and Wolff (1989).

8. The original statement is: "$E \Rightarrow G$ is incorrect," where $E=$a well-educated citizenry and $G=$economic growth. The inverse of the original statement is $\sim E \Rightarrow \sim G$, where $\sim$ indicates "not." That statement is correct and thus the converse of the original statement, $G \Rightarrow E$, is as well.

9. See notes to Figure 1.1 for the precise years and definition of net secondary school enrollment. A similar figure appears in Goldin (2000) but uses 1990 per capita GDP and 1990 gross enrollment rates. The net rates more accurately reflect schooling for a fixed age group.

10. Krueger and Lindahl (2001) show that measurement error and omitted variables bias are concerns in interpreting the impact of education on growth in cross-country regressions.

11. GDP per capita (in 2000 dollars) was $4,596 in 1900, $5,904 in 1920, $8,086 in 1940, and $14,382 in 1960. See sources and notes to Figure 1.1. Because income inequality of currently low income countries is generally greater than it was in the United States in 1900, the measure probably overstates median incomes in currently poor countries relative to the United States in 1900.

12. Enrollment rates in the UNESCO data for the United States are similar to official U.S. enrollment rates since 1950 as well as those calculated for this work (see the enrollment data in Chapter 6).

13. The outlier country is Equatorial New Guinea, which has had large offshore oil discoveries since 1995. Although oil revenues have been substantial, the wealth dividend has not been widely distributed.

14. A further reason to consider the South-East quadrant as the bad education quadrant is that research, starting with Barro (1991), has documented for the second half of the twentieth century that countries with lower rates of investment in schooling (particularly as measured by secondary schooling rates for young people) have lower subsequent rates of economic growth conditional on their starting per capita income.

15. The South-West and North-East quadrants are not interesting without considerably more data. Nations in the South-West part of the graph have both incomes and schooling rates that are lower than given by the comparison point, and

those in the North-East have both incomes and schooling rates that are higher. Whether or not they are too low or too high would require a model of the relationship between education and income.

16. The 15 nations, in ascending order of net enrollment rates, are: Zimbabwe, Bangladesh, Ecuador, Indonesia, Philippines, Cape Verde, Moldova, Bolivia, Azerbaijan, Albania, Tajikistan, Jamaica, Jordan, Egypt, and Armenia. Some of these nations were once part of the Soviet Union and had high rates of education relative to their income levels, but for the other nations there are more complicated reasons for their persistent poverty.

17. The median per capita income of the 53 low-income nations in Figure 1.1 was $2,601 (2000 $). Among 10 of the European nations in Figure 1.7, both the median and the mean per capita incomes in 1955 were about $8,600 (2000 $). The Penn World Table data are in 1955 dollars but are consistent across nations; a multiplicative factor of 4.98 was used (from U.S. data) to convert GDP/capita into year 2000 dollars for consistency with the low-income nation comparison.

18. Among all low-income countries in the data set, 36 percent had enrollment rates below 0.2, 28 percent were between 0.2 and 0.4, and 36 percent exceeded 0.4. If only male youths are considered, 31 percent had enrollment rates below 0.2, 31 percent were between 0.2 and 0.4, and 38 percent exceeded 0.4.

19. Acemoglu and Zilbotti (2001) discuss why the spread of modern technologies creates strong economic incentives for the less-developed parts of the world to educate their people when technologies are not easily refitted for less-skilled, low-educated workers.

20. Summers (1994) provides a forceful statement concerning the economic and social benefits arising from the education of girls in developing countries.

21. In later chapters we consider the impact of education on distribution and why schooling advances took place.

22. Changes in the educational attainment of the workforce can differ from the trends by birth cohort for the U.S.-born because immigrants enter and birth cohorts differ in size, among other factors.

23. Educational attainment was first asked in the U.S. Census in 1940, and thus we must infer educational attainment at age 35 for older cohorts (those born prior to 1905) based on their educational attainment reported at older ages in the 1940 Census. Because U.S. federal population censuses are decennial, we do not observe all the younger cohorts at exactly 35 years of age. Thus, we use a regression adjustment procedure described in the notes to Figure 1.4 to estimate educational attainment at age 35 for each birth cohort.

24. A discernible acceleration can be observed with the 1940s birth cohorts, in large part due to Vietnam War college draft deferments.

25. These calculations are based on the 2005 Current Population Survey Merged Outgoing Group samples. If one includes the foreign-born, the educational gap between white non-Hispanics and Hispanics expands to 2.8 years (or four times the white-black gap) for those born in the 1970s.

26. See OECD (2006, table A1.3a), which aggregates tertiary-type A (four-year) and B (two-year) education. The four nations that exceeded or equaled the U.S. college degree rate (39 percent) for 25 to 34 year olds in 2004 are Belgium (41 percent), Ireland (40 percent), Norway (39 percent), and Sweden (42 percent). The eight others with rates greater than 30 percent are Denmark, Finland, France, Iceland,

Luxembourg, the Netherlands, Spain, and the United Kingdom. Canada, Japan, and Korea had rates far exceeding that of the United States.

27. Maddison (1987, table A-12) presents data on educational attainment for the United States and several European nations that differ substantially from those presented here. Maddison, whose data come from a 1975 OECD publication, claims that in 1950, for example, the U.K. 15- to 64-year-old population had 3.27 years of secondary school, on average; the German population had 4.37 years; and the U.S. population had 3.4 years. These numbers are graphed in Nelson and Wright (1992, figure 6), as "average years of secondary education." But the Maddison data are *not* comparable across countries. They neither refer to a fixed type of school nor to a fixed age of pupils. They assume, for example, that "secondary school" began after grade 4 in Germany but after grade 6 in the United Kingdom. They also implicitly assume that everyone in the United Kingdom in 1950 completed eight years of school and that all schooling beyond grade 6 was in a secondary school or a college. When schooling by age of student is used (as in Table 1.1), levels of attainment in Great Britain are found to be vastly below those in the United States for at least the first two-thirds of the twentieth century. The OECD currently publishes data that are comparable by country.

28. A good example is provided by Cohen and Hanagan (1991), who compare early to mid-twentieth century schooling in Birmingham, England, and Pittsburgh, Penn., both industrial towns with heavy industry and little demand for youthful labor. Elementary schools were established early in both cities, but in Birmingham the shift to high school came very late (in the 1960s) whereas it arrived in Pittsburgh in the 1920s.

29. Real per capita income in the United States in 1940 was actually less than that in seven of the ten European nations for which GDP data exist for 1955. The seven are Denmark, Luxembourg, the Netherlands, Norway, Sweden, Switzerland, and the United Kingdom. The three that had lower incomes are Austria, Belgium, and France. The list excludes some of the poorer nations in Western Europe such as Greece, Italy, Portugal, and Spain.

30. Enrollment in U.S. secondary schools in 1940 was 73 percent. An unweighted average of enrollment in full-time general and technical schools for the ten European nations is 25 percent.

31. More formally, consider a two-period model, where agents invest in general training (formal schooling) or in specific training (an apprenticeship) during period 0. All work full-time in period 1. Technology is given by f_i in period 0 but has probability p of changing to f_j by period 1. The agent who invests in an apprenticeship (specific training) earns $f_i(a_i)^0$ in period 0 and $f_i(a_i)^1$ in period 1 independent of whether technology changes. Because apprenticeships involve training, $f_i(a_i)^0 < f_i(a_i)^1$. The agent who invests in formal schooling (general training) earns $-C$ in period 0, where C = the direct cost of schooling, and $f_i(s)^1 = f_i(a_i)^1$ in period 1, if technology does not change. The formally-schooled person earns $f_j(s)^1 > f_j(a_i)^1$ if technology does change. The agent, therefore, should choose the training type dependent on whether $p > [f_i(a_i)^0 + C]/[f_j(s)^1 - f_i(s)^1]$. That is, the rule would be to invest in formal schooling if the probability of technical change exceeds the ratio of the costs to the benefits of the general training. General training is more valuable the higher the probability of technical change (or a geographic move), the lower the cost of formal schooling, and the larger the gain from general schooling conditional on the technical change.

32. Demographer Everett Lee (1961) emphasized that migration "was and is a major force in the development of American civilization and in the shaping of American character" (p. 78), and noted that he and Henry Shyrock made independent calculations showing that the high migration rates in the 1960s extended back to at least 1850 (p. 79).

33. On U.S. and European migration rates since the 1960s, see, for example, Eichengreen (1992) who shows that the mobility rate within the United States was two to three times that within European nations from the 1960s to the early 1980s. Hughes and McCormick (1987), using longitudinal data for the 1970s and 1980s, show that the U.S. migration rate for manual workers was about four times the U.K. rate and that job-related migration for that group in the United States was more than ten times the U.K. rate.

34. See Ferrie (2005). Counties in the two nations were of approximately the same size. Adult men were more than 30 years of age.

35. See, for example, Schultz (1964) on the role of education in how individuals respond to economic change. Galor and Moav (2000) formalize Schultz and show the conditions under which the rate of technological change, and not skill bias, increases the relative demand for skill.

36. Elbaum (1989) argues that the growth of formal education in the United States led to the breakdown of apprenticeships; that is, the causation runs from increased education to the cessation of apprenticeships. Although possible, the greater geographic mobility in a country with enormous land availability was more consistent with formal education than with apprenticeships.

37. The school year increased during the early part of the twentieth century, particularly for the elementary years and in rural schools. Nonetheless, we do not adjust for days, although such a correction would probably increase the growth in our measure of educational attainment, particularly for cohorts born in the early twentieth century. The reasons for not adjusting the estimates are several. All cohorts born after 1900 attained at least grade nine, on average. Thus the marginal school year would not have been affected since the length of the school year was already high for the upper grades by the 1910s. Furthermore, the adjustment would be imprecise. We also do not adjust for aspects of educational quality such as teacher certification, school facilities, and curriculum. See Chapter 6 for more on these issues with respect to secondary schools and see Denison (1962) and Card and Krueger (1992a) for analyses of how the growth in school quality affected growth of the human capital stock and returns to schooling.

38. The conclusions we draw from Table 1.2 are almost identical to those from a similar set of calculations that weight the labor force by hours worked.

39. Educational attainment of the overall U.S. workforce comes from Table 1.2 and that for the U.S.-born uses the same sources as does Table 1.2. We define the U.S.-born as those born in one of the current 50 U.S. states plus Washington, D.C. For consistency across data sets, we treat the foreign-born children of U.S.-born parents and those born in U.S. territories and possessions as foreign born. The estimate of the foreign-born share of the U.S. workforce in 1915 is the average of 1910 and 1920 from the 1910 and 1920 IPUMS. It should be noted, and is discussed in more detail in Chapter 8, that the education distribution of recent immigrants is bimodal with respect to educational attainment. There are many with little education and there are also many with a considerable amount of schooling.

40. The 1940 Census has been shown to overstate the fraction of older Americans who completed high school when compared with administrative records during the years when the individuals would have graduated from high school (see Goldin 1998).

41. Fogel (1964) dispelled the notion that any single innovation, such as the railroads, could have been the impetus to economic growth as had been claimed by Rostow (1960).

42. The key assumption is that the price paid to each factor of production—a wage, profit, or rent—is equal to the value of its marginal product (that is, the marginal contribution to production).

43. Among the earliest to point out the role of an augmented stock of labor was Schultz (1960), who noted, "there are many indications that some, and perhaps a substantial part, of the unexplained increases in national income in the United States are attributable to the formation of . . . [human] capital" (p. 571). Denison (1962) made a large number of adjustments to the labor input in addition to education. Jorgenson and Ho (1999) and Gordon (2000) have refined the analysis further.

44. The wage of each education group is adjusted for differences across the groups in experience and demographic variables.

45. Jones (2002) argues that the standard growth accounting framework understates the growth contribution of human capital since it does not include the indirect effect on capital investment from the higher incomes generated by increased human capital. Jones' alternative framework implies a 1 percent increase in human capital per worker boosts output by a full 1 percent. In contrast, Bils and Klenow (2000) argue that standard growth accounting *overstates* human capital's causal contribution to growth to the extent that increased schooling endogenously responds to other sources of improvements in productivity.

46. Jorgenson and Ho (1999), using a slightly different methodology, provide estimates of the educational quality growth of the U.S. workforce since 1948, and Aaronson and Sullivan (2001), using a methodology close to ours, provide estimates for the post-1960 period. Our estimates of the growth of the educational productivity of the workforce are quite similar to these estimates.

47. The educational productivity growth estimates for the native-born workforce use the same sources and methods as for the overall workforce in Table 1.2, but differ in limiting the underlying samples to U.S.-born workers. The difference between the change in educational productivity for the overall workforce and the U.S.-born workforce is the immigration impact. The immigration impact includes the combined impacts of both legal and illegal immigration to the extent possible. The U.S. censuses and Current Population Surveys attempt to include illegal immigrants and adjust the sampling weights accordingly.

48. Using the alternative growth accounting framework of Jones (2002), which includes the implied impact on physical capital investment from the higher incomes generated by increased education, our estimates imply the contribution of education to labor productivity growth was 0.48 percent per year (or 20 percent of the overall increase) for the full 1915 to 2005 period.

49. The change in output per capita differs from the change in output per hour because hours per capita declined over the period examined. In practice, GDP per capita trends can diverge from standard labor productivity measures since GDP in-

cludes government output while labor productivity measures only cover the business sector. Our estimate of real GDP per capita growth from 1915 to 2005 is based on data for 1915 to 1960 from *Historical Statistics, Millennial Edition* (2005), table Ca9–19, series Ca11; and for 1960 to 2005 from the *Economic Report of the President 2006* (2006), table B31.

50. The rising share of women in the workforce slightly lowered measured labor quality. Changes in the age composition of the workforce had effects that varied by subperiod. As children increasingly remained in school until their late teens, the share of youth in the labor force declined and that contributed to faster growth of labor quality from 1915 to 1940. The entrance of the large baby boom cohorts into the labor force from 1960 to 1980 decreased labor force quality because they were young. As a result, the period from 1960 to 1980 saw an unusually large increase in educational attainment, but unusually small improvement in overall labor quality growth. As baby boomers acquired work experience, the corresponding increase in labor force quality from 1980 through 2000 offset the unusually small increase in educational attainment. See DeLong, Goldin, and Katz (2003) for the details of this analysis.

51. Nelson and Phelps (1966) provide a seminal conceptual analysis of the indirect effect of education on economic growth through its impact on hastening the diffusion of new technologies.

52. Economic growth models that explicitly incorporate the effect of education within the workforce on the rate of technological progress include Romer (1990) and Jones (1995).

53. See, for example, Doms, Dunne, and Troske (1997) and Bresnahan, Brynjolfsson, and Hitt (2002). Earlier work showed that more educated farmers were earlier adopters of new crop varieties and thrived in changing economic and technological environments (Schultz 1964; Welch 1970).

54. Jones (2002) estimates the increasing R&D intensity can account for 49 percent of the growth in U.S. output per worker from 1950 to 1993.

2. Inequality across the Twentieth Century

1. See, for example, Krugman (1990), Prestowitz (1988), and Reich (1991).

2. We illustrate productivity trends using output per hour in the non-farm business sector from the U.S. Bureau of Labor Statistics (series PRS85006093 from http://www.bls.gov/lpc/home.htm). Trends are similar using GDP per hour or overall business sector output per hour.

3. See, for example, Baumol, Blackman, and Wolff (1989) and Gordon (2004). According to Gordon, the growth of GDP per hour worked declined in Europe by 2.52 percent per year, from 4.77 percent per year during 1950–73 to 2.25 percent per year during 1973–95, as compared with a decline of 1.29 percent per year for the United States, from 2.77 to 1.48.

4. Although inequality measures vary and there is a vigorous debate over which are the best, the sharp rise in inequality of family, household, and individual economic resources since 1980 is a robust finding across a wide range of data sets, inequality metrics, and measures of resources. Mishel, Bernstein, and Allegretto (2005, 2007) provide a useful summary.

5. In the year 2000, the 90–10 ratio of income per adult equivalent in the United States was 5.5, but was 3.0 in Sweden, 3.3 in Germany, 3.5 in France, 3.9 in

Canada, 4.5 in Italy, and 4.6 in the United Kingdom (Brandolini and Smeeding 2006; figure 2.1).

6. See Burtless and Jencks (2003) on some of the broader social and political consequences of high and rising recent U.S. inequality.

7. See Dew-Becker and Gordon (2005) on the changing distribution of the benefits of U.S. productivity growth.

8. A focus on the distribution of annual incomes may be misleading if some of the variation is driven by transitory shocks and individuals have good access to credit markets that allow them to smooth consumption. Although a substantial share (perhaps one-third) of cross-section variation in annual incomes in standard household data sets represents measurement error or relatively transitory income shocks, the evolution of inequality of more permanent measures of income (averaging incomes over multiple years or even an entire labor market career) generates the same trends as the ones we will focus on using annual income measures (Bradbury and Katz 2002; Gottschalk and Danziger 1998).

9. This income measure is used to determine the official U.S. poverty rate. The U.S. Census Bureau does not include in their definition of families individuals living on their own and unrelated individuals in a household, but the trends are quite similar for household inequality measures including all individuals.

10. The Gini coefficient ranges from 0 to 1 with a higher value indicating greater inequality. The Gini coefficient can be defined in the following manner. One begins by ordering income units (households, families, or individuals) from the poorest to the richest and then plotting the cumulative share of total income received by income units up to that percentile (on the vertical axis) against each unit's percentile point in the income distribution (on the horizontal axis). This functional relationship is known as the Lorenz curve. The Gini coefficient measures how much the Lorenz curve deviates from the line of perfect equality (the 45 degree line or the diagonal) for which all income units have the same income. The Gini coefficient is given by the area between the Lorenz curve and the line of perfect equality expressed as a share of the total area below the line of perfect equality (which is $\frac{1}{2}$). See Atkinson (1983) on the Gini coefficient, the Lorenz curve, and alternative measures of inequality.

11. The unusually large jump in the Gini coefficient from 0.404 in 1992 to 0.429 in 1993 partially reflects a change in Census survey methodology which increased incomes at the very top end of the distribution. The basic finding of rising inequality since the late 1970s versus stable inequality before is robust to changes in the methodology. See, for example, Mishel, Bernstein, and Allegretto (2005, figure 1I.)

12. The use of broader CPS measures of household resources that adjust for taxes and the valuation of in-kind transfers does not change the qualitative findings presented here, nor does adjusting household resources for family size. See Cutler and Katz (1992) and U.S. Census Bureau (2005a).

13. Piketty and Saez (2003, 2006) use tax return data and find that the share of income accruing to the top 10 percent of tax units declined from 33.0 percent in 1947 to 31.9 percent in 1973, and then increased to 44.3 percent in 2005. Moreover, these changes were largely driven by the top 1 percent.

14. On consumption inequality trends, see Cutler and Katz (1991, 1992) and Attanasio, Battistin, and Ichimura (2004). On trends in long-run earnings and income inequality, see Kopczuk, Saez, and Song (2007) and Gottschalk and Danziger (1998).

15. Piketty and Saez (2003) even find that changes in recent decades in the incomes at the top 1 percent of the distribution are dominated by changes in labor market earnings. Burtless (1999) discusses the contribution of changes in the inequality of labor market earnings to rising family income inequality.

16. Katz and Autor (1999) and Autor, Katz, and Kearney (2005a, 2007) provide detailed summaries of the evidence and literature on the changes in the U.S. wage structure over the last four decades.

17. It is occasionally argued that the large increase in wage inequality of recent decades has little or no effect on living standards because the U.S. economy generates enormous churning and economic mobility. But the evidence from multiple data sources shows no increase in earnings mobility in the United States in the past three decades and possibly even a decrease (Gottschalk and Moffitt 1994; Haider 2001; Kopzuk, Saez, and Song 2007). Recent increases in cross-section (single-year) wage inequality have, therefore, translated into increased *permanent* or lifetime inequality.

18. The gap was 0.21 log points (24 percentage points) in 1979 and increased to 0.49 log points (63 percentage points) in 2005 using weekly wage data for full-time, full-year workers, 25 to 34 years old, from the March CPS.

19. We use full-time, full-year (FTFY) workers, who are defined as those working at least 35 hours per week and 40 or more weeks per year. The March CPS provides individual-level data for the period covered on prior year's annual earnings, weeks worked, and hours worked per week. Changes in the questions on weeks and hours worked make information on FTFY workers the most comparable.

20. Lemieux (2006a) and Autor, Katz, and Kearney (2007) similarly document the growing "convexification" of U.S. returns to schooling since 1980 with the returns to post-secondary schooling (especially post-college training) rising sharply and returns to lower schooling levels increasing modestly.

21. The 1940 Census data cover labor income received in 1939.

22. On the "Great Compression," see Goldin and Margo (1992), who coined the term.

23. See, for example, Autor, Katz, and Krueger (1998) and Murphy and Welch (1993).

24. The material on skill ratios in the 1920s and 1930s is from Goldin and Margo (1992).

25. Kuznets' (1953) data extend from 1913 to 1948.

26. See also Goldsmith (1967) and Goldsmith, Jaszi, Kaitz, and Liebenberg (1954), who revised and extended Kuznets' estimates. Budd (1967, table 1) reports Gini coefficient summary measures of the family income distribution based on Goldsmith's data, indicating a decline from 0.49 in 1929 to 0.47 in 1935/36 to 0.44 in 1941.

27. These data are from Piketty and Saez (2003). The tax data also show that the top decile share of wage income increased only slightly from 1929 to 1939 and that the top 1 percent share of wage income was actually lower in 1939 than in the late 1920s. The wage income shares of top earners (the top decile and 1 percent groups) subsequently declined sharply in the 1940s.

28. See, for example, Bell (1951), Keat (1960), Lebergott (1947), Ober (1948), Williamson and Lindert (1980), and Woytinsky (1953).

29. Keat (1960), Ober (1948), and Woytinsky (1953) report such ratios.

30. Bell (1951) creates a distribution of wages by industry where each wage is an average for an occupation. He does this for different years and measures percentage changes at points in the distribution. Bell reported only his conclusions and gives no data. Lebergott (1947) had earlier done about the same thing for two years, 1900 and 1940, and looked at the degree to which there was convergence in wages for specific industries. He chose occupations in each industry that did not change significantly during the 1900 to 1940 period.

31. Williamson and Lindert (1980) construct a long-run series similar in spirit to Ober's and, in fact, use Ober's series for the critical 1907 to 1920 period. Although their series shows compression in the 1940s, it does not reveal any persistent narrowing from the late nineteenth century to the 1940s. The finding is due entirely to an error made in copying Ober's 1920 data point, the year of an important splice to the National Industrial Conference Board data (see Williamson 1975, table 11). The 1920 Ober data point should have been 166 but was, apparently, mistakenly copied as 186. When that error is corrected, the series has virtually identical properties to Ober's original series. The premium to still decreases during World War I, recovers somewhat in the early 1920s but never regains its high pre-war level, and then decreases again in the 1940s.

32. The Bureau of Labor Statistics publication *Employment and Earnings* is a recent version of the series.

33. On the relationship between earnings and hours worked in the past and present, see Costa (1998). There are two industries in the c.1940 data that have the distribution of weekly, in addition to hourly, earnings: soap and shipbuilding. For the soap industry, the distribution of weekly earnings is more compressed than that for hourly wages. The reverse occurs for shipbuilding, but the weekly earnings distribution for shipbuilding in c.1940 is not as dispersed as it was in 1890.

34. There are minor differences between the two years in coverage. The 1890 data exclude piece-rate workers whereas those for c.1940 include them. Men are less affected by this exclusion than women in 1890. The industries having the highest fraction of male production workers paid by the piece in 1890 are furniture and silk. Product lines across the half-century changed in some industries. "Soap and candles" in 1890 becomes soap in c.1940; silk in 1890 becomes "silk and rayon" in c.1940. The two tobacco industries change their most important products between the two years. In 1890 "cigars and cigarettes" is mainly cigars and thus we compare it with cigars in c.1940, while in c.1940 "chewing, smoking, and snuff," is mainly cigarettes and we compare it with the 1890 category of "chewing, smoking, and snuff," which excludes the less important cigarette category.

35. For more information on these industries, see Goldin and Katz (2001a, appendix table 1).

36. For 1890, the source is the U.S. Census Office (1895b). The hand-trades (e.g., carpentering, plumbing, plastering, blacksmithing) are subtracted from the 1890 total to make the data comparable with the later definition of manufacturing. For 1940, see U.S. Bureau of the Census (1942).

37. These measures refer to log wage differences at various points in the distribution. We use the usual convention that, for example, 90–10 is the log wage at the 90th percentile minus the log wage at the 10th percentile

38. The "flouring and grist mill products" industry is the one exception across the board, and possibly for good reason since the industry changed radically after

1890. In 1890, there were almost 18,500 flour-mill establishments in the United States. With the diffusion of reduction milling and the invention of methods to grind hard spring wheat, enormous economies of scale resulted (James 1983). Each flour mill, in the earlier era, employed just a few high-paid workers, whereas after the concentration of the industry, the fraction of less-skilled mill employees increased. Note that fully 16 percent of the male employees in the industry were white-collar workers, many of whom were probably owner-operators (see Goldin and Katz 2001a, appendix table 1). The 90–10 measure in 1890 inclusive of the nonproduction workers was 2.94 and it exceeds that in c.1940 of 2.69.

39. The evidence presented in Table 2.1 concerns changes in the dispersion of wages of male production workers *within* detailed manufacturing industries. A full analysis of changes in the overall dispersion of production workers in manufacturing requires knowledge of changes in the dispersion of mean industry wages for detailed manufacturing industries. The available evidence suggests no widening of inter-industry wage dispersion in manufacturing over the period studied. For example, Cullen (1956) finds that inter-industry wage dispersion among 84 manufacturing industries narrowed from 1899 to the mid-1930s, widened in the late 1930s, and narrowed again in the 1940s. Cullen's estimates indicate that overall inter-industry wage dispersion, as measured by the inter-quartile range, was quite similar in the 1899 to 1904 and 1937 to 1939 periods. Accounting for changes in inter-industry wage dispersion is unlikely to alter our conclusions concerning the substantial compression of the wage distribution among manufacturing production workers from 1890 to 1940, as well as the further compression in the 1940s.

40. The weights are the production-worker share of the industries in 1940. Of the nine industries, two (lumber, tobacco: cigarettes) experienced no compression in the 1940s and one (flouring) may not have for the 1890 to 1940 period. For the remaining six that did experience a wage compression in the 90–10 in both periods, the weighted means for the log differences are 24.7 log points for 1890 to 1940 and 14.0 log points for c.1940 to the early 1950s. See Goldin and Margo (1991) for a discussion of the data used for the nine industries for c. 1940 to the early 1950s.

41. On this point, see the work of Jerome (1934).

42. See also Douglas (1930), who presents the earliest series on the wages of "ordinary white-collar workers," by which is generally meant most clerical employees (e.g., clerks, typists, stenographers, secretaries, bookkeepers) and lower-level managers, but not sales workers.

43. Goldin and Katz (1995, table 1). Clerical workers are defined here in three groups: (1) bookkeepers, cashiers, and accountants; (2) clerks, except those in stores; and (3) stenographers, typists, and secretaries.

44. The percentage changes are taken from the log point values. The two series are spliced using the overlap at 1939, assuming that the difference is a factor of proportionality. The estimate for females uses the log of the average wage ratio from 1890 to 1914; that for males uses the log of the average wage ratio from 1895 to 1914.

45. Earnings data also exist for ministers of various Protestant denominations and for public school teachers. The series for ministers also decreases, relative to production workers, before 1940, but the factors causing that decrease are probably different from those for other white-collar groups since the demand for religious

training decreased. The series for teachers is subject to various forces, such as the increased demand for high school instructors in the era of the high school movement.

46. The Boothe-Stigler data are for land-grant institutions and refer to 9- to 10-month salaries. See also the notes to Table 2.3.

47. Note in Figure 2.8 that the earnings of professors relative to wage and salary earners in manufacturing rise considerably in the depths of the Great Depression but then resume their former level. This feature of the series is characteristic of other wage series for skilled relative to unskilled workers.

48. The ratio of the earnings of full to assistant professors was virtually constant from 1910 to 1960. We divide the professor earnings by those for all (wage earnings) manufacturing workers. The production worker series, used in the clerical work comparison, does not exist for all years. Note that the manufacturing worker series includes clerical workers in the manufacturing sector.

49. The relative decline in engineering salaries is less apparent in the data for "beginning engineers" (Table 2.3, cols. 1 and 2) than it is for "all engineers" (Table 2.3, col. 5).

50. See Goldin and Margo (1992, table 7) on the Interstate Commerce Commission (I.C.C.) series for skilled and unskilled railroad workers and also for the National Industrial Conference Board (N.I.C.B.) series relating to the hourly pay of skilled and semiskilled workers in manufacturing relative to unskilled workers in manufacturing.

51. The decrease in the 1970s is the major instance in which the returns to college education do not track changes in the wage structure generally.

52. Goldin (1999) estimates the returns to education in precisely that manner.

53. State censuses all but disappeared in the 1930s and did not reappear with the improved post-Depression economy because of extensions to the federal census and other expansions in federal data collection.

54. Iowa had more than 1,000 towns with populations of fewer than 1,200 persons in 1915.

55. These data are for 1914 and reflect the contemporaneous graduation rate among youths approximately 17 years old and that for enrollment among youths about 14 to 17 years old. Even though Iowa was ranked tenth in graduation, its graduation rate in 1914 was just 19 percent; its contemporaneous enrollment rate was 31.5 percent.

56. See Goldin and Katz (1999b, table 4).

57. By "returns to education" we do not mean the internal rate of return, but, rather, the coefficient on years of education in a (log) earnings regression. That is, the usual assumptions of Mincer's (1974) framework apply—that there are no direct costs of education to the individual and that all persons are in the labor force for the same number of years independent of educational attainment. See Card (1999) on the problems of "ability bias" and other issues involved in providing a causal interpretation of such estimates of the returns to education.

58. The material on Iowa counties is from Goldin and Katz (1999b).

59. We use the 1940 Census occupation codes.

60. The Census years are 1940, 1950, and 1960, but income is for the previous year. In the discussion, we will often use the federal Census year for convenience, similarly for the 1915 Iowa State Census.

61. The 1915 data are restricted to non-farm male workers. The 1950 Census restricted to Iowa produces a rather small sample.

62. Not surprisingly, educational returns also decreased between 1915 and 1960 (Table 2.6, rows 1 and 3).

63. Income from farming was the most important source of self-employment income in the 1915 Iowa data. In comparing the farm income data from the Iowa state census with that on gross agricultural income from the agricultural census, we have concluded that the 1915 Iowa data are, by and large, net income measures.

64. There were 24 colleges and universities in Iowa in 1897 of which two were under state control. The rest were small sectarian liberal arts colleges. See the data sources in Goldin and Katz (1999a).

65. See Chapter 7 and Goldin and Katz (1999a) on the evolution of U.S. colleges and universities. Many of the "older" institutions that had been staffed by a handful of faculty were transformed into more modern colleges by expanding in size and having greater specialization in teaching.

66. The evidence is from Bishop (1989).

67. See Taubman and Wales (1972).

68. The estimates differ slightly for the high school calculation but not at all for the college calculation.

69. These estimates use the 1915 Iowa State Census sample and the 1940 IPUMS.

70. "The present assault upon capital is but the beginning. It will be but the stepping-stone to others, larger and more sweeping, till our political contests will become a war of the poor against the rich—a war constantly growing in intensity and bitterness" (*Pollock v. Farmers Loan* 158 US 601, 1895).

3. Skill-Biased Technological Change

1. See Autor, Katz, and Kearney (2005b) and Lemieux (2006b) on the recent patterns of the evolution of U.S. residual (within-group) wage inequality.

2. Skill-biased technological change refers to any introduction of a new technology, change in production methods, or change in the organization of work that increases the demand for more-skilled labor (e.g., college graduates) relative to less-skilled labor (e.g., non-college workers) at fixed relative wages.

3. See Bresnahan and Trajtenberg (1995) on general purpose technologies and their contributions to economic growth.

4. See Tinbergen (1974, 1975) and Freeman (1975) for pioneering analyses of how the evolution of the wage structure depends on a race between technological advance and access to education.

5. See Cline (1997) for a comprehensive evaluation of the international trade explanation for changes in the U.S. wage structure in the 1980s and 1990s.

6. The percentage college wage premium (adjusted for demographics) can be derived by exponentiating the log college/high school wage differential shown in the final column of Table 3.1, subtracting 1, and then multiplying by 100. The levels of the log college wage premium from the Census and CPS are not fully comparable due to differences in the construction of hourly wages in the two surveys. Thus, we add the 2000 to 2005 CPS change in the log college premium to the

2000 Census log college premium to get a 2005 log college premium that can be compared to the 1950 Census log college premium.

7. See Bound and Johnson (1992), Katz and Autor (1999), and Katz and Murphy (1992) for more detailed expositions of this framework.

8. Changes in institutional factors or norms of wage setting that lead to deviations from competitive labor market outcomes could have played a role, although the basic logic of the framework would still hold, as would the implication that the relative demand function shifted outward as long as firms remain on their labor demand curves. It is possible that institutional factors, such as unions, produced employment levels off the demand curve and that declines in union strength could have led to a reduction in the relative wage and employment of less highly educated workers even in the absence of demand shifts against them.

9. Autor, Katz, and Krueger (1998) find that growth in the employment and wage bill shares of more-educated workers from 1960 to 1996 is dominated by within industry changes using data on U.S. three-digit industries. Dunne, Haltiwanger, and Troske (1996) show that the growth in the nonproduction worker share in U.S. manufacturing has been dominated by within plant changes since 1970. Borjas, Freeman, and Katz (1997) illustrate that between industry labor demand shifts from international trade explain only a modest portion of the rise in the demand for more-skilled U.S. workers from 1980 to 1995.

10. Doms, Dunne, and Troske (1997) provide a detailed plant-level analysis of the correlates of the adoption of new technologies in U.S. manufacturing in the 1980s and 1990s.

11. Autor, Katz, and Krueger (1998) document strong positive correlations of skill upgrading with computer investments, increases in capital intensity, R&D investments, and increased employee computer usage for U.S. industries. Allen (2001) demonstrates a positive relationship between the employment of scientists and engineers and the employment of more-educated workers. Machin and Van Reenen (1998) find positive effects of R&D intensity of the growth of both nonproduction employment and high-education employment for an industry panel for seven OECD countries.

12. On the banking sector, see Autor, Levy, and Murnane (2002) and Levy and Murnane (1996). Levy, Beamish, Murnane, and Autor (1999) examine auto repair, and Bartel, Ichniowski, and Shaw (2007) study the valve industry.

13. Bresnahan, Brynjolfsson, and Hitt (2002) study these issues combining a detailed survey of senior human resource managers on organizational practices and labor force characteristics with detailed information on information technology investment for U.S. companies in the mid-1990s.

14. Bresnahan (1999) and Autor, Levy, and Murnane (2003) posit such an organizational complementarity between computers and workers who possess both greater cognitive skills and greater people skills.

15. Friedberg (2003) and Valletta (2006) use questions on computer use at work from a series of CPS Computer and Internet Use supplements to document the growth of U.S. employee computer usage.

16. See, for example, Berman, Bound, and Griliches (1994) and Autor, Katz, and Krueger (1998).

17. The approach, following Katz and Murphy (1992) and Autor, Katz, and Krueger (1998), is to relate the college–high school wage differential to the relative

quantities and demands for these education classes expressed as "equivalents." College equivalents are given by college graduates plus half of those with some college; high school equivalents are those with 12 or fewer years of schooling plus half of those with some college. A calculation of demand shifts for college graduates requires an estimate of the degree of substitutability of college and non-college workers in production (including the substitutability for consumers between products that are more- and less-intensive in college equivalents). These substitution possibilities are summarized by a key parameter known as the aggregate elasticity of substitution. Our measures of demand shifts for college equivalents are calculated using our preferred estimate of the aggregate elasticity of substitution between college and high school equivalents (1.64). In Chapter 8, we develop this framework in more detail, extend the analysis to cover 1915 to 2005, provide new estimates of the aggregate elasticity of substitution, and explore the sensitivity of the results to different assumptions about the parameter.

18. Autor, Katz, and Kearney (2006, 2007) show that the slowdown in relative demand growth for college workers since 1990 reflects changes in the composition of skill demand growth. Demand growth has remained strong for those with post-college education and for the highest-earning college graduates, has accelerated for in-person service jobs employing the least-educated workers, and has slowed for "middle skilled" workers including many with a four-year college degree or some college.

19. The direct estimation of a supply-demand model of the college wage premium using annual data from 1963 to 2005 leads to the same conclusion, that reasonably stable rapid secular growth in demand combined with a slowdown in relative supply growth after 1982 can largely explain the evolution of the college wage premium during the last four decades (Autor, Katz, and Kearney 2007). But such an annual time-series analysis suggests some acceleration in the relative demand for college equivalents in the 1980s combined with a deceleration in the 1990s. We extend this time-series analysis to 1915 to 2005 in Chapter 8. A similar model extended to allow for multiple age and experience groups also helps explain the larger rise in the college wage premium for younger workers since 1980 through the slowdown in the growth of relative skill supplies, both overall and between birth cohorts (Card and Lemieux 2001).

20. See, for example, Nelson and Wright (1992), who state that "there is no reason to believe that the [U.S. manufacturing] labor force was particularly well educated by world standards" (p. 1947). Although the manufacturing labor force was less well educated than the average worker with the same demographic characteristics, Nelson and Wright do not consider the fact that many of these workers were well educated by the standards of the industrial world. The main reason that they do not consider this fact is that they rely on Maddison's (1987) data on years of secondary school education across various countries. As we demonstrated in Chapter 1, these data are incorrect and vastly overstate educational attainment in Great Britain and other European nations.

21. These tabulations use the 1900, 1920, and 1940 Census IPUMS and include males 14 years and older with gainful occupations for 1900 and 1920 and in the labor force for 1940. Blue-collar workers for the purposes of these tabulations include those in craft occupations, operatives, and laborers (excluding farm laborers).

22. Our substantive findings are similar if we look, instead, at the educational attainment of all workers, all male workers, or all blue-collar workers, rather than restrict attention, as we do, to 18- to 34-year-old blue-collar males. The age limitation is imposed because the educational attainment reported by older Americans in the 1940 Census appears overstated (see Goldin 1998).

23. The sample is restricted to the currently employed. In 1940 more than 10 percent of those in the labor force were unemployed and another 4 percent were on work-relief.

24. The list would probably include many others in the batch and continuous-process group if the 1940 Census tabulated finer categories of industries, e.g., distilled liquors and pharmaceuticals. Industries are defined as high-technology if a large percentage of their total labor force were engineers, chemists, and other scientific personnel, similar to currently used definitions.

25. In 2000, 28 percent of the U.S. male labor force 25 to 34 years old had 16 or more years of schooling. All figures are based on tabulations from the 1940 and 2000 IPUMS for the labor force, which includes the unemployed and, in 1940, those on emergency work relief. The comparison is robust to including only the employed.

26. The 1940 IPUMS separately identifies 61 manufacturing industries that, for the most part, correspond to current 3-digit SIC industries and some of the larger 4-digit industries.

27. The results differ trivially if years of education instead of the percentage graduating from high school were used.

28. The standard deviation of the share of high school graduates among young, male, blue-collar workers across manufacturing industries is 0.086, as compared with 0.080 for the adjusted industry coefficients. A more extreme adjustment uses the mean industry residuals from an analogous regression but excluding the industry dummies. In that case, the standard deviation of the share of high school graduates is 0.071.

29. See Burdge (1921, table 24-L, p. 339).

30. The group interviewed included employed boys between the ages of 16 and 18 in cities with more than 25,000 people. Those who were not employed in the metal trades were in a variety of industries, such as wood, clothing, food, textiles, and leather, and also in transportation and construction. Of course, the majority of boys who did some high school were in the clerical trades.

31. Data on employment by detailed (4-digit) industries from the 1909 to 1929 censuses of manufactures indicate that the majority of employees in five 1940 Census industry categories (beverages, dairy products, grain-mill product, paints and varnishes, and petroleum refining) worked in industries classified by Chandler (1977) as using continuous-process or batch production methods. Consistent with our framework, these industries employed a disproportionate share of more-educated blue-collar workers in 1940: 36.0 percent of young, blue-collar workers in these continuous-process and batch production industries had 12 or more years of schooling as compared with 27.1 percent in the remainder of manufacturing. It should also be noted that some of the more-educated industries remained rather artisanal in their methods of production (e.g., jewelry).

32. The industry categories in the 1940 Census are broader than those in the earlier censuses of manufactures. We aggregated the earlier data to conform to the

1940 categories for our analysis of the determinants of industry variation in education levels (see the data appendix to Goldin and Katz 1998).

33. The education data are adjusted for differences in age composition, urbanization, and geographic distribution of employment. The regression results are quite similar when we use the actual industry-level high school graduate share of 18- to 34-year-old, blue-collar males (unadjusted for differences in age composition and geography) in 1940 as the dependent variable.

34. These industries were chosen because oleomargarine production used a continuous-process technology, whereas lumber and timber were mainly factory produced. The ratio of capital to wage-earners in lumber and timber was $1,693 but it was $5,871 in oleomargarine in 1909.

35. See Du Boff (1979).

36. For various reasons, the measurement of horsepower driven by generated electricity is imprecise. The generated electricity variable is estimated using the procedure described in Du Boff (1979, appendix A), although see Jerome (1934) for another method. The problem is that motors powered by generated electricity were often rated above their actual use, which is limited by the horsepower of the prime movers, whereas those powered by purchased electricity can be run at or above their rating.

37. See Jerome (1934), Nelson (1987), and Nye (1990). In iron and steel "the proportion of common laborers was cut approximately in half from 1910 to 1931. The evidence is unmistakable," notes Jerome, "that recent progress has eliminated unskilled labor to a much greater extent than other grades" (1934, p. 63). In all the industries mentioned, the use of conveyors, traveling cranes, jitneys, carriers, industrial trucks, and other handling devices reduced the relative demand for unskilled labor. The changes, moreover, were evident as early as 1916. We find, from the 1910 and 1940 Census IPUMS, that laborers as a share of total manufacturing employment declined from 23.6 percent in 1910 to 14.3 percent in 1940.

38. Electricity played a complex role in increasing the relative demand for skilled workers. Although Nye (1990, pp. 234–235) concludes that electricity increased the relative demand for skill, he also describes opposite effects. "As the electrified factory evolved it required a different mix of labor and management . . . more middle management; more engineers and technicians; fewer artisanal workers; and a more complex grading of worker skills, with many more semiskilled laborers . . . and far fewer unskilled workers. . . . Boy mule drivers in coal mines, carriers in tire factories, or shovelers of raw materials in steel mills saw their work taken over by electric locomotives, conveyors, and cranes . . . as a few skilled men using expensive machines did work formerly performed by a mass of the unskilled." The increase in purchased electricity use, moreover, decreased the need for prime movers and the skilled labor that serviced them. See also Du Boff (1979) and Devine (1983) on the transition from mechanical to electric drive and the introduction of group and unit drive motors.

39. The adjustment in Table 3.4 for the growth of the industry will account for some of this factor. The newness of the capital stock concerns the replacement of the shafts and pulleys of the older system of power with separate machines (unit drive) associated with electricity. Note that either generated or purchased electricity could have accomplished the same transformation. Data on the average horsepower of motors suggest that firms with purchased electricity switched more

completely to unit drive (smaller motors) whereas those generating their own appear to have used group drive more.

40. See, for example, the descriptions of positions in electrical machinery, glass, medicinal manufacturing, paint and varnish, and the printing trades in U.S. Department of Labor (various years, 1918–1921).

41. Electrical Merchandising (1922).

42. The earliest nationally representative sample with data on earnings and education is the 1940 IPUMS, which contains earnings data for 1939. The coefficient on years of schooling for the blue-collar sample is 0.083 (s.e. = 0.0013) in a log (full-time equivalent) weekly earnings equation that includes potential experience and its square. The analogous regression for young, male, ordinary white-collar workers (white, 18–34 years old, sales and clerical occupations) in manufacturing is 0.091 (s.e. = 0.0028). The blue-collar and ordinary white-collar samples contain 27,942 and 4,892 observations, respectively. The estimated return in the blue-collar sample is 7 percent when a full set of state dummies are also included.

43. The standard error on the 0.065 coefficient is 0.0012.

44. Our 1915 Iowa State Census sample contains 3,134 white, male, non-farm, blue-collar workers, 18 to 34 years old. The mean of schooling in this sample is 8.05 years, and 19.5 percent attended some high school or college. The estimates of returns to schooling in Iowa in 1915 are similar for a broader sample of young, male, blue-collar workers that include workers in service occupations as seen in Table 2.5 of Chapter 2.

45. Within-industry variation over the 1909 to 1929 period leads to similar conclusions to the between-industry regressions reported in Table 3.6. There are positive and significant relationships between the change in the (log) average earnings of production workers from 1909 to 1929 and changes in capital intensity and electricity use (conditional on controls for changes in the other covariates included in the Table 3.6 regressions).

46. The wage premia use the 1929 coefficients. The actual difference in wages is 33 percent. Oleomargarine had a capital-to-labor ratio in 1919 of $8,759 and 69.4 percent of its horsepower came from purchased electricity; the numbers for lumber and timber are $3,028 and 27.5 percent, respectively.

47. Another interpretation is that the wage differentials reflect premia for identically skilled individuals working in more capital- and electricity-intensive industries in which there was greater worker bargaining power and managerial discretion (e.g., Slichter 1950). That may well be the case, but the strong correlation between wages in the 1909 to 1929 period and education in 1940, by industry, provides evidence that the compositional effect matters. There is, as well, a relationship between the high-education industries given in Table 3.3 and the percentage of the industry's 1910 labor force that was "machine related" (e.g., machinist, electrician).

48. Chandler (1977).

49. Goldin and Katz (1998, table 6).

50. See, for example, Braverman (1974) and Piore and Sabel (1984).

51. Acemoglu (1998, 2002) emphasizes a market-size effect whereby growth in the share of more-educated workers produces a larger market for more skill-intensive technologies and creates economic incentives for the R&D sector to produce more skill-biased innovations. Galor and Moav (2000) focus on the increase in supply of R&D workers from a more-educated workforce. These approaches

imply that the large secular increases in U.S. educational attainment over the twentieth century and the growth of new organizational forms should have generated an accelerating rate of increase in the relative demand for the more skilled.

52. U.S. Census Office (1895a); U.S. Bureau of the Census (1933).

53. Data on the operating establishments for U.S. manufacturing industries since 1958 are based on the Annual Survey of Manufactures and the censuses of manufactures. Data from 1958 to 1996 are from the NBER Manufacturing Productivity Database (Bartelsman and Gray 1996). Data for 1996 to 2004 are from the U.S. Census Bureau (2005b).

54. The reason that the ratio of the wage bill between skilled (S) and unskilled (U) labor would not change is that the definition of $\sigma = 1$ along a relative labor demand curve is that $(w_S \cdot L_S / w_U \cdot L_U)$ is a constant, where w = wage and L = labor.

55. Since the relative earnings of nonproduction workers declined from 1890 to 1929 and increased from 1960 to 1999, the rate of growth of the relative demand for nonproduction workers would be even greater in the earlier, relative to the later, period if this elasticity of substitution were below 1. Of course, the sign pattern could reverse if the elasticity of substitution were sufficiently larger than 1. The elasticity of substitution between skilled and unskilled labor is likely to be lower for an individual sector (manufacturing) than for the overall economy. The economy-wide (aggregate) elasticity of substitution includes consumer substitution possibilities across the products of different sectors (e.g., manufactured goods as opposed to services). Thus, our preferred estimates for the aggregate elasticity of substitution, in the 1.5 to 2 range, mean that a sector-level elasticity of substitution for manufacturing of 1 is plausible. See Katz and Autor (1999) on these aggregation issues and on the role of the elasticity of substitution between more- and less-skilled workers in the estimation of relative labor demand shifts.

56. See Goldin and Katz (1995) and Autor, Katz, and Krueger (1998). For example, the share of manufacturing employment in the top five (two-digit) industries by education (petroleum, chemicals, electrical machinery, printing and publishing, and scientific instruments) expanded from 10 percent to 16 percent during the 1910 to 1940 period (Goldin and Katz 1995, table A4).

57. Goldin and Katz (1998, table 7).

58. We probe the issue in more detail in Chapter 8.

59. The full framework is contained in Goldin and Katz (1998).

60. The term "artisan" is used here to mean a worker who produces virtually the entire good in a production process containing almost no division of labor.

61. James and Skinner (1985) divide industries in 1850 into two categories: "skilled" (e.g., woodworking) and "unskilled" (e.g., clothing). They find that in both skilled and unskilled industries raw materials were the relative complements of physical capital, although the effect was greater in the skilled sector. More importantly for the capital-skill complementarity hypothesis is that skilled labor was the better substitute for capital in its sector than was unskilled labor in its. Thus an increase in capital (or raw materials) would have decreased the relative demand for skilled labor. Cain and Paterson (1986) do not consider skill differences but find, analogous to James and Skinner, that capital and raw materials were relative complements and that both together substituted for labor. Williamson and Lindert (1980), however, assume capital-skill complementarity and generate, in their model, rising inequality with capital deepening during the nineteenth century.

62. See Hounshell (1984) on gun making. According to Sokoloff (1986), some initial deskilling, for example in shoemaking, involved little capital, and no mechanization. It was, rather, of the Smithian pin-factory variety. Landes (1972) takes an opposing view. Braverman (1974), among others, argues that industrialization and mechanization served to deskill a host of artisanal trades and to reduce the relative earnings of craftsmen.

63. See Atack (1987) and Sokoloff (1984) on the transition from the artisanal shop to the factory in the nineteenth century. Both make the important point that the transition was slow in some industries and it depended not just on technological change in manufacturing (often in the organization of work) but also on decreases in transport costs. In some industries (e.g., boots and shoes, clothing, furniture, leather, meatpacking, tobacco) a significant minority of value added was produced in artisanal shops (<7 employees with no power source) even as late as 1870, while in others (e.g., saddlery), the majority of value added came from artisanal shops in 1870.

64. Recall that the term "batch" refers to production "in a batch," generally used for liquids (e.g., liquors), semi-solid liquids (e.g., oleomargarine), or molten metals (e.g., steel, aluminum). It is not to be confused with another usage, the production of items in batches (e.g., clothing pieces) for later assembly.

65. Braverman (1974, p. 146) quotes Eli Chinoy on automobile production: "Final assembly, for example, had originally been a highly skilled job. Each car was put together in one spot by a number of all-around mechanics." See also Hounshell (1984).

66. The full model is contained in Goldin and Katz (1998). Atack, Bateman, and Margo (2004, 2005) explore several of the empirical implications of the model.

67. Jerome (1934) illustrates this point for many industries.

68. A walk through most any factory today—one that assembles autos or their parts, makes high-grade steel, or fabricates just about anything except clothing—will reveal few production operatives but many capital-maintenance workers. In one auto assembly plant we visited, an engineer proudly reported that any human welder we saw would soon be replaced by robots.

4. Origins of the Virtues

1. The enormous racial divide in access to schooling and in educational resources, from the founding of the republic to at least the 1970s, has been the major exception to "egalitarianism" in U.S. educational institutions.

2. See Easterlin (1981) and Lindert (2004), for example. Later in this chapter we explore the enrollment data of the early U.S. censuses and raise questions about these and the attendance data for selected states.

3. We distinguish between "schooling," for which we mean the contemporaneous level of enrollment, attendance, or graduation, and "education," for which we mean the average years (and quality) of schooling attained by a population.

4. Watson (2006) documents the rise in U.S. residential segregation by economic status since 1970.

5. The U.S. Office of Education did not collect information on the number of school districts until 1932 when it reported that there were 127,531 (*Historical Statistics*, series H 412). It is likely that there were more in 1900. Even though some

school districts were extremely large, for example the New York City school district enrolled almost 560,000 school children in 1900, most youths lived in small towns and rural areas in 1900. Only 22 percent of 5- to 14-year-olds lived in cities with more than 25,000 people, and 27 percent resided in cities with more than 10,000. Fully 60 percent lived in rural areas or towns with fewer than 1,000 persons. U.S. Office of Education, *Annual Report* (1900–01), p. 1547; U.S. Bureau of the Census (1975), series A 57–72; IPUMS for the 1900 U.S. census of population.

6. Lindert (2000, 2004) emphasizes the role of decentralization in spurring educational advances in the nineteenth century. According to Lindert, the United States, Canada, and Prussia had decentralized systems, while Great Britain and Scandinavian countries were highly centralized, which allowed national elites to gain control of educational decisions. Even though Prussia is deemed "decentralized" by Lindert, Ringer (1979, p. 32) notes that "a fairly homogeneous national system of education emerged even before 1870, especially at the secondary and university levels." The imposition in 1812 of state "leaving exams" in Prussia underscores the centralization of educational control. See also Fishlow (1966b, p. 435) on the growth of centralization in Europe relative to the United States in the late nineteenth century. According to Fishlow, local sources in England provided 75 percent of primary school income in 1876 with half coming from private sources, but by 1900 Parliamentary grants exceeded 50 percent of income and private fees had been eliminated. France, which had been highly centralized for both fiscal and curriculum matters, further centralized in the late nineteenth century. In 1877, the French central government provided 25 percent of public primary school revenue but by 1900 it accounted for 80 percent. Fishlow notes a similar trend in Prussia with respect to its school-tax law.

7. Burdge (1921) reports the findings of a 1918 New York State survey of 16- to 18-year-old males in the state. Among the employed boys who graduated high school and were from cities with more than 25,000 residents, 92 percent indicated they wanted to work in white-collar occupations and 82 percent were currently employed in such jobs. In contrast, among those who completed no years of high school 57 percent desired white-collar jobs and 46 percent were currently employed in them. The 1915 Iowa State Census provides similar results. Among 18- to 24-year-old males living in Iowa's larger cities, 82 percent of high school graduates were white-collar workers whereas just 21 percent of those without any high school were. For women in the same age group, the fractions are 95 percent for high school graduates and 41 percent for those with no high school (see Appendix B).

8. The First Amendment (the Establishment Clause) says: "Congress shall make no law respecting an establishment of religion, or prohibiting the free exercise thereof." The Supreme Court decision, *Zelman v. Simmons-Harris* (no. 00–1751, June 27, 2002), which upheld the use of public vouchers by parents to send their children to religious schools in Cleveland, may have important consequences in general for the use of vouchers by denominational institutions and, more important, for the nation's public school systems.

9. We define "common school" below. By "free" schooling we mean education for which the marginal cost to the user is zero.

10. "Wisdom, and knowledge . . . depend on spreading the opportunities and advantages of education in the various parts of the country . . . it shall be the duty of legislatures and magistrates, in all future periods of this commonwealth, to cherish

the interests of literature and the sciences, and all seminaries of them; especially the university at Cambridge, public schools and grammar schools in the towns" (Constitution of the Commonwealth of Massachusetts, chapter V, section II). For state constitutions (and their changes over time), see John Wallis's, University of Maryland and the NBER State Constitutions Project, which at the time of this writing was located at http://www.stateconstitutions.umd.edu/index.aspx

11. See Kaestle (1983) for a fine introduction to the literature on education in the period of the early republic and Rudolph (1965) for some of the primary sources.

12. Benjamin Rush, "Plan for the Establishment of Public Schools" (1786), in Rudolph (1965, pp. 4, 6).

13. Benjamin Rush, "Thoughts Upon the Mode of Education Proper in a Republic" (1786), in Rudolph (1965, p. 19).

14. Kaestle and Vinovskis (1980, chapter 5).

15. On local and state taxation, see Stewart (1914, pp. 77–92), who provides much of the evidence cited in this section.

16. Fernandez and Rogerson (2003) analyze the impact on educational resources and equity of alternative school finance systems with an emphasis on the degree of centralization of school finance decisions.

17. Hoxby (1996, 1999) discusses efficiency-equity tradeoffs, the importance of competition among school districts in efficiency, and the role of sorting among school districts within a Tiebout framework.

18. On the role of property taxes, and the assumptions needed to obtain different incentive effects from the property tax as opposed to an income tax, see Hoxby (1996).

19. Among the states entering the Union after the original 13, only Maine, Texas, Vermont, and West Virginia did not receive land grants. Three of them were carved out of pre-existing states and Texas owned its own land. See, for example, Cremin (1951), p. 119.

20. Cubberley (1934, orig. pub. 1919; chapter IV). Randall (1844, p. 83) provides annual data from 1815 to 1843 on the amount paid from the state treasury, the amount received by the districts, and the amount paid by individuals on rate bills.

21. The term "common school revival" was used by the U.S. Commissioner of Education in 1900 to describe the fervent campaigning for free schooling in the 1840s (see Fishlow 1966a).

22. The definition of "rural" used here is any place with fewer than 2,500 people (IPUMS of the 1850 federal census of population).

23. Landes and Solmon (1972) provide evidence of the ineffectiveness of state compulsory schooling laws in the late nineteenth century. Compulsory schooling laws were not effective until the early twentieth century and by then most of the children who would have been constrained by the laws were already in school (Goldin and Katz 2003). We address the role of state compulsory education legislation from 1900 to 1940 in Chapter 6.

24. Kaestle and Vinovskis (1980, table A2.2).

25. Of the top ten cities by population in New York in 1850, seven (ranked by size: New York City, Brooklyn, Buffalo, Rochester, Troy, Syracuse, and Utica) had abolished the rate bill by 1853.

26. This paragraph relies on Cubberley (1934, orig. pub. 1919, p. 200).

27. Aurner (1914, pp. 21–22, 47). All of Iowa's schools became free in 1858.

28. Fishlow (1966a, table 1) uses mainly U.S. decennial census documents from the censuses of schools (as opposed to an actual census of individuals). The "school census" was continued in the 1850 to 1870 Social Statistics of the Census, although the censuses also asked individuals whether they attended school for at least one day in the previous year and also their occupation.

29. The data given are for enrollment, not attendance, and may be overstated for various reasons. The work of Kaestle and Vinovskis (1980, table A2.5; Vinovskis 1972) on antebellum Massachusetts schools implies that the attendance rate c.1840 for 5- to 19-year-olds was around 43 percent, far lower than the enrollment rate. They report average daily attendance rates among *all* children and youth under 20 years old as 37 percent in Massachusetts during the 1840 to 1880 period. We use their estimates for the attendance rate of 0- to 4-year-olds and the fraction of the population in each age group from Census of the United States (1841, pp. 8–9) to obtain the estimate of 43 percent for the 5- to 19-year-old group. Even if enrollment rates are overstated, there is no reason to believe that they are more overstated in New York State than in Massachusetts.

30. Computed from the IPUMS of the 1850 and 1860 population census.

31. We have not corrected the figures for differences in urbanization and other possible intermediating factors, but these would not lead us to reject the conclusion that the rates were similar between the two states. As is noted in Table 4.1, many of the larger cities of New York had free schooling long before the state abolished rate bills.

32. Data for the Midwest are from Fishlow (1966a, p. 49) and are from the school census.

33. This is also the point made by Fishlow (1966a). According to Fishlow, increased enrollment in the 1840s and 1850s occurred, in many states, prior to the abolition of the rate bills. Only after enrollment increased did free schooling pass at the state level. Fishlow does mention that Louisiana was the only state in the South to adopt a free school law before the Civil War (in 1847), and that it experienced the third greatest gain in the South in (white) enrollment from 1840 to 1850. The two other states (North Carolina and Tennessee) with large enrollment gains had increases in school revenue due to the distribution of revenue in 1837 by the federal government from the sale of land surplus (p. 52).

34. We demonstrate a similar result with regard to state compulsory education and child labor laws in Goldin and Katz (2003). See also Chapter 6.

35. For a complete history of state educational offices, see Cubberley and Elliott (1915). New York, in 1812, appointed the first state officer to supervise schools, but the office was abolished in 1821. Although in the 1820s some secretaries of state were asked to act as the state superintendent of schools, it was not until the 1840s that various states began to appoint individuals to a separate office of state superintendent of schools. In that sense Mann's position was a first in the history of state educational offices and was, according to Cubberley (1934, orig. pub. 1919), the first real state board of education in the United States.

36. See Kaestle and Vinovskis (1980), chapter 8 on Mann's political problems with the Massachusetts state legislature.

37. Mann (1891, vol. 3, pp. 94–95). Also see Vinovskis (1995), chapter 5, who is critical of Mann's empirical methodology.

38. See, for example, Bowles and Gintis (1976). Field (1979) uses cross-section evidence from Massachusetts to suggest that the Irish influx led industrialists to lengthen the school term for social control and other related reasons.

39. For the details on the referendum for taxation in Indiana and for free schools in New York State, see Cubberley (1934, orig. pub. 1919, chapter 6).

40. Mann, as well as many others throughout U.S. history who have been associated with the separation of church and state, espoused nonsectarian schools but not secular teaching. According to Mann: "Our system . . . earnestly inculcates all Christian morals; it founds its morals on the basis of religion; it welcomes the religion of the Bible. . . . But here it stops" (Mann 1891, vol. 4, pp. 222–340). See also Nord (1995), who notes that Unitarians allowed that schools could teach religious morals without reference to a particular theological system.

41. See Michaelsen (1970) and Glenn (1988). The third article of the Declaration of Rights of the Massachusetts Constitution states "the people of this commonwealth have a right to invest their legislature with power to authorize and require . . . towns, parishes, precincts . . . to make suitable provision . . . for the institution of the public worship of God, and for the support and maintenance of public Protestant teachers of piety, religion, and morality." With the change in this article, the Congregational order was disestablished.

42. Not until 1963 did the U.S. Supreme Court, in *Abington Township v. Schempp* (374 U.S. 203), prohibit bible reading and the recitation of the Lord's Prayer in public schools.

43. Cubberley (1934, orig. pub. 1919, p. 238). See Michaelsen (1970) and Stokes and Pfeffer (1964) on the narrowly defeated Blaine amendment voted in 1876 (63 percent of the House voted for it; 39 percent voted affirmatively on the Senate version). The House version of the Blaine amendment read: "No State shall make any laws respecting an establishment of religion . . . ; and no money raised by taxation in any State for the support of public schools . . . shall ever be under the control of any religious sect" (Stokes and Pfeffer 1964, p. 434). The Senate version was similar. With the defeat of the Blaine amendment, a congressional law (Enabling Act) was passed in 1876 requiring that all newly admitted states establish a provision in their constitution against the public support of sectarian schools (Michaelsen 1970, p. 68). Because all states had amended their constitutions or enacted legislation to prevent public funding for sectarian schools it was deemed reasonable that newly admitted states be required to do the same.

44. This story has been well told by several historians, among them Kaestle (1973) and Ravitch (1974).

45. Kaestle (1973, chapter 2) remarks that school fees for "common pay schools" in New York City in the mid-1790s were low enough that all but the very poorest could attend. The fees he reports were about $2.50 (or 20 shillings) per child per quarter. Annual family income would have been $250 for a laborer who worked 250 days and around $350 for a house carpenter (Adams 1967, table 1 for Philadelphia or New York City, see table 16). If a family had three school-aged children each of whom went to school for two quarters a year, the fees would have amounted to 4 to 6 percent of annual income but a considerably larger fraction of income net of necessities.

46. Kaestle (1973, table 19).

47. A rate bill that covered the entire cost of education was, in stark contrast, publicly provided but privately funded education.

48. See Kaestle (1983) on educational changes from the colonial period to that of the Revolution when "Many elementary schools in the North admitted girls for the first time . . . although access was often limited and segregated" (p. 28). Similarly, Kaestle and Vinovskis (1980) argue that much of the advance in school enrollments in the years from 1800 to 1830 occurred because of the increase in the schooling of girls (pp. 24–26).

49. As we noted before, the data from the U.S. decennial population census appear to give too high an estimate of the fraction of youths in school full-time but there is no reason to believe that the upward bias was greater for girls than it was for boys.

50. Whether or not mid-nineteenth-century common schools were largely sex segregated is another matter. Some have argued that most common schools were intended to be sex segregated (e.g., Vinovskis and Bernard 1978), whereas others have convincingly demonstrated that boys and girls, especially in the less settled areas, learned together in practice (Tyack and Hansot 1990). Places that had a sufficiently large population, on the other hand, could afford to educate girls and boys in separate facilities.

51. U.S. Commissioner of Education (1895, p. 786).

52. Larger cities would later retreat from complete coeducation with the establishment of special secondary schools for vocational skills that were gendered.

53. U.S. Commissioner of Education (1895, p. 799).

54. See Easterlin (1981), for example.

55. Ringer (1979, p. 34).

56. These data are similar to those used by Fishlow, although we are able to use the micro-data that were not available when he did his work.

57. "School" included public and private schools of various types, including common, grammar, high schools, academies, colleges, and seminaries. Sunday schools (from 1850 to 1880) and evening schools (from 1860 to 1880) were explicitly excluded from the census definition of a "school." The U.S. decennial census has asked a similar question on school attendance in every subsequent census. A question on educational attainment was first asked in the 1940 Census.

58. The census asked the question on occupation of males older than 14 years in 1850, of all youths older than 14 years in 1860, and of all youths older than 10 years in 1880. The question on occupation apparently was asked of everyone in 1870.

59. For ease of presentation, Figure 4.3 does not contain the data for 1870.

60. Public high schools were still rare outside the largest cities in 1880.

61. Compulsory education laws accomplished this, but only Massachusetts had such a law until the mid-1860s. The decline in "infant schools" in New England might also be a factor.

62. The reduction is greater considering just boys. For example, in 1880 51 percent of all 16-year-old males in the Midwest were enrolled in school but just 28 percent were in school full-time.

63. Cubberley (1934, orig. pub. 1919, p. 259); U.S. Office of Education (1906), pp. 1855–1863; Census of the United States, 1830 (1832), pp. 16–19.

64. The Massachusetts law was not seriously enforced. As a rough approximation to the 500 families rule, we use all towns with more than 600 white males older than 20 years.

65. See Labaree (1988) for a history of the school's first century.

66. The New York City high school was built with funds from the sale of stock and the stockholders reserved the right to place their own children in the school.

67. Cubberley (1934, orig. pub. 1919, p. 262).

68. We discuss the education of older youths in common schools in Chapter 6 using the 1915 Iowa State Census.

69. Only a small literature exists on the subject of nineteenth-century academies. See, for example, Brown (1899), Cubberley (1934, orig. pub. 1919), Sizer (1964a), and Tyack (1967, chapter 10).

70. Many of the older academies that currently exist in the East were founded in the late eighteenth century. Phillips Academy, for example, was established in 1778. Many of those in the Midwest that still exist were established during the "academy movement."

71. Kandel (1930) recounts the histories of academies in Illinois, Indiana, and Michigan, which existed until public high schools were established. Elsewhere, academies remained after the establishment of a public high school because the public high school did not offer a classical course, as was the case in Groton, MA (Katz 1968).

72. "By the 1820s and 1830s, a growing network of reformers sought to . . . lobby for a common system of tax-supported public high schools. . . . school reformers began to argue that public high schools alone should provide advanced instruction for the talented few" (Reese 1995, p. 17).

73. The decision, *Charles E. Stuart et al. v. School District No. 1 of the Village of Kalamazoo*, 30 *Michigan* (1874), is reprinted in Cubberley (1970, orig. pub. 1934, p. 240). On the Kalamazoo high school and the opposition to it, see Reese (1995, pp. 76–79).

74. See Dunbar (1960) concerning why *Kalamazoo* had so far-reaching an impact. The best reason given is that Justice Cooley of the Michigan Supreme Court argued that the high school was an integral part of the complete system of public education established by the state legislature.

75. Easterlin (1981).

76. See also the many citations, from 1834 to 1880, in Reese (1995, p. 96, fn. 52).

5. Economic Foundations of the High School Movement

1. See, for example, Easterlin (1981). The pre–Civil War U.S. data include slaves in the population and thus the enrollment data would be considerably higher without their inclusion. Canada had schooling levels that were close to those in the United States and throughout this volume we often mean North America when referring to the United States or America.

2. On the franchise, mass education, and the causes of both, see Engerman and Sokoloff (2005) and Sokoloff and Engerman (2000) who emphasize early levels of factor endowments, such as ratio of land-to-labor as well as the existence of slavery. Acemoglu and Robinson (2000) model the franchise as a commitment mechanism for governments, and Acemoglu, Johnson, and Robinson (2001, 2002) see early colonial rule as critical in later development.

3. Lindert (2004) reports that elementary school students (public and private) per 1,000 children 5 to 14 years of age increased in England and Wales from 657 in

1890 to 742 in 1900. The rate in Sweden, according to Lindert's data, increased largely during the 1870s but never reached a level above 700 in the nineteenth century (Swedish children, it should be noted, begin school at an older age than most others). In France the increase was uniformly steady from 1850 to 1880 when the figure stood above 800. In Prussia, the educational leader prior to the 1840s, the rate was already quite high in the mid-nineteenth century but was 768 in 1900, not much higher than it was earlier. The United States in 1900 had attained a rate well over 900. Therefore, even though the education gap between the United States and much of industrialized Europe narrowed considerably in the late nineteenth century, it did not close. It should be noted that there are serious problems of comparability across countries and even within countries since the starting and ending ages of elementary school students varied considerably.

4. The phrase the "second transformation" of education comes from an insightful article by Trow (1961).

5. A simple model of the process is the following. Consider a two-period model of the decision to invest in education (say at the secondary school level). An individual with no education beyond the elementary years earns w_n in both periods and a person with more education earns w_h in the second period. The direct cost of education to the student's family (e.g., tuition, room, and board) is C. The decision to invest in education is: $[(w_h/w_n) - 1]/(1+r) \geq [(w_n + C)/w_h]$, where the discount rate is r. That is, invest in education when the discounted benefits exceed the direct and indirect costs relative to the wage for individuals with no further education. In the nineteenth century the ratio (w_h/w_n) was high and some parents sent their children to academies, that is, private secondary schools. But costs (C) were high at the academies so that only wealthier families (those with a low enough r) would educate their children there. The equilibrium level for (w_h/w_n) was, consequently, high. Demand and supply shifted out at the same rate across the latter part of the nineteenth century, maintaining the wage ratio by skill. But when demand for skill shifted out sufficiently to justify the large fixed costs to building and maintaining a community high school, the majority in the district would demand a public high school (typically without tuition to residents) and C would plummet. The supply of skill would then increase and the wage ratio by skill would fall (as we observed it did after 1914 in Chapter 2) to maintain the equality of the equation. Thus, the skill ratio (w_h/w_n) was high for a long time before the high school movement.

6. Occupational data are from Edwards (1943). Agricultural occupations varied in their demand for formal schooling, and by the early twentieth century farmers in certain parts of the nation were in the forefront in seeing value from education to the farm community.

7. Most of the women in the blue-collar group were working as seamstresses, milliners, and machine operators. The higher level of education for these women than of comparable males may have been due to the fact that some women desired to attain a teaching credential and may have taught in the past.

8. A small additional group stated that they went to college even though they did not attend a high school. Some had attended the preparatory institute of the college or had been tutored at home.

9. See Michaels (2007) on the role of increased complexity in production processes on the growth of demand for office workers in the early twentieth century.

10. See *Historical Statistics* series P 4 and 5, also cited in Goldin (2000, p. 564). Manufacturing excludes the "hand and neighborhood" industries, such as those working in blacksmith shops, carpenters, plasterers, painters, and others in the building trades.

11. We found in Chapter 3 that growth in the relative demand for nonproduction workers in manufacturing was as rapid from 1890 to 1929 as it was late in the twentieth century (1960 to 1999).

12. Not until 1870 did the printed volumes of the U.S. population census use the term "accountant" to describe an occupation, and even then the occupation listed was "book-keepers and accountants in a store." In 1870, banks, express companies, insurance offices, railroad offices, and telegraph offices hired "clerks and book-keepers," but, apparently, not accountants. It was not until 1890 that the U.S. Census used the occupational grouping "bookkeepers and accountants" in the trade and transportation sector to describe such work regardless of where the individuals were employed (U.S. Census Office 1897, p. 304).

13. The Eleventh Annual Report of the Commissioner of Labor (U.S. Bureau of Labor 1897) reported on almost 150,000 employees gleaned from firm payroll records in establishments that hired a disproportionate number of women and youths (for example, textiles and glassware in manufacturing, dry goods, and insurance in trade). Detailed occupations were enumerated separately.

14. Strom (1992) claims that the outflow of experienced personnel during World War I led to the greater use of office machines which saved on skilled labor in the 1910s. See Yates (1989) on the proliferation of simple filing systems in the late nineteenth century.

15. For a fairly complete listing of all office equipment available for sale in 1924, see Office Equipment Catalogue (1924). See also Morse (1932, p. 272) for a chronology of business machine and related inventions.

16. The 1922 figure is from the Federal Board for Vocational Education (1922). There were at least 100 separate office and clerical occupations enumerated in the original surveys of a 1940 Women's Bureau Bulletin housed at the National Archives. The surveys were conducted in five cities, although only Philadelphia firms were selected for the sample. See Goldin (1990), data appendix.

17. Strom (1992, p. 283) discusses the skills that businesses demanded in the early twentieth century from ordinary clerical workers.

18. Edwards (1943). The group in 1890 consists of three groups: bookkeepers, cashiers, and accountants; clerks (but not clerks in stores, which Edwards estimates and subtracts); and stenographers and typists.

19. See the important work of Rotella (1981) on women's employment in the clerical sector.

20. The O. Henry story is "Springtime a la Carte" from his collection *The 4 Million* (1906). The Dos Passos novel *1919* (1932) is the second in his trilogy *U.S.A.* Sinclair Lewis' *The Job: An American Novel* (1917) is part of his feminist trilogy which includes his better-known novel *Main Street: The Story of Carol Kennicott* (1920).

21. Some went to night school instead. Dorothy Richardson (O'Neill 1972, orig. pub. 1905), in a popular autobiographical story of dubious veracity, wrote of her life as a lamentable, low paid factory girl who enrolled in night school and became a "prosperous" stenographer.

22. Carlton (1908, p. 133). Carlton also advised that: "Not many can become a skilled worker to-day who does not understand the scientific principles underlying his trade, who does not understand why certain methods are preferable to others, who is not able to act upon his own initiative in cases of emergency . . . a trade cannot be properly learned without a school . . . In the machine-building trades it is almost indispensable" (p. 236).

23. Also see the discussion in Chapter 3.

24. See Kocka (1980) for an insightful account of the etymology of the terms "blue-collar" and "white-collar" worker. The terms are an American invention, for only in America would the words be needed. Working class or manual workers in Germany and elsewhere in Europe were obvious by the grime on their hands and their trademark blue overalls. In the United States, on the other hand, manual workers often showered at work and changed out of their work clothes before going home. In egalitarian America, the manual worker could become whomever he wished. The terms "white-collar" and "blue-collar," therefore, had more meaning in the less class-oriented American society.

25. Both were Midwestern firms; Deere Tractor was in Moline/Davenport, Iowa, and National Cash Register was in Dayton, Ohio.

26. National Cash Register Company (1904) is a compilation of comments from the Mosely Industrial and Educational Commissions, which were brought to the United States in 1902 to observe American industry. Quotations are from pp. 28–29 and p. 34. The reference by a British observer was to "Deer Plough Works." National Cash Register was a pioneer in industrial welfare and was a progressive company that did not hire any worker below 17 years of age (p. 14). Its product market position—it had an exceptionally large share of its industry—may have given it the ability to be selective in its hiring and to engage in progressive employment policies.

27. Wagoner (1966, p. 86).

28. On mechanical engineers, see the insightful work by Calvert (1967, p. 70).

29. Elbaum (1989) discusses the decline in apprenticeship programs in the United States relative to those in Europe and Britain. His main thesis is that increased education in America led to the breakdown of formal apprenticeship programs. See also Douglas (1921).

30. General Electric Company (1924) and National Cash Register Company (1919). Although the information about General Electric comes from a 1924 booklet, their apprenticeship program was in existence from 1903.

31. Iowa Department of Public Instruction, *Biennial* (1905, p. 143); attributed to college president O. H. Longwell, *The Twentieth Century Farmer.*

32. On the diffusion of hybrid corn, see Ryan and Gross (1950) and Schultz (1964) on the role of education in agriculture more generally. Evans (1926) claims that across 769 farms in Tompkins County, NY, farmers with a high school education were making nearly twice as much as those with an eighth grade education, but it is not clear whether the size of the farm was held constant in the analysis.

33. Gabler (1988, p. 67).

34. At age 14 in 1845 Andrew Carnegie became a messenger in a telegraph office of the Pennsylvania Railroad. In eight years he became the private secretary and personal telegrapher of a superintendent of the railroad and a bit later was awarded the position of superintendent. Thomas Edison became an apprentice telegrapher

in 1863 at age 16 and worked for six years as an itinerant telegrapher. Some of Edison's earliest patents concern the telegraph.

35. Gabler (1988, p. 67).

36. David Sarnoff, founder of NBC, is an example of an extraordinary individual who was "discovered" because he was a superb telegrapher. As a very young man, Sarnoff was a radio operator for the Marconi Wireless Telegraph Company and soon became the operator of a powerful radio station from which, in 1912, he picked up the distress signal from the sinking Titanic. The Marconi Company rapidly promoted him. Sarnoff later demonstrated his genius at promoting new technology when he pioneered the use of radio to broadcast sports with the Dempsey-Carpentier fight in 1921.

37. See Goldin (1994) on the impact of immigration on the wages of urban unskilled workers and artisans.

38. For example, in 1920 3.5 percent of all manufacturing workers (both production and nonproduction) were British born, but 11.4 percent of jewelers were. For the data on separate nativities, the British and Italians are in the Northeast United States and Scandinavians are in the Midwest. Data are restricted to white males between 18 and 64 years of age. Our source is the 1920 IPUMS of the federal population census.

39. We use the IPUMS of the 1920 population census for these calculations.

40. See also the discussion in Chapter 3 of young men during the World War I years from Burdge (1921). The manual workers with more education were disproportionately in the metals industries, whereas those with less education were in wood, cloth, and leather.

41. The actual returns were probably higher since the lower educated and lesser skilled faced higher unemployment and nonemployment.

42. The original source used in Margo (2000, pp. 25–30) is an extensive group of payrolls, known as *Reports of Persons Hired*, of the civilian employees of Army forts, which were located in various states, some close to urban areas and some in more remote settings.

43. These ratios are based on the data on monthly earnings of white-collar workers (clerks) and daily wages of common laborers from Margo (2000, table 5B.4) for 1826–1830 and 1856–1860. We have multiplied the daily wages of common laborers by 26 to convert them into monthly earnings.

44. Average annual income for a male (ordinary) clerk in 1895 was $1,097 and was $1,099 in 1914 (see Goldin and Katz 1995, table 2). The deflator necessary to convert 1914 dollars to 1895 dollars is 0.8403 (*Historical Statistics, Millennial Edition*, table Cc1). Weekly earnings for the lowered skilled are $8.45 in 1895 and $10.78 in 1914. Multiplied by 52 weeks yields $439 for laborers in 1895 and $561 in 1914. Our sources are *Historical Statistics*, series D 778, for lower-skilled labor, and full-time weekly earnings from Coombs (1926), corrected for a transcription error in the 1895 figure. Coombs used series that did not include "laborers," but instead used the lowest paid workers in various industries.

45. Iowa Department of Public Instruction (1903, p. xv) noted that: "In Iowa several thousand [ungraded] rural schools follow a course of study as regularly and completely as do town schools. Pupils furnishing the course of study are granted [eighth grade] diplomas of graduation . . . and are admitted to the first year of high school," if there was a high school for them to attend. But the report went on to

mention that in places without a high school, two additional years of common school (beyond the usual eight) were often provided. In 1912 Iowa passed an education law mandating that the upper grades be taught only in schools with more than one teacher, even though the school report of that year noted: "A number of rural districts have set up a claim of furnishing high school facilities in the one-room schools" (Iowa Department of Public Instruction 1911/12, p. 23).

46. The total number of classical academies for the entire United States given in the 1870 social statistics is slightly more than 1,500 (U.S. Census Office 1872). These data appear fairly accurate. The number for New York State is not much different than that given in New York State Regents reports for about the same year (see New York State Regents 1869). We estimate the number of public high schools that existed in 1870 from the dates of their establishment as given in "Education Report, 1904" (U.S. Office of Education 1906, table 43). Not all high schools that existed during the 1903–1904 scholastic year gave an establishment date, and we employ the assumption that the establishment years of those without dates were distributed identically to those with dates. Of the 7,230 public high schools listed in the 1903–1904 report, about 7 percent were established before 1870. Some others, but probably not many, may have existed in 1870 but closed their doors before 1903. Note that it does not seem possible to use the public high school data in the 1870 social statistics since the number of reported high schools in some states (e.g., New York State) is far too low and the number in others (e.g., Ohio) seems too high. It would appear that some states included in their high school data elementary schools that taught older students and other states included them under the grammar school heading.

47. These are estimates based on the aggregate data from the U.S. Office or Bureau of Education and the authors' reworking of these data using disaggregated data.

48. The academies supported by the New York State Regents were incorporated and had boards of trustees. They were, in consequence, larger institutions. The funds granted by the Regents came from the Literary Fund, which was divided equally among the eight state senate districts and then on a per student basis within each district. Each academy received a subsidy only for students who took an approved classical course of study. Kandel (1930) notes that Illinois, Indiana, and Michigan also provided state support to academies.

49. In drawing up this list of courses we consulted New York State Regents (1841, 1869). Almost all the courses we list were offered as early as 1840. By 1868 several were added, including electricity, magnetism, mechanics, statics and dynamics, principles of teaching, domestic economy, psychology, drawing, mapping, and calisthenics.

50. See, for example, Cubberley (1934, orig. pub. 1919), Kandel (1930), Riordan (1990), and Sizer (1964a).

51. See U.S. Census Office (1853, 1864, 1872). The data were supposed to be collected at the county level but in some cases (e.g., New York State) the data were recorded at the township level, as is obvious from the census manuscripts.

52. The manuscripts for various states are available on microfilm from the National Archives. Manuscripts are not available for all states and in some cases are not available for all counties in a state. We thank Robert Margo for providing microfilms for Arkansas, Indiana, Iowa, Massachusetts, Michigan, New York, Pennsylvania,

and Texas, although not all years are included for each state. National Archives and Records Administration, Record Group 29, Records of the Bureau of the Census, Social Statistics, Seventh, Eighth, and Ninth Censuses. The films for New York are from the New York State Library.

53. Riordan (1990) cites a figure of more than 250,000 students in about 6,000 academies. Sizer (1964a) and Kandel (1930) use the 1850 Census data from an article in *American Journal of Education* by Henry Barnard. There is no indication that the published census reports were consulted, although Kandel correctly notes that "If these figures are accurate, one in every seventy of the white population attended an academy—a proportion not even yet realized in most European countries, and a little more than half of the proportion in high schools at the present time [1930]" (p. 418).

54. U.S. Census Office (1853); emphasis has been added.

55. The number of (white) 15 to 18 year olds in 1850 was approximately 1.703 million (*Historical Statistics*, series A 123, assuming an equal number by single year of age for 15 to 19 year olds). The most compelling evidence on the matter is that the ratio of public school (meaning elementary or common school) pupils to academy pupils was extremely low in the South. The ratio was just 5.5 in the South, but 31.5 in the Midwest. Furthermore, the manuscripts of the 1850 Census of social statistics reveal that many southern counties did not list a single common or elementary school but did record an academy and its students.

56. Some private secondary students may have been included in "day and boarding" schools, when these schools contained students in both the elementary and upper grades. The group will also exclude students in the large number of small and short-lived private schools that gave training primarily in commercial subjects and in the arts and music.

57. The figure is derived by assuming that academies and public high schools were four years and dividing by an estimate of the 15- to 18-year old population. Note that the estimate will not change much if a different set of ages is used. The figure would be larger if students in commercial and music and art schools were included. Although these types of schools were often included in the academy group, they can be reasonably omitted from the group of academic secondary schools. It should also be noted that students older than 14 years in rural areas often remained in common schools even though the educational gains from doing so were far less than had they been able to attend a high school. See Chapter 6.

58. *New York Daily Times*, Apr. 7, 1853 and Sept. 7, 1852.

59. *New York Daily Times*, Sept. 11, 1857.

60. Some public high schools in New York State and Pennsylvania charged tuition in 1870, which is odd. New York State passed its free schooling act in 1867 and Pennsylvania did in 1834 (see Chapter 4). In New York State schools in Brooklyn (Kings county) and Yonkers (Westchester county) listed tuition, as did those in Bedford, Jefferson, and Westmoreland counties in Pennsylvania. One possibility is that before the *Kalamazoo* (1874) decision (see Chapter 4) the legality of spending public funds on secondary schools was in question.

61. Extremely small and extremely large values for total revenue per student were eliminated. Medians are computed based on the mean per township or county and most (49/68) had between one and three academies.

62. Long (1960) gives $2.67 as the daily wage of machinists in 1870 and $1.52 for that of laborers. We multiply by six days a week and fifty weeks per year. Office

workers are assumed to have twice the earnings of laborers, as per the results for 1860 in Margo (2000).

63. These figures are for a slightly earlier period, c.1860, in New York State from advertisements in the *New York Daily Times*.

64. The transition from academy to high school can be traced in the New York State Board of Regents volumes. Towns that had academies in one year and public high schools in the next must have used the academy building.

65. The same was probably true for other states. As Kandel (1930) notes: "The story of secondary education in Illinois follows the same lines as in Indiana— private institutions, incorporated by the legislature, and aided from time to time out of state or local funds until the development of the public high school deprived them of the reasons for their existence" (p. 411).

66. "The public high school in the nineteenth century was mainly an urban invention. The Boston high school in 1821 was built to complement the Latin school. It was to give students mechanical and mercantile skills for those professions" (Tyack 1967, p. 354)

67. For a timeline of compulsory education laws see, for example, Steinhilber and Sokolowsi (1966).

68. Margo and Finegan (1996), using the 1900 IPUMS and exploiting information on birth month, find an impact on 14 year olds but only in states having both compulsory education and child labor laws.

69. See Goldin and Katz (2003). Some of the child labor laws, such as those that mandated continuation schools, were effective in keeping youths in school since they increased the costs of working.

70. The population is as of the 1910 U.S. Census.

71. Iowa Department of Public Instruction, *Biennial, 1912/13* (1914, p. 35), on the campaign for consolidated school districts.

72. Ueda (1987).

73. California Superintendent of Public Instruction (1910, p. 26).

74. Iowa Department of Public Instruction (1893, p. 25).

6. America's Graduation from High School

1. By secondary or high school we mean grades nine to twelve regardless of the type of school. In Appendix B we discuss how we treat enrollments in various types of schools, including junior high schools. To obtain the graduation rate, we divide the number of graduates by the number of 17-year-olds and to obtain the enrollment rate we divide the number of enrollees by the number of 14- to 17-year-olds. We use these ages because of the apparent under-enumeration of 18-year-old males.

2. Krug (1964) is an excellent source on the high school movement. See also Herbst (1996) and Trow (1961); Reese (1995) and Vinovskis (1985) on its origins; and Labaree (1988) and Ueda (1987) for insightful examples.

3. California, Office of Superintendent of Public Instruction (1908/10).

4. North Carolina State (1910). The report went on to note: "In 1898 there were only 14 public high schools in North Carolina that reported to the U.S. Commissioner of Education. . . . In 1908 there were 100 public high schools . . . and 37 private schools." That is, even though the high school movement was detected in North Carolina, it was very much in its infancy.

5. About 130,000 school districts existed nationwide in the 1920s, many of which were fiscally independent (Gordon 2000; 127,531 existed in 1932, the first year the Office of Education reported the data). But most were common school districts that did not have a resident student population that could justify a high school. The number of districts that could have maintained a public high school in the early 1920s would have been much smaller. In Iowa, for example, there were about 5,000 school districts in the 1920s, but only about 1,000 independent city, town, and village corporations that could have had a high school. The other 4,000 were school districts in the open country.

6. See Meyer, Tyack, Nagel, and Gordon (1979) on nation-building and the roles of collective ideology and economic world view.

7. Enrollment and graduation statistics by sex are available only for the public sector.

8. The New England states did not have as high a standing in terms of all defense contracts, but they were among the highest in contracts to light industry, such as textiles, clothing, and boots and shoes, which routinely hired youth. On total defense spending by states, see Miller (1947).

9. The data include those in two-, three-, or four-year public and private high schools, in the final year of junior high, and in the preparatory departments of colleges and universities. They do not generally include students attending common schools beyond eight years, although in some states they may. Students in the preparatory departments of colleges and universities have been omitted from all other series we know of despite the fact that they accounted for about one-third of all private school students in the 1910s.

10. The Office of Education, also known as the Bureau of Education, was the forerunner to today's Department of Education. It was established in 1867 as the Department of Education and became the Office of Education in 1869, an agency of the Department of the Interior, where it remained for 70 years. It was known as the Bureau of Education for those years, but was formally renamed the Office of Education in 1929. In 1939 it became part of the Federal Security Agency and was included in the new agency of Health, Education and Welfare (HEW) in 1953. The Department of Education became a separate cabinet-level agency in 1980.

11. The title was initially *Digest of Educational Statistics*.

12. See Chapter 9 for a discussion of state graduation requirements.

13. We use the number of 14- to 17-year-olds for the enrollment rate and the number of 17-year-olds for the graduation rate, but it would not matter if we used 15- to 18-year-olds and 18-year-olds or just about any other combination as long as it is four years for attendance and one year for graduation. This procedure is accurate as long as a substantial fraction of youth were not routinely left back or attended school without intending to graduate. We address this issue in the discussion of African American youth in the South.

14. The data for the period after 1970 are discussed in Chapter 9.

15. The census divisions and the (48) states (plus the District of Columbia) they include are: *New England:* Connecticut, Maine, Massachusetts, New Hampshire, Rhode Island, Vermont; *Middle Atlantic:* New Jersey, New York, Pennsylvania; *South Atlantic:* Alabama, Delaware, District of Columbia, Florida, Georgia, Maryland, North Carolina, South Carolina, Virginia; *East South Central:* Kentucky, Tennessee, West Virginia; *West South Central:* Arkansas, Louisiana, Mississippi, Oklahoma, Texas; *East North Central:* Illinois, Indiana, Michigan, Ohio, Wisconsin;

West North Central: Iowa, Kansas, Minnesota, Missouri, Nebraska, North Dakota, South Dakota; *Mountain:* Arizona, Colorado, Idaho, Montana, New Mexico, Nevada, Utah, Wyoming; *Pacific:* California, Oregon, Washington.

16. The U.S. unemployment rate was 23.6 percent in 1932 (*Historical Statistics,* series D 86).

17. Data are from *Historical Statistics,* series A 176. *De jure* segregated schools also existed in parts of other states, as in certain Kansas cities (including the well-known example of Topeka), but they were not legally mandated by the laws of the state.

18. Data on the enrollments in segregated high schools in the 17 states of the South were first published by the Office of Education in 1930 and were ended in 1954 after the Supreme Court's ruling in *Brown v. Board of Education,* even though *de jure* segregation was not ended until a series of court orders forced desegregation plans in various urban areas. See Appendix B for more detail.

19. In 1916, for example, about half of all African American secondary school students in the South attended private schools and half of the remaining youth attending schools in the public sector were educated in the secondary departments of normal schools and colleges. There were in 1916 three times more private than public high schools (216 compared with 64) and virtually all the public high schools were in the larger cities of the border states. See Anderson (1988) who, along with Caliver (1933b), discusses the Slater Fund, whose mission was to provide secondary education to rural blacks. Although the Slater Fund established 355 of these schools by 1933 throughout the South, they termed them "county training schools," rather than high schools, to avoid public scrutiny.

20. Hall (1973, p. 156). The schools established by the Slater Fund are not included in the total because they were not called high schools. See also Caliver (1933a) on secondary schools for blacks in the South.

21. The population per square mile in 1930 was 44.1 for Iowa and 49.7 for Georgia (*Historical Statistics,* series A 196). The fraction of males involved in agriculture was 43.3 percent for Iowa and 48.5 percent for Georgia (Lee, Miller, Brainerd, and Easterlin 1957, table L-4). The high school graduation rate for whites in the South is given for 1930 since the data on black high school enrollments begins then.

22. The seven states are Alabama, Arkansas, Georgia, Louisiana, Mississippi, North Carolina, and South Carolina.

23. In the aggregate, the tiny towns had 380,000 people.

24. The Iowa *Biennial Report* on public education lists all incorporated towns in three groups: cities, towns, and villages with more than 3,000 persons, with between 1,500 and 3,000, and with fewer than 1,500 (Iowa Department of Public Instruction 1914). We use the aggregate data on total and urban population from *Historical Statistics,* series A 195–209.

25. Becker and Murphy (1988) make a similar point. They go one step further and suggest that the intergenerational loan was paid back in the form of social security. We, on the other hand, are conceptualizing the intergenerational loan as being shifted within the community from one group of grandparents to the next.

26. See Epple and Romano (1996), who analyze the level of support, and Fernandez and Rogerson (1995), who investigate the existence of public education.

27. Alesina, Baqir, and Easterly (1999) show in a majority-voting model that an increase in the polarization of preferences concerning spending on public goods (formally an increase in the median distance from the median) reduces the amount

of public goods provision. They find, using a cross-section of U.S. cities c. 1990, a negative relationship between spending on "productive" public goods (schooling, roads, and libraries) and the city's degree of ethnic fragmentation.

28. In fact, the earnings of white-collar workers were far more similar across the United States in the 1909 to 1919 period than were the earnings of production workers. The coefficient of variation of city-level mean clerk wages is smaller than that for production workers in a sample of 227 non-southern cities in 1919. Similar patterns are apparent in 1909 and 1914. See Goldin and Katz (1995) for a description of the wage data, which come from the U.S. census of manufactures.

29. We have also divided non-Catholics into two other groups: non-hierarchical religions that encourage the reading of the Bible by the laity (e.g., Lutherans, New England Protestants) and non-hierarchical ones that do not (e.g., most evangelical religions). Only the percentage Catholic is of statistical and economic significance. Race is another important factor in U.S. educational history, but given the large percentage of blacks living in the South during the 1910 to 1940 period, there is little systematic relationship between percentage non-white and graduation rates once measures of income and wealth and a South dummy are included in the state-level regressions.

30. Portland, Oregon (1920, p. 26).

31. The estimates in Table 6.1, columns 1 to 4 are unweighted, but these results are not very sensitive to weighting by state population. Columns 5 and 6 are weighted by the population of 17-year-olds in each state because unweighted estimates of models to explain the change in graduation rates from 1928 to 1938 are greatly influenced by two extreme outliers (DE and NV). Thus we present the more robust, weighted estimates.

32. The fractions of the population that are urban, foreign-born, and Catholic are all strongly collinear, and each of these variables is also collinear with the fraction of workers employed in manufacturing. Similarly, per capita wealth, income, agricultural income, and automobile registrations are all collinear. We use a subset of each of these groups in the regressions.

33. Today's elderly can, and do, escape the higher taxation that comes with more and better quality education. In the period we are examining, the elderly generally did not, or could not, move from places with more expensive educational public goods. Grandparents who lived in towns and villages at the turn of the twentieth century often boarded their grandchildren who lived on farms to enable them to attend high school. This interpretation is consistent with the findings in Hoxby (1998) concerning the changing impact of the elderly on school expenditures across the twentieth century.

34. If the (log) per capita wealth were omitted from Table 6.1, column 2, the role of automobiles per capita would greatly increase. A shift from the state at the 25th percentile to that at the 75th percentile would increase the graduation rate by 8 percentage points (27 percent of the mean graduation rate) in 1928.

35. The strong positive impact of automobile registrations per capita on graduation rates is robust to the inclusion of controls for population density, percentage urban, and access to improved roads. Mroz, Rhode, and Strumpf (2006) also find a large and robust positive effect of automobiles per capita on public school spending per child at the county level for 1931–1932.

36. Lindert (1994, 1996), in two cross-country studies of the twentieth century, finds that greater equality fosters more social spending (e.g., transfer programs) and that a greater percentage of Catholics lowers it.

37. We have also estimated a state fixed-effects model for high school graduation rates (not shown) that pools data from 1910, 1920, and 1930. We find results similar to the levels regressions in Table 6.1, columns 1, 2, and 3. Auto registrations per capita and the percentage older than 64 years remain strongly and positively related to the state graduation rate. Percentage Catholic and the manufacturing employment share variables have coefficients similar to those in the cross-section regressions but are not precisely estimated due to the persistence of cross-state differences in these variables.

38. The only variable to change signs in the difference regression, compared with the levels regression, is that regarding the elderly.

39. Agricultural income (natural log of) per agricultural worker in 1920 is used here instead of the (log) wealth variable. The results are virtually unchanged if (log) wealth in 1922 is used.

40. The historic 1944 Education Act increased the age of compulsory education (the school leaving age) in England, Scotland, and Wales from 14 to 15 in 1947. The fraction of those leaving school at age 14 declined from 57 percent in 1945 to less than 10 percent in 1948 (Oreopoulos 2003). More important, perhaps, was that secondary education became fully funded after 1944, whereas it had previously been on a tuition basis and was inaccessible to those of modest means except for the small share awarded merit scholarships.

41. For a timeline of U.S. compulsory education laws, see, for example, Steinhilber and Sokolowsi (1966). Goldin and Katz (2003), from which much of this section derives, contains detailed information on the major aspects of compulsory schooling and child labor laws for all states from 1910 to 1939.

42. David Tyack (1974), in a widely cited volume on educational history, has stated: "Attendance in high schools increased [from 1890 to 1918]. . . . The curve of secondary school enrollment and graduation continued to soar: in 1920, 61.6 percent of those fourteen to seventeen were enrolled . . . in 1930, the [figure was] 73.1 percent. . . . As these statistics suggest, during the first two decades of the twentieth century compulsory schooling laws were increasingly effective" (p. 183). Many other well-regarded historians, such as Troen (1975), have also accorded compulsory schooling and child labor laws a large role in the increased school enrollment and attendance of teenaged youths during the Progressive Era.

43. Emmons (1926, p. 134) contains a summary of the required attendance each week in continuation schools. Of the 23 states having a mandatory continuation school law in 1925, eight required up to 8 hours per week, nine required 4 hours, and six were at the 5- or 6-hour level. See Hogan (1985) for a description of continuation schools in Chicago.

44. Further details about the construction of these variables can be found in Goldin and Katz (2003). We follow Acemoglu and Angrist (2000) in the construction of "child labor school years," but we lag the school entrance age law by eight years to better capture rules relevant for youths of secondary school age at *t* when they were of school entrance age (about eight years before). We also follow the approach in our measure of "compulsory school years."

45. We report robust standard errors clustered by state to account for serial correlation in the residuals. All regressions are weighted by the contemporaneous number of 14-year-olds in the state. The regression results are not sensitive to weighting. Our work is similar to that of Lleras-Muney (2002) and Schmidt (1996), but the explanatory variables differ.

46. But, as shown in Goldin and Katz (2003), the effect of "compulsory school years" on secondary school enrollment rates is eliminated by the inclusion of state trends and the effect of "child labor school years" is much weakened, although the impact of the continuation law variable remains about the same.

47. See Goldin and Katz (2003) for a more complete empirical analysis, which also demonstrates modest impacts of compulsory schooling and child labor laws on the educational attainment (years of completed schooling) for the affected birth cohorts (born from 1896 to 1925).

48. U.S. Commissioner of Education (1906) contains a listing of the approximately 7,200 public schools that reported to the Commissioner of Education with information on the numbers of secondary students by sex, whether they were in a college preparatory program, the number of teachers by sex, and the date of establishment, among other information. It should be noted that around one-quarter of these schools also contained elementary school students and thus they were not necessarily dedicated high schools. It is likely that the actual number of public schools giving high school instruction exceeded the number in this study by about 20 percent, although the undercount would have been the greatest for small schools. (On the reporting of schools to the U.S. Office of Education before the 1930s, see Appendix B, Table B.4, which shows that the undercount was about 15 percent even as late as the early 1920s.)

49. We have coded information for all high schools listed in five states (GA, IA, IN, MA, PA) in a 1903 federal survey (U.S. Commissioner of Education 1906). On average, 25 percent of all schools contained elementary school students. The fraction was highest in Georgia (45 percent) and Indiana (34 percent) and lowest in Iowa (9 percent).

50. In some states legislation defining what constituted a high school directly followed the passage of a "free tuition" act, which required a school district without a high school to pay tuition to a neighboring district. Standards for a high school ensured the paying district that a high school education was being provided. In other cases, the state university crafted such legislation with the state to ensure that high school graduates, who were guaranteed an education by the state, were properly prepared.

51. These data refer to full-time school attendance rates from the 1910 Census of Population and are reported in more detail in Goldin and Katz (1999b, table 2). Full-time school attendance means a youth attended school at some point after September 1 of the previous year and was not working (reported no gainful occupation in the census). Similar patterns by size of place are apparent in 1920 and remain even after including controls for race, ethnicity, parental background variables, and region.

52. Census division dummy variables are also included but are not shown in the table. We include the fraction Catholic in part to account for the private (parochial) schools that drew students away from the public schools since the dependent variable measures the attendance rate in *public* schools. Yet even when we use the total

attendance rate for 16- to 17-year-olds (available in the U.S. Population Census) as the dependent variable, which includes attendance at *both* public and private schools, we find a similar result (see Goldin and Katz 2005). In all estimations, the fraction Catholic in the population has a strong and negative effect on the schooling rate. Moreover, it is a more powerful variable than is the fraction foreign-born or the fraction native-born of foreign-born parentage.

53. We cannot distinguish between the number of teachers hired by existing schools and those recently employed by the new schools and thus we will include all change in the number of teachers per school as part of the "intensive" margin. The adoption of something known as the "platoon system," whereby students spent part of their day in large, monitored groups either outside or inside a study hall, could also have been used. The platoon system would have allowed a school to accommodate an increased number of students.

54. In 1910–1911, just 41 percent of public elementary school graduates in the Bronx entered a high school, whereas the figure was 71 percent in Queens, 56 percent for Manhattan, and 65 percent for Brooklyn.

55. In 1923 the 280 cities that contain enrollment information had 34 percent of the nation's public secondary school youths, and in 1933 they had 36 percent, correcting for the missing information for 28 cities using 1927 data. Cities with more than 25,000 persons accounted for 36 percent of the entire U.S. population in 1920 and 40 percent in 1940 (*Historical Statistics*, series A 57–72). Not surprisingly because of demographics and the relationship between school enrollment and city size, these larger cities had a smaller fraction of secondary school students than of the total population

56. Of the 28 cities that did not report in 1933 but reported in 1927, just 2 were outside the South, and of the 47 that failed to report in 1937 but reported in 1927, just 3 were not in the South. It is not clear why so many cities in the South did not return the Office of Education surveys in the 1930s.

57. See Tyack, Lowe, and Hansot (1990) on the role of the Depression.

58. Some of the students were in the preparatory departments of universities that were publicly funded. The term "private" here simply distinguishes these students from those in the public high schools. However, most of the students in the preparatory departments of colleges and universities were in privately controlled schools.

59. In 1880, for example, there were about 26,000 students in preparatory departments of which about 12,000 were training to go to college. At the same time there were a total of 116,000 students of all types in colleges and universities in the United States. The vast majority of preparatory students in 1880 were in the southern, central, and western states (*Annuals* 1880, p. cxxxi).

60. The number of youths enrolled in private schools decreased from 1930 to 1934, the only years when the absolute number declined.

61. The data we have compiled on secondary school enrollment and graduation rates (see Appendix B) give breakdowns by sex for public schools only. We have inflated the public rates by sex by the ratio of the private to the public graduation rates for both sexes. Boys may have had higher rates of enrollment and graduation in private schools and the preparatory departments of colleges and universities than did girls; if so, we will have somewhat biased the totals in favor of girls. But that is apparently not the case. We find a similar female advantage of 5 percentage points on

average in the high school graduation rate for the 1892 to 1911 birth cohorts (those who would finished high school from 1910 to 1929) using the data on adult educational attainment from the integrated public use micro-samples of the U.S. population censuses for 1940 to 1970. The census data on educational attainment include both public and private schools.

62. These data come from our secondary school graduation data by states (see Appendix B) and have been weighted by the number of youths in the population. The data are biennial except for 1910 to 1913 when they are annual, and thus the 1910s have one more observation compared with the 1920s.

63. Vermont Superintendent of Education (1900, p. 36).

64. Iowa Department of Public Instruction (1893, p. 25).

65. *Annuals* 1904.

66. Thorndike (1907, p. 246).

67. By chief, we mean that a plurality of the students and in many areas a majority of the students—as we will demonstrate—planned to continue with their education.

68. Ironically, the famed *Committee of Ten* report, issued in 1893 in the early dawn of the high school movement, advised an academic curriculum for U.S. secondary schools that would allow all youth to enter college. The report, commissioned by the National Education Association and chaired by Charles Eliot (president of Harvard), looked back in time, rather than forward. See Herbst (1996, chap. 9) and Sizer (1964b).

69. Smith and Ward (1984) also make this point using data on male educational attainment by birth cohort from the federal population census. We also find, using the census data on the adult educational attainment of U.S.-born males, that 52 percent of the male high school graduates at the beginning of the high school movement (those in the 1886–1890 birth cohort) reported having attended at least some college as compared with 43 percent of male high school graduates in the late high school movement period (those in the 1916–1920 birth cohort). It is only with the birth cohorts of the late 1930s and early 1940s that the share of male high school graduates eventually attending college reaches and then surpasses in subsequent cohorts the level achieved by the 1886–1890 birth cohort.

70. The data on the continuation of high school graduates have been the focus of various educational historians. Krug (1962) corrects errors made in the early 1900s that led some to believe that the vast majority—far above the majority—of high school graduates continued to college. That impression came from an incorrect comparison of the number of college students with the number of high school graduates. Some college students had never graduated from high school but were, instead, home tutored or went to the preparatory divisions of colleges. Krug's evidence is consistent with that given here.

71. Another check on the survey data could come from comparing the actual number of high school graduates with the number of first year students in the nation's college and universities. But first college degrees were often given in professional programs, such as the law and medicine, in this early period and these students were not separated from those attaining a further college degree.

72. Population data for cities are for 1910.

73. Ravitch (2000) argues that the high school curriculum was altered and diluted in the 1920s not because schools changed from institutions that educated

youths to attend college to those that awarded terminal degrees, but rather because the good intentions of progressive reformers went awry. The truth is that a larger fraction of high school graduates went to college in the 1910s than subsequently and that the fraction for the entire nation had been so high that many high schools were largely preparing students to attend college, even if some never went. Whether or not the progressive reformers failed their charge is another story. See also Angus and Mirel (1999) on the subject.

74. Historians of education will connect these changes to the shift of the National Education Association from the precepts of the *Committee of Ten* report in 1893 to the *Cardinal Principles* statement in 1918 (National Education Association 1918). But the 1918 report reflected what was already happening in secondary schools throughout the nation. Although it is viewed as an influential document, it reinforced trends toward the teaching of a variety of non-academic courses that were already taking place.

75. The usual distinction was that the Classical course included the study of Latin, whereas the English course substituted a modern language.

76. New York State academies that received state funds (and were therefore chartered by the state) had to report their curriculum. For one view of vocationalism in American history—what it has meant and why it has not succeeded—see Grubb and Lazerson (2004).

77. Washington State (1922), p. 294.

78. The conclusions in this and the following paragraphs are drawn from several sources. These include the high school curriculum for the Ottumwa, IA, high school for the years from 1900 to 1929 (Ottumwa, IA, various years); that for Davenport, IA, for 1917 to 1934 (Davenport, IA, various years,); and those for Iowa's small cities in 1924, cited below.

79. The Ottumwa data are from the *Annual Directory* for the city schools (see Ottumwa, IA, various years, 1900 to 1929). The 35 separate courses do not include separate math courses since they were not listed; science shifted from three to six separate subjects, history from one to three, and English from one to five.

80. The data are from Iowa (1925). The 20 towns and small cities were randomly selected.

81. An important example is provided by English literature and rhetoric, which were listed as separate subjects in 1915 but grouped together in 1922. It is likely that these courses were combined in some schools so that a student enrolled in one of the courses would have automatically been listed in the other. The sharp change from 1915 to 1922 and equal numbers for the subjects in 1915 suggest that explanation.

82. The curriculum survey was next taken in 1949.

83. The calculation assumes that all academic subjects met for the same number of hours per week. These data come from *Historical Statistics, Millennial Edition*, table Bc115–145, compiled by Claudia Goldin. We assume, for reasons given in the text, that the combined total for English literature and rhetoric in 1915 was 100, which means that every student took either subject or both combined. During the later period, 1922 to 1934, approximately equivalent data exist.

84. See Davis (1927, table 53) on courses taken by public high school students in 1925 in various states. A far larger fraction took English in grades 9 and 10 than in grades 11 and 12. A far higher fraction took beginning algebra and beginning geometry than the upper level math courses.

85. See Appendix B on the construction of the secondary school statistics. The basic sources are *Biennials* (various years).

86. The actual figure depends on how the courses are distributed across the grades. The total decrease was from 76.9 percent of all students to 56.2, or 21 percentage points. As a fraction of the initial amount, this is 27 percent. The lower bound estimate for the decrease within grades uses the assumption that all lower grade students take as much math as possible consistent with the total numbers. Within grades 9 and 10 the decrease would be 7 percentage points and the decrease for grades 11 and 12 would be 6 percentage points. The weighted average is a decrease of 6.6 percentage points or 8.6 percent relative to the initial 76.9 percent.

87. For detail on the various academic and non-academic subjects see *Historical Statistics, Millennial Edition*, table Bc115–145.

88. The Smith-Hughes Act, which federally funded vocational education, was passed in 1917. But the act did not contain any provision for the use of these funds for commercial education and therefore was probably not a causal factor in the enormous increase in commercial education in public high schools in the early 1920s. See McClure, Chrisman, and Mock (1985).

89. Students in commercial schools include both day and night students. The data on public and private commercial enrollment come from *Annuals* 1892–1893 (vol. 2, p. 2020); *Annuals* 1899–1900 (p. 2470); U.S. Office of Education (1920, table 1); Bolino (1973); Proffitt (1930, table 1); and Weiss (1978, tables 1–3).

90. Most junior high schools went from seventh to ninth grades. Of the 133 cities with junior high schools in 1923 (out of a possible 280 cities), 79 percent went from seventh to ninth grades and the rest varied from 1 to 4 years between sixth to tenth grades. Of the 195 cities with junior high schools in 1927 (out of a possible 287 cities), 88 percent included grades seven to nine. See Appendix C for sources.

91. See Table 6.4 for the number of public school teachers of the high school grades. The data on elementary school teachers are from the same source. The lower bound for high school teachers excludes junior high school teachers in 1930, whereas the upper bound includes them. (Junior high school teachers in the lower two grades are added to the elementary school teacher data.) These data are only for public school teachers, and since the public sector rose at the relative expense of the private sector, the total increase would have been smaller. Using estimates of the number of both public and private school students yields a figure of 7.6 percent average annually (*Historical Statistics*, series H 424 and H 429). If the teacher to pupil ratio did not change, then the number of public and private school teachers of the high school grades would have increased by that amount, or about the midpoint of the public school teacher data calculation.

92. This calculation uses the balanced panel of 215 cities (see Table 6.4 and Appendix C for details on the sources).

93. Data on secondary school teachers, their salaries, sex composition, and credentials do not exist for most states and municipalities in their biennial or annual reports. Although salary information is given, the data generally aggregate males and females. The data for Kansas and California are, therefore, unique. See Frydman (2001).

94. The fraction female decreased at the end of the 1920s in Kansas' larger cities and in the villages, for which the World War I data are unavailable.

95. See Goldin (1990, 1991) on marriage bars.

96. Two of the cities have missing data for 1922, but on average they were no different from the mean in 1920, when the aggregate figure was 70 percent. The fraction of high school teachers with a college or university degree increased to a far greater degree in the rest of Kansas. Although just 49 percent had graduated from a four-year college or university in 1912, more than 60 percent did in 1920. Most of the remainder had graduated from a "normal" school, generally a two-year institution.

97. The data set is from Oregon (1923) and contains 1,616 teachers, principals, and superintendents in Oregon high schools.

98. We infer graduation from college from the listing of a college and a year of graduation.

99. The four-year colleges and universities of Kansas in 1922 graduated around 475 women and 630 men. In 1930 they graduated around 1,000 women and 1,100 men. The number of high school teachers in Kansas increased by about 1,000 during the years 1922 to 1930 and the fraction with four-year degrees increased from around 60 percent to probably around 80 percent. Even though there was rapidly increasing demand for teachers with a college degree, the supply appears to have been increasing rapidly as well.

7. Mass Higher Education in the Twentieth Century

1. As the *New York Times* reported on December 11, 1944, this was true even among veterans who had not graduated from high school: "Very few veterans want to return to high school. They feel that their experience, even if they didn't finish high school, makes college the next step."

2. Initially during debate on the bill, the education section was not as central as it would become. The take-up rate for education benefits was underestimated and the potential for unemployment in the postwar domestic economy was overestimated. The first section of the act is a guarantee of 52 weeks of unemployment benefits at $20/week for returning veterans. Roosevelt signed the bill into law on June 22, 1944, just two weeks after D-day. The bill had been in the planning stages ever since 1942.

3. See Stanley (2003) for more of the compensatory view and Bound and Turner (2002) for more of the greater impact of the bill.

4. The fraction of young Americans with some college was high enough by the end of the twentieth century to be considered a mass movement, but the fraction who obtained a four-year college degree was far lower. Although the fraction soared for females, the fraction of males with a college degree has not expanded much from cohorts born in the late 1940s to those born in the mid-1970s (see Figure 7.1).

5. For a fuller discussion of the reversal of the college gender gap, see Goldin, Katz, and Kuziemko (2006).

6. The figure is exclusive of specialized and independent professional and religious schools. There were about 2,000 four-year schools including the specialized and independent group. The fraction of students enrolled in publicly controlled institutions vastly exceeded that for the privately controlled, as we will demonstrate in a later section of this chapter.

7. Data are from WebCASPAR [Computer Aided Science Policy Analysis and Research Database System] of the NSF [National Science Foundation] and NCES [National Center for Education Statistics], which uses the IPEDS [Integrated Postsecondary Education Data System] data.

8. A comprehensive list of undergraduate degree granting institutions in England was compiled from the Higher Education and Research Opportunities in the United Kingdom (HERO) website, http://www.hero.ac.uk/uk/home/index.cfm. The websites of the individual institutions gave the establishment year and the year the institution was granted undergraduate degree granting status, which is the one used here. Institutions in Germany were similarly compiled using a comprehensive listing and individual websites: http://www.mit.edu:8001/people/cdemello/geog .html. The population adjustment is for 15- to 19-year-olds in or around the relevant dates. The figure for Germany includes both East and West in 1950. The U.S. figures used in this example for four-year colleges and universities are 550 in 1900, 1,150 in 1950, and 1,400 in 2005. The 1900 and 1950 figures are conservative estimates and exclude independent professional schools and theological institutes.

9. These data, and others in this section, use colleges and universities recorded in the 1922–24 and 1932–34 *Biennials* and various college guides, such as the College Blue Book (1933). The *Biennial* for 1938–40 was the last to present data on separate institutions. The institutions used here exclude independent professional and religious colleges. See Goldin and Katz (1999a) for details. There were 26 religions listed as establishing at least one institution of higher education. By religious groups we mean an established religion, (e.g., Lutheran), rather than a subset the group (e.g., Norwegian Lutheran).

10. The funds to establish the University of Georgia came in the form of land script, which set the stage for the mechanism used to establish the federal land grant institutions.

11. The Ohio Company was required, by Congress, to set aside two townships for the establishment of a university.

12. The institutions for 1860 and 1900 are those existing by the 1990s. We find similar results using a data set of institutions existing by around 1930. Small private institutions failed in far greater numbers than did public institutions and larger private institutions. It is probably the case, therefore, that the fraction private was somewhat larger in the past than these data sets reveal.

13. The graph begins with the 1785 to 1790 five-year interval for ease of viewing since only 28 institutions had establishment dates from 1636 to 1784. We have excluded all independent specialized schools, such as military, medical, legal, and religious schools, from the 1992 list.

14. The peak establishment dates are also the same for all institutions in existence in 1934, rather than in 1992.

15. The rankings are from the *US News and World Report* website, "America's Best Colleges, 2006," and weigh undergraduate education indicators heavily. Brandeis is a special case. It was established, in large measure, because Jewish academics and students had long been discriminated against, because large numbers of Jewish scholars took refuge in the United States during the war, and because the Jewish community had amassed funds to found a great university.

16. The three non-U.S. institutions in the top 20 are Cambridge, Oxford, and Tokyo. These statements rely on rankings produced at Shanghai Jiao Tong Uni-

versity for 2005 and cited by *The Economist*, http://ed.sjtu.edu.cn/rank/2005/ARWU2005_Top100.htm. The rankings are based on a weighted average of research output, major research awards, and academic citations (with some portion scaled by the size of the institution), and have been criticized for placing disproportionate emphasis on the sciences. The *Times Higher Education Supplement* (owned by TSL Education, London) produces a similar list and awards the United States 12 of the top 20 spots, and Britain and Australia a combined total of 5 spots.

17. The data on science and engineering are from a 2003 NSF-NIH Survey of Graduate Students; all others are from WebCASPAR, which uses the IPEDS data.

18. Nobel Prize fields used are Chemistry, Physics, and Medicine and are limited to winners born in the United States. If we also include those who immigrated as young children, the pre-1936 figure becomes 41 percent and the post-1935 figure is 11 percent. Many of those who received their Ph.D. after 1955 and studied in Europe were funded by the U.S. government under the auspices of the National Science Foundation or through the Ford Foundation with U.S. funds. A substantial fraction studied at the Niels Bohr Institute in Copenhagen, Denmark. For Nobel Prize information and the biographies searched, see http://nobelprize.org/.

19. See Kane and Rouse (1999) on the history of community colleges and their role in the U.S. higher education system.

20. We use an estimate of 550 for the number of four-year institutions in 1900, 780 for 1933, and 1,400 for 2000. Institutions of higher education exclude independent theological and professional schools and schools that were highly specialized in other dimensions. The number of undergraduates, graduates, and professional students is from our data sets for 1897, 1933, and the early 1990s.

21. In the data given here we exclude all independent teacher-training institutions and two-year colleges, as well as students in the preparatory departments of higher educational institutions. See Goldin and Katz (1999a), data appendix for further information.

22. We have collected data from U.S. Office of Education, *Annuals* and *Biennials* for various years (1897, 1923, and 1933) when rich data at the institutional level exist.

23. We do not have good measures of total research funds at either the institutional or state levels, but we do have expenditures for "organized research separately budgeted" for public and private institutions (not including teachers colleges, normal schools, and junior colleges). In 1933, 2.4 percent of all educational and general expenditures for privately controlled colleges went for "research," defined in this manner, and 9.3 percent did in the publicly controlled sector. Interestingly, the highest private percentage by state was New Jersey, because of a state-supported but privately controlled institution—Rutgers University. Source: *Biennial, 1932–34*, table 22.

24. The early 1990s data (a five-year average) are from WebCASPAR from which were excluded junior colleges, normal schools, independent teaching colleges, independent professional schools, and independent theological institutes. Students include all those enrolled. The reason that the relative number of students in public versus private institutions of higher education has remained fairly constant is that the number of public institutions, relative to private ones, has increased.

25. Institutions are ranked by the number of bachelor's degrees granted in 2001, excluding the University of Phoenix (an institution catering to working adults with a substantial on-line program). The sole private institution in the top 25 is Brigham Young University in Utah. Data are from WebCASPAR.

26. See the discussion in Veysey (1965).

27. We use all students, other than those in preparatory departments, because many individuals in professional schools were obtaining their first post-secondary degree in the early period.

28. Starr (1982) contains an excellent discussion.

29. Only in the case of medical schools is the causation unambiguous. The Flexner Report caused independent medical schools to be merged with existing universities or led to the creation of new medical schools by universities.

30. On professional training, see Abbott (1988). As late as 1934 only 10.7 percent of the 122 law schools listed in *The College Blue Book* (1933), a contemporary guide to colleges, required four years of college; 67.2 percent required one semester to two years of college, and 9.0 percent required just a high school diploma.

31. See, for example, the discussion in Kevles (1979).

32. The data come from Kaplan and Casey (1958, table 6).

33. These ideas rely heavily on Rossiter (1979) and Ross (1979).

34. Our sample consists of all national learned societies existing in the United States in about 1980, when Kiger (1982) wrote his last volume on the subject, and those that are current members of the American Council of Learned Societies.

35. Hofstadter and Hardy (1952, p. 31) write that "by 1910 the American university as an institution had taken shape," and Veysey (1965) discusses how various factors, such as the rise of the research university and the increase in vocational subjects, had become accepted facts of higher education by 1910. All changed as the scientific method, practically oriented courses, the "lecture method" of teaching (Handlin and Handlin 1970), and specialization in a host of dimensions swept the world of knowledge (Bates 1965; Kimball 1992; Oleson and Voss 1979).

36. O'Neill (1971, table 3) presents enrollments (in credit hours) in the public and private sectors from 1930 to 1968 and finds little upward trend until around 1952. However, her data include teacher-training institutions, and once that adjustment is made, are consistent with our data.

37. There were 275 teachers colleges offering undergraduate degrees in 1941, a historic peak. But their numbers slid rapidly to 218 in 1949, the last year teachers colleges were enumerated separately and just as many of the public institutions were being recreated as state colleges and universities. For these data and those on two-year institutions, see *Historical Statistics, Millennial Edition*, table Bc510–522.

38. Of the nine states in the New England and Middle Atlantic regions, only Vermont and Pennsylvania had a state college in 1860. On the establishment of state universities, see Brubacher and Rudy (1958).

39. Even some of the nineteenth-century academies established in New York State received state funds to train common school teachers.

40. Rosenberg and Nelson (1996) discuss the role of state and private universities in producing "local public goods" before the 1940s and the shift to defense and health-related work later. Another reason offered for state funding of university research, as in Jaffe (1989), is that it has localized positive spillovers that increase eco-

nomic growth and industrial development in the state, even if the research does not directly relate to the current industries.

41. The fraction of engineers working for the government in 1940 would be even greater if the figure included those working in the private sector under government contract. Note that 1940 was before the large increase in defense spending.

42. Of the 28 state institutions founded before the Civil War broke out, 19 were in the South (including the border states).

43. *Historical Statistics*, series Y 684–685, F 1.

44. The material that follows is drawn from Goldin and Katz (1999a), the longer version of that paper as given in the bibliography, and Goldin and Katz (2001b).

45. We use the standard BLS consumer price index (CPI) deflator as given in *Historical Statistics, Millennial Edition* (table Cc1) and the BLS website.

46. See Goldin and Katz (1999a NBER longer version, figure 7).

47. The date we use here is 1994.

48. The state of residence is generally that of their parents. The percentage of students attending college who enrolled in their state of residence was 76.4 percent in 1897, 75.6 percent in 1923, and 80.3 percent in 1931. Hoxby (1997) uses these data to track the evolution of a national market in education.

49. See Goldin and Katz (1999a, table 3).

50. The positive impact of later statehood may originate, in part, in the greater generosity of the federal government over time in granting lands to new states for institutions of higher education. But the relationship holds even if we exclude the original thirteen states plus Maine and Vermont. See Quigley and Rubinfeld (1993) for a discussion of the relationship between state higher education more recently and the year of statehood.

51. The data are from Zook (1926) and include, by sex, all residents of each state who were attending college or university anywhere in the United States. Colleges exclude junior colleges and teacher training institutions, but include professional and technical institutes.

52. *Digest of Education Statistics 2005*, table 328. Total revenue excludes that from hospitals and auxiliary services.

53. We include required student fees in the tuition measure.

54. Public-sector tuition generally refers to the flagship universities, but some of the data include all state universities. When the data sets overlap, as from 1964 to 2004, the numbers are virtually identical for the flagships and all the state universities combined averaged by state. Private-sector tuition is for universities and is "sticker price" rather than actual cost.

55. For the details of the California Master Plan and the division of students among the various schools before the plan was adopted, see California Liaison Committee (1960).

56. State and institutional data from WebCASPAR. These data contain Fall enrollments for all students in institutions of higher education based on the highest degree given by the institution and whether it is a four-year or two-year institution.

57. *American Men of Science* included individuals each of whom was a member of a scientific society, including the National Academy of Sciences and those of a more specialized nature, and who responded to a survey; see Cattell (1938, 1960). Individuals were selected because of their contributions to the advancement of

pure science. We sampled all the listings of individuals whose last names began with the letter "b" and categorized them on the basis of their current employer. There were 1,501 usable listings in 1938 and 2,541 in 1960.

58. The ratings of graduate programs from the 1930s to the 1960s reveal similar findings. See, for example, American Council on Education (1934) and Cartter (1966). We have examined the ratings for the larger programs in the sciences and engineering. Among the top departments there is some movement toward the public sector.

59. The 16 largest fields for the 1928/32 and 1969 comparison are botany, chemistry, classics, economics, engineering, English, geology, German, history, mathematics, philosophy, physics, political science, psychology, sociology, and zoology. They were 72 percent of all Ph.D.s in 1928/32 and 53 percent in 1969 (82 and 65 percent, respectively, without education doctorates). The fields for the 1969 and 1993 comparison are biochemistry (including molecular biology), chemistry, economics, engineering (chemical, civil, electrical, mechanical), English, history, mathematics, music, philosophy, physics, political science, psychology, and sociology. They were each about 50 percent of all Ph.D.s produced in either 1969 or 1993 and were 60 percent excluding education doctorates (*Digest of Education Statistics 1970* and *1997*). Departmental rankings are from American Council on Education (1934); Roose and Anderson (1970); and Goldberger, Maher, and Flattau (1995). Disciplinary rankings by department for 1928/32 were grouped and each department in the grouping was given the mean ranking of the group.

60. Enrollment data by type of institution were downloaded from WebCASPAR.

61. See, for example, Geiger (1990, p. 2), from which most of this paragraph derives.

8. The Race between Education and Technology

1. The phrase "the best poor man's country" was initially used in the eighteenth century to describe economic conditions in Pennsylvania but was later used to describe the entire northern part of America. See Lemon (1972, p. 229, fn. 1), who took the title of his book on the early history of southeastern Pennsylvania, *The Best Poor Man's Country*, from several contemporary comments about the region. The ideas are similar to those in Tocqueville's *Democracy in America* (1981, orig. pub. 1832).

2. In his two-volume treatise, *The American Commonwealth* (1889), Bryce often commented on Tocqueville's observations.

3. For example, Bryce considered neither slavery nor the urban poor.

4. On the trend in the wealth distribution from 1776 to the 1920s (1776, 1850, 1860, 1870, and 1920s), see Wolff (1995) and the compilation of wealth data in Nasar (1992). Piketty and Saez (2003, 2006) contain data on the incomes of the top 1 percent of the distribution from 1913, the beginning of the U.S. income tax, to the early 2000s and for the top 10 percent from 1916.

5. Douglas (1930, p. 367, italics added).

6. Douglas (1926, p. 719). Paul Douglas was born in 1892 and would have been in his mid-twenties just as the returns to various skills began to be reduced and the wage distribution started to narrow. He was 34 years old when he wrote about the non-competing groups that had previously existed.

7. Our estimates of the decrease in the pecuniary returns to a year of education are robust to the level of schooling as well as to the age and sex of the individuals.

8. Long and Ferrie (2007) find that intergenerational mobility in the United States was higher than that in Britain in the nineteenth and early twentieth centuries but that it is fairly similar in more recent decades.

9. Growth is given by productivity trends using output per hour in the non-farm business sector from the U.S. Bureau of Labor Statistics (series PRS85006093 from http://www.bls.gov/lpc/home.htm).

10. The black-white schooling completion gap narrowed from 3.84 years for those born in 1885 (25 years old in 1910) to 1.35 years for those born in 1945 (25 years old in 1970), based on tabulations from the 1940 and 1970 IPUMS. The cross-state standard deviation of mean years of schooling narrowed from 1.60 years for those born in 1885 to 0.62 years for those born in 1945. On the evolution of racial and regional differences in school resources, see Card and Krueger (1992a, 1992b) and Margo (1990); on regional income convergence, see Barro and Sala-i-Martin (1991); and on racial income convergence, see Donohue and Heckman (1991).

11. Lemieux (2006a) finds that 60 percent of the increase in overall wage inequality (using the variance of log wages) from 1973 to 2005 is accounted for by the expansion in educational wage differentials, especially the rise in the return to post-secondary schooling.

12. We use the term "dropout" for individuals who did not graduate from high school even though some individuals, early in our period, did not drop out since there was no four-year high school in their locale.

13. Differential effects of changes in the prices or quantities of other production inputs (e.g., capital and energy) on the demands for different types of labor are subsumed into λ_t and θ_t. The total factor productivity parameter A_t implicitly includes technological progress and physical capital accumulation.

14. Heckman, Lochner and Taber (1998) find that relaxing the assumption of predetermined relative skill supplies and using youth cohort size and military requirements to instrument for relative skill supplies yield estimates similar to those from ordinary least squares for the aggregate elasticity of substitution between college and non-college workers for U.S. national time series data. Ciccone and Peri (2005) instrument for relative skill supplies in state-level panel data for 1950 to 1990 using measures of state compulsory schooling and child labor laws. Their instrumental variable estimates of σ_{SU} cluster around 1.5, almost identical to our implied estimates of σ_{SU} given in Table 8.2.

15. We follow two complementary approaches to measuring skill supplies in efficiency units. The first approach, following Autor, Katz, and Krueger (1998), starts with information on the total wage bill in each skill group and on composition-adjusted prices (wages) based on our estimated educational wage differentials. The wage bill (prices×quantities) for each skill group is then adjusted for changes in wages (prices) to get a pure composition-adjusted quantity (supply) measure. The details of the approach are given in the notes to Table 8.1. The second approach, following Katz and Murphy (1992), starts with hours worked and wages for detailed age-sex-education groups. Efficiency units are then computed by weighting the hours worked of each age-sex-education group by the relative wage of that age-sex-education group in a base period. The first approach can be

thought of as a chain-weighted price index that adjusts raw labor inputs into efficiency units, whereas the second approach uses a fixed-weighted price index. For comparability with the existing literature, we use the wage bill–based approach for measuring broad long-run supply and demand shifts by skill groups in Tables 8.1 and 8.3. We also follow the literature in using the fixed-weighted approach for time series regressions to explain the evolution of educational wage differentials (Tables 8.2 and 8.4) and for the decomposition of skill supply shifts into different components (e.g., U.S.-born vs. immigrant labor inputs in Tables 8.5 and 8.6). We have checked the robustness of the findings in each case by using the alternative approach to measuring skill supplies. The answers are similar in all cases using both approaches.

16. The wage and skill supply data are actually for the years 1914, 1939, 1949, and 1959, but for simplicity of presentation we will refer to these dates as 1915, 1940, 1950, and 1960, which are the years of the censuses (state and federal) from which these data were collected. See Acemoglu (2002) for a related time series analysis of the college wage premium and the relative supply of college skills using data for 1939 to 1996 (1939, 1949, 1959, and 1963 to 1996).

17. Our empirical specification and measurement choices follow Katz and Murphy (1992) and Autor, Katz, and Kearney (2005a, 2007). The empirical findings are similar for alternative measures of the skilled-unskilled wage premium, such as a fixed-weighted average of wages of all workers with some college or more to all workers with no college. The basic results are also robust to the use of different relative supply measures (such as workers with any college versus those with no college) and to adding controls for cyclical factors (such as the unemployment rate).

18. See also Autor, Katz, and Krueger (1998) and Autor, Katz, and Kearney (2005a, 2007).

19. See Goldin and Margo (1992) for a detailed analysis of these factors in the 1940s wage compression.

20. On union wage developments in the 1970s and early 1980s see Mitchell (1980, 1985). On the role of institutions in the growth of wage inequality in the 1980s see DiNardo, Fortin, and Lemieux (1996).

21. Autor, Katz, and Kearney (2006, 2007) discuss the "polarization" of the U.S. labor market since 1990, by which they mean that the two ends of the distribution are doing better than the middle. The top is doing well, the middle is doing poorly, and the bottom is doing fairly well. Their explanation is that demand is soaring for those who have analytical and people skills and is strong, as well, for those who have lower-skilled jobs in the service sector. Computers substitute for routine manual and cognitive tasks, thus reducing demand for many high-end jobs taken by high school graduates and low-end jobs taken by those with any college. But new information technologies complement the non-routine analytic and interactive tasks of those with post-college training and have relatively little impact on non-routine manual tasks of many lower-skilled service sector jobs. The growth of international outsourcing (also known as offshoring) appears to have had similar impacts on labor demand. See also Autor, Levy, and Murnane (2003) and Levy and Murnane (2004).

22. See Card (2001) on the high returns to college for marginal college enrollees whose college attendance decisions are impacted by changes in public tuition and geographic access to college.

23. The 1970s contain similarities to the 1940s, as we noted in the text, in the overshooting of the reduction in the college wage premium due to institutional factors. Thus the 1950s and the 1980s contain increases in the college wage premium that overshot market forces because of the erosion of the institutional factors that had protected lower- and middle-wage workers in the 1940s and 1970s.

24. We use the entire 1940 to 1960 period rather than the two subdecades for the reasons provided in the text. The college wage premium in the 1940s, it would appear, decreased more than justified by fundamentals and the increase in the 1950s brought it back to its equilibrium value.

25. The rapid relative demand growth we estimate in Table 8.1 for college workers from 1980 to 1990 may have been due to the computer revolution but may also be an overshooting due to institutional factors (declines in both union strength and the real minimum wage).

26. We use the wage differential between those with exactly a high school degree (12 years of schooling) and those with eight years of schooling. Those margins are the most relevant ones for measuring the full returns to high school in the first half of the century, because the majority of workers had eight or fewer years of schooling in 1915. In contrast, almost no U.S.-born workers today have fewer than nine years of schooling (under 1 percent in 2005) and the more meaningful margin is between those with a high school degree and high school dropouts (those with 9 to 11 years of schooling). Empirically, the distinction does not matter much for the time series path of the high school wage premium or for our analytic conclusions. These two measures of the high school wage premium are compared in Appendix Table D.1.

27. The large and significant coefficient on the interaction of the high school relative supply and the post-1949 dummy variable should be contrasted with that for the college wage premium analysis for which there is virtually no impact of adding a similar term (Table 8.2, col. 5).

28. The specifications in Table 8.4 that do not allow for a break in the elasticity of substitution in 1949 (cols. 1, 2, and 3) produce the implausible result that there was essentially no trend increase in the demand for high school graduates relative to dropouts during the pre-1950 period.

29. The decrease in the wage premium from 1940 to 1950 was even larger than from 1915 to 1940, but institutional factors of the 1940s make analyzing the longer 1940 to 1960 period more sensible.

30. The 21 percent figure is an average from the 1910 and 1920 U.S. population censuses.

31. On immigration restriction in the early twentieth century, see Goldin (1994).

32. These estimates are based on tabulations from the 2005 CPS MORG sample for those aged 18 to 65 years in the civilian workforce.

33. Immigrants made up 15.6 percent of employment in 1915 Iowa (Table 8.5) but were 21 percent of employment for the entire United States. The data on educational attainment in 1940 of older immigrant birth cohorts (those who arrived by 1915) and the U.S.-born in the same cohorts confirms that the contribution of immigration to skill supply gaps for the United States in 1915 is well approximated by our direct estimates for Iowa.

34. The derivation of the "immigrant contribution" is provided in the notes to Table 8.5.

35. Our implicit assumption that immigrants and the native-born are perfect substitutes within education groups may slightly overstate the impact of immigration on the wages of the U.S.-born. Estimates of the wage impacts of immigration also tend to be smaller in local labor market analyses than using our approach of looking at skill supplies at the national level. See Borjas (2003); Borjas, Freeman, and Katz (1997); Card (2005); and Ottaviano and Peri (2006) on alternative approaches and estimates of the impact of immigration on recent U.S. labor market outcomes.

36. The slowdown in the U.S. high school graduation rate will be discussed in Chapter 9.

37. We use a methodology analogous to that described in Tables 8.5 and 8.6 for decomposing overall relative skill supply growth into immigrant and native-born components.

38. We use the following four groups to measure the white-collar wage premium with the 1910–30 change in the log wage premium and the weight for each group given in parentheses: male clerks (−0.379, 0.3), female clerks (−0.229, 0.2), associate professors (−0.247, 0.25), and starting engineers (−0.143, 0.25). The rationale for the weights is that white-collar work was about 50 percent clerical at the time and males were about 60 percent of clerical workers. See Goldin and Katz (1995, tables 1 and 10).

39. Our assumption about initial conditions—that high school graduates were 4 percent of the workforce in 1890—is a compromise between historical administrative data on the high school graduation rate (which increased from 2 percent in 1870 to 3.5 percent in 1890) and the higher estimate for 1890 of a 6.3 percent share of high school graduates implied by "backcasting" household survey data from the 1915 Iowa Census and 1940 IPUMS. Modest changes in the assumed 1890 initial conditions do not substantially change our conclusions.

40. See Goldin and Katz (1995, table 8) for further details on the methodology and a discussion of how accounting for immigration and differential mortality by education could affect our series on the high school graduate share of the labor force for 1890 to 1940 derived from administrative data.

41. An alternative approach is to use data on educational attainment by birth cohort from the 1915 Iowa State Census and the 1940 Census. Estimates of the high school graduate share of the labor force using this method are shown in column 2 of Table 8.7.

42. One reason to prefer the administrative data is that high school graduation rates probably advanced faster in Iowa than in the rest of the United States in the late nineteenth and early twentieth centuries. The census-based estimates of the high school graduate share in column 2 of Table 8.7 are much higher than the administrative-based estimates in every year from 1890 to 1930. See Goldin (1998) on the overstatement of high school graduation rates of older cohorts in the 1940 census.

43. It will be recalled that the inverse of the elasticity of substitution, $-1/\sigma_{SU}$, is $\partial \log(w_S/w_U)/\partial \log(S/U)$, the slope of the inverse relative demand curve.

44. If we assume, instead, an elasticity of substitution of 2 (our preferred estimate between high school graduates and dropouts in the early period), then we conclude that demand grew at 1.9 percent annually (38.5 log points) from 1910 to 1930, which also implies some demand acceleration after 1910.

45. Our census-based estimates of the labor force share of high school graduates (col. 2 of Table 8.7) imply that immigration accounts for a 0.9 percentage point increase in the high school graduate share of the labor force from 1910 to 1930 as compared with a 10.1 percentage point contribution from the U.S.-born.

46. More precisely, the growth in the relative supply of high school graduates increased by 58.4 log points from 31.5 log points for 1890–1910 to 89.9 log points for 1910–30. The rising high school graduation rate of the U.S.-born accounts for 53.7 log points of the acceleration and declining immigration explains the remaining 4.7 log points.

9. How America Once Led and Can Win the Race for Tomorrow

1. The 26 nations also include some OECD partner countries and were chosen because of available administrative data. See OECD (2006, table A2.1).

2. The OECD estimates of the U.S. graduation rate using the administrative records is 75 percent whereas the completion rate using 25- to 34-year-olds is 87 percent in 2004. The main difference is that some high school dropouts in the United States later earn a high school equivalency certificate (known as the GED).

3. These estimates cover the civilian, non-institutional population and are based on tabulations from the 1980 to 2003 October CPS provided to us by Judith Scott-Clayton (personal communication). The share of new high school graduates continuing on to college in the fall following graduation has also increased from 49 percent in 1980 to 64 percent in 2003 (*Digest of Education Statistics* 2005, table 181).

4. Turner (2004) provides a comprehensive analysis of recent U.S. trends in the persistence of students in college and time-to-degree.

5. See OECD (2006, table A3), includes four-year colleges or tertiary type-A programs.

6. See OECD (2006, table A1.3a), includes both OECD members and partners.

7. See OECD (2006, table A1.5).

8. See U.S. Department of Education (1998). Even the TIMSS did not completely equalize the age and educational attainment of students. For example, even though all students were in their final year of secondary school, that was grade 13 or 14 for some and grade 12 for others, such as U.S. students.

9. See U.S. Department of Education (2004). Out of 29 OECD countries, the U.S. ranked 24th in both mathematics literacy and problem solving for 15-year-olds in the 2003 PISA.

10. Mishel, Bernstein, and Allegretto (2005, table 7.10) find that the United States had the largest increases in wage inequality (90–10 wage gap) for both men and women among OECD nations from 1979 to 2000.

11. Card and Lemieux (2001), Gottschalk and Joyce (1998), and Katz, Loveman, and Blanchflower (1995) examine the role of cross-country differences in skill supply growth for changes in educational wage differentials. See Abraham and Houseman (1995) and Dustmann, Ludsteck, and Schönberg (2007) on Germany.

12. See Acemoglu (2003), Blau and Kahn (2002), Freeman and Katz (1994), and Kahn (2000) on the interplay of market forces and institutional factors in accounting for cross-country differences in the evolution of wage inequality.

13. Controversy exists concerning the accuracy of conventional measures of the high school graduation rate based on administrative data on diplomas. For example,

Mishel and Roy (2006) find that estimates based on individual-level longitudinal samples linked to transcript records, such as the National Education Longitudinal Study (NELS), show somewhat higher high school graduation rates than do the administrative data estimates for youth cohorts expected to graduate in the early 1990s. As we will discuss below, household surveys such as decennial censuses and CPS show somewhat larger increases since 1970 in the share of young adults with at least a high school degree than do administrative data on high school degrees.

14. For a summary of the history of the GED see Quinn (1993).

15. The data for 1961 come from U.S. Department of Health, Education, and Welfare (1961–62) and are for the 33 reporting states weighted by the total number of graduates (regular and GED). The 1971 and 1995 national data are from the *Digest of Education Statistics;* see also *Historical Statistics: Millennial Edition* table Bc265–272.

16. There are several possible reasons why the number of individuals receiving the GED plummeted after the late-1990s. One is that prisons no longer funded GED courses. Another is that welfare recipients had to find jobs and could not receive aid while studying for the GED. Perhaps most important of all, various changes to the GED in the early 2000s made it a more difficult examination to pass.

17. The fraction older than 24 years was 38 percent in 1994 and was 40 percent in 1991. In 2003, however, it was 27 percent, probably because of regulations that served to decrease test taking by single mothers and the incarcerated population. Data are from American Council on Education (1992, 1995, 2005).

18. We add GED recipients to the number of diplomas received in June of the following year. After 2000 the fraction of certificates earned by individuals older than 34 years old declined and was less than 11 percent in 2003. The higher the cutoff age, the more years we need to compute the statistic, and the fewer years we can report.

19. Cameron and Heckman (1993) and Heckman and LaFontaine (2006) analyze differences between GED recipients and conventional high school graduates in their labor market experiences. On the labor market returns to the GED for dropouts see Tyler (2004). On the noncognitive and cognitive skills of GED recipients see Heckman and Rubinstein (2001).

20. Based on tabulations from the 1970 Census IPUMS and the 2005 CPS.

21. The 2000 Census IPUMS data indicate that about half of the immigrants who arrived in the United States when of high school age obtained high school degrees or GEDs by the time they were 20 to 22 years old.

22. These estimates and statements are based on tabulations from the 1970 and 2000 Census IPUMS. The overall Hispanic share increased from 4.5 percent in 1970 to 14.5 percent in 2000 of the total U.S. population and from 18.5 percent in 1970 to 45.6 percent in 2000 of the U.S. foreign-born population.

23. In addition, as the share of the foreign-born increases, school systems are often strained to teach language skills to recent entrants and resources are diverted to the children of the foreign-born.

24. Ethnic and racial identifiers were not, until recently, part of the administrative records on high school diplomas collected by the U.S. Department of Education.

25. The pre-1990 Census data are presumed to include GED recipients with regular high school graduates and they explicitly do so starting in 1990. A change in the coding of educational attainment occurred with the 1990 Census when ac-

tual degrees were requested. We use standard definitions for a high school graduate in both periods. For the pre-1990 censuses we define high school graduates as those with 12 or more years of completed schooling. Beginning with the 1990 Census, we define high school graduates as those who indicate their highest level of schooling is at least a high school degree or a GED. It is likely that the pre-1990 data overstated the high school graduation rate by about 2 percentage points for 20- to 22-year-olds. Even with the adjustment the growth in the high school graduation rate for young adults remains modest from 1970 to 2000.

26. An analogous decomposition by foreign-born status using the 1970 and 2000 Census IPUMS shows that the growth in the foreign-born share of 20- to 22-year-olds from 9.2 percent in 1970 to 14.7 percent in 2000 served to reduce the growth in the high school graduation rate by 0.9 percentage points. The high school graduation rate for U.S.-born 20- to 22-year-olds increased by 3.9 percentage points from 80.4 percent in 1970 to 84.2 percent in 2000. We have also examined the effect of changing the Hispanic share on the growth of the high school graduation rate for 20- to 24-year-olds in the U.S. civilian, non-institutional population from 1973 to 2003 using the October CPS for 1973 to 2003. The high school graduation rate for 20- to 24-year-olds increased by 2.9 percentage points from 83.7 percent in 1973 to 86.6 percent in 2003. The growing Hispanic share from 1973 to 2003 reduced the growth in high school graduation rate by 3.1 percentage points for 20- to 24-year-olds in the October CPS. In fact, the high school graduation rate for non-Hispanics aged 20 to 24 years increased by 5.7 percentage points from 85.0 percent in 1973 to 90.7 percent in 2003.

27. See Borjas (1994) for an historical analysis of the rate of intergenerational convergence in educational attainment of U.S. immigration groups. Card, Di-Nardo, and Estes (2000) find that rate of assimilation in educational attainment for U.S.-born children of immigrants was as rapid in the 1990s as in the first-half of the twentieth century.

28. Tabulations from the 1940 to 2000 IPUMS and the 2005 CPS indicate that mean years of schooling at age 30 for the U.S.-born increased from 8.49 years for those born in 1900, to 10.90 years for those born in 1925, to 13.16 years for those born in 1950, to 13.92 years for those born in 1975. These estimates of mean years of schooling at age 30 by birth cohort use a methodology analogous to that described in the notes to Figure 1.4 of Chapter 1.

29. Card (1999, 2001) provides critical and comprehensive reviews of recent studies using such "natural experiments" to estimate the returns to schooling. More recently, Oreopoulos (2007) finds that post-1970 changes in state compulsory schooling laws to raise the minimum school-leaving age above 16 years modestly but statistically significantly raise the educational attainment of affected cohorts and generate substantial labor market returns to such increased schooling.

30. See Lleras-Muney (2005) on education and health; Lochner and Moretti (2004) on education and crime; and Milligan, Moretti, and Oreopolous (2004) on education and citizenship.

31. See, for example, Currie and Moretti (2003).

32. In 1914 about 23 percent of all students in the high school grades in Iowa were non-resident, tuition-paying pupils (Iowa Department of Public Instruction 1914).

33. We exclude military academies, such as West Point.

34. See Hoxby (1999) on the potential efficiency benefits of a decentralized educational system with local control and local property tax finance.

35. See Hoxby (1996, 2001) on the possible equity-efficiency trade-offs in shifting from local to state school finance and on how the specific design features of school finance equalization can lead some plans to level up spending and others to level down spending. Murray, Evans, and Schwab (1998) find that court-ordered school finance equalization plans from 1971 to 1996 typically served to reduce within-state spending inequality by leveling up spending with increased aggregate school spending financed through higher state taxes. Hoxby and Kuziemko (2004) provide a case study of the adverse impacts of the Texas "Robin Hood" scheme for school finance reform enacted in 1994. Hoxby (2001) and Card and Payne (2002) provide contrasting estimates of the impacts of school finance reforms on student outcomes.

36. We do not deflate by local price indices and this would probably reduce past disparities.

37. The data on expenditures on K–12 students by state have been com-piled from the various *Annuals* and *Biennials* (Alicia Sasser, personal communication).

38. The coefficient of variation (in percent) in K–12 per student school expenditures across all states was around 50 in 1900. It fell to around 40 by 1920 and continued to decrease to 1930 when it was around 35. Although it increased somewhat during the Great Depression, it then continued its decline to around 25 by 1950 when these data conclude. Differences existed by region, but the general path is about the same.

39. We have included all students in grades K to 8 in the elementary school group, even if some of them were in junior high schools. Similarly, we have included all those in grades 9 to 12 in the high school group.

40. On recent estimates the variation in per student spending across school districts, see Murray, Evans, and Schwab (1998, table 2).

41. Of course, the provision of generous vouchers could lead to the entry of new nondenominational private schools. Analyses of the impacts of large-scale voucher programs in other nations include Angrist, Bettinger, and Kremer (2006) for Colombia; Hsieh and Urquiola (2006) for Chile; and Fiske and Ladd (2000) for New Zealand. Analyses of recent U.S. voucher programs include Rouse (1998), Howell and Peterson (2004), and Krueger and Zhu (2004).

42. As of 2006, nine states have had court challenges to their voucher systems: California, Colorado, Florida, Maine, Missouri, Ohio, Pennsylvania, Vermont, and Wisconsin. Maine, Ohio, Vermont, Utah, Wisconsin, and the District of Columbia have state-funded vouchers (federally funded in the case of the District of Columbia) that can be used at any private school. The Florida system was struck down by the state supreme court. A few other states offer tax credits or deductions for education expenses.

43. In many states in the early part of the twentieth century, local districts tested students at the end of their common school years or eighth grade to make certain that they were ready to attend the district high school. But that testing was different from the tracking in Europe.

44. I. M. Kandel, one of the "deans" of U.S. educational history remarked that,

the American tradition of free secondary education . . . is today the most important guarantee of that equality of opportunity, the strongest pillar on which democracy rests. The principle of giving every boy and girl a chance was accepted in this country before other countries, not excluding the most advanced, had even established the right of every child to a free elementary education. . . . National interest, it is now universally recognized, demands the same common education for all, whether in the same school or not depends upon strength of class consciousness in a country . . . the American practice still stands alone as the only genuine attempt to give every young person an equal chance to the right to make the best of himself. (Kandel 1934, pp. 13–21)

45. See Counts (1926). Jessen (1928), in a report for the U.S. Bureau of Education, reports somewhat less strict requirements around the same period. The differences may come from the interpretation of the state laws or from different aggregation of the subjects. According to Jessen's table, 27 states had three or fewer subjects required other than ninth grade science, 41 required English, and 14 required algebra or geometry, although an additional 10 had the requirement of some mathematics. We use Counts as an upper bound on state requirements because our point is that states did *not* set high standards for graduation.

46. The New York State Regents' examination, which began in 1865, was only for elementary school students until 1878 when it became a test for high school students, as it is today (Marks 1989).

47. See, for example, Dee and Jacob (2006).

48. Kessler (2007) uses variation across states in changes in GED passing standards in the 1990s to examine how the difficulty of attaining a GED affects conventional high school graduation rates. He presents suggestive evidence that higher GED passing standards led to small increases in the share of students staying in high school and earning a conventional high school degree.

49. Heckman and Krueger (2003) offer different perspectives on the sources of problems in the U.S. education and training systems and on the effectiveness of alternative human capital policies.

50. Many of these inadequately prepared students eventually end up doing some post-secondary schooling. Bettinger and Long (2005) report that in 2001 an astounding one-third of all entering college students was required to take remedial courses in reading, writing, or mathematics.

51. Corcoran, Evans, and Schwab (2004) analyze changes in teacher quality in recent decades and the role of changing labor market opportunities for women.

52. Krueger (2003) assesses the evidence.

53. See Krueger (1999) on an evaluation of STAR. Attempts to reduce class size for all students are unlikely to hold teacher quality constant, at least in the short run. Universal reductions in class size require more teachers for a given number of pupils and may require the hiring of less qualified teachers. Richer schools are likely to outbid their poorer counterparts for the most qualified teachers. Thus, broad-based attempts to reduce class size might have little benefit for students from low-income backgrounds.

54. Hanushek (2002) and Hoxby (2003).

55. The impacts of recently adopted city- and state-level school accountability policies are discussed in Jacob (2005) and Hanushek and Raymond (2005). On improved approaches to evaluating teacher effectiveness see Kane, Rockoff, and Staiger (2006).

56. See Jencks and Phillips (1998), Fryer and Levitt (2004), and Neal (2006).

57. On the effectiveness of charter schools see Hanushek, Kain, Rivkin, and Branch (2006). Recent U.S. public and private school choice options are assessed in Carnoy (2001), Cullen, Jacob, and Levitt (2006), and Hoxby (2003).

58. See Akerlof and Kranton (2002), Austen-Smith and Fryer (2005), and Fryer and Torelli (2006).

59. New York City is currently undertaking policy experiments along these lines for children in low-income families and troubled schools. Jackson (2007) finds that a Texas program providing substantial financial incentives for teachers and students in low-income high schools to increase participation in Advanced Placement (AP) courses significantly increases the SAT scores and college matriculation rates of minority students.

60. This point is emphasized by Carneiro and Heckman (2003) and Heckman and Masterov (2007). U.S. Census Bureau (2006) reports the share of children living in poverty increased from 15.1 percent in 1970 to 17.6 percent in 2005 using the official U.S. poverty line. And U.S. Census Bureau (2007) reports the share of children not living with two parents increased from 14.8 percent in 1970 to 32.7 percent in 2005.

61. See Cunha, Heckman, Lochner, and Masterov (2005).

62. See, for example, Currie (2001) and Ludwig and Miller (2007).

63. Kane (1999, table 3–1, p. 60) estimates that mean real net tuition for public two-year and four-year colleges increased by 107 percent and 106 percent, respectively, from 1980–81 to 1994–95. In contrast, real family incomes were essentially stagnant for the bottom 40 percent of families from 1980 to 1995. The U.S. Department of Education (2007) reports that the mean real college price net of grants (in 2003–4 dollars) increased for two-year public colleges by 15 percent from $7,300 per year in 1989–90 to $8,400 per year in 2003–04 and for four-year public colleges by 30 percent from $9,600 in 1989–90 to $12,500 in 2003–4. Real median family income only increased by 9 percent from 1990 to 2004.

64. See Kane (1999) and Dynarski (2002, 2005).

65. See, for example, Carneiro and Heckman (2003) and Ellwood and Kane (2000).

66. Carneiro and Heckman (2002) provide a critique of the evidence that credit constraints affect the post-secondary schooling investments of low-income youth. See Cadena and Keys (2006) on debt aversion and the borrowing behavior of college students.

67. See Bound and Turner (2007) for evidence on how increased cohort-crowding at the state level lowers government resources per student in higher education and reduces college-going and completion rates.

68. These estimates are from Scott-Clayton (2007) and are based on the October CPS.

69. Tabulations from the 2003–4 National Postsecondary School Aid survey reported by Scott-Clayton (2007) show that 88 percent of college students who work indicate the need to pay tuition, fees, or living expenses as the main reason for employment during school. Stinebrickner and Stinebrickner (2003) use quasi-experimental variation in hours of work assigned to college students in a financial aid program and find that increased work hours harm academic performance in college.

70. Avery and Kane (2004) and Dynarski and Scott-Clayton (2006) give detailed assessments of the complexity of the U.S. college financial aid system and the adverse impacts that the system has on the college going of low-income youth.

71. Watson (2006) details the rise in neighborhood segregation by income since 1970. On the relationship between concentrated poverty and poor educational and labor market outcomes see Jencks and Mayer (1990)

72. Recent evaluations of housing mobility policies provide evidence that moves to lower-poverty areas may improve the educational and social outcomes of disadvantaged youth, particularly for girls. See, for example, Rosenbaum (1995) and Kling, Liebman and Katz (2007).

73. See Heckman and Lochner (2000).

74. See Krueger (2003) for a review of research on such approaches and Kemple and Scott-Clayton (2004) on the impacts of Career Academies. The residential-based Job Corps program, which serves mostly poor, urban dropouts aged 16 to 24 years, has consistently produced high social returns through increased labor earnings and reduced criminal activity.

75. See Heckman and Masterov (2007) for cost-benefit analyses of expanded educational investments with a focus on policies targeting young children.

76. Piketty and Saez (2007a) find that the income share of the top 1 percent of households increased from 8.2 percent in 1980 to 17.4 percent in 2005. And the growth of the top 1 percent share was dominated by households in the top 0.1 percent of the income distribution.

77. See Piketty and Saez (2007b).

78. See Furman, Summers, and Bordoff (2007) on such a strategy.

79. See, for example, DiNardo, Fortin, and Lemieux (1996) and Levy and Temin (2007).

80. See Freeman and Katz (1994) and Freeman (2007).

Appendix B

1. The Office [Bureau] of Education, the forerunner to today's Department of Education, was established in 1867 as a department of education and became the Office of Education in 1869, an agency of the Department of the Interior, where it stayed for 70 years. It was known as the Bureau of Education for those 70 years, but in 1929 it was formally renamed the Office of Education. In 1939 it became part of the Federal Security Agency and was, in 1953, included in the new agency of Health, Education and Welfare (HEW). The Department of Education became a separate cabinet-level agency in 1980. We call all of these agencies the U.S. Office of Education.

2. Summer school and night school are generally omitted.

3. See also *Historical Statistics, Millennial Edition.*

4. The "Statistical Summary of Education, 1937–38," is chapter 1 of the *Biennials* 1936–38.

References

Note: In the listing of authors we have omitted "State of." Thus "State of California Superintendent" is listed as "California Superintendent." For many state and federal publications, e.g., California, Office of the Superintendent of Public Instruction, *Biennial Report*, or U.S. Office of Education, *Biennial Report of the Commissioner of Education*, we reference the volume by the years covered by the report rather than by the date of publication. In the midst of writing this book the new *Historical Statistics of the United States: Millennial Edition* (2006) was released. We reference it for some series but have kept references to the older version, *Historical Statistics of the United States (1975)*, when the series did not change. Because we use both often, we reference them as *Historical Statistics, Millennial Edition and Historical Statistics* rather than by their authors. Similarly, we reference the Digest of Education Statistics by name rather than author.

Aaronson, Daniel, and Daniel Sullivan. 2001. "Growth in Worker Quality." *Economic Perspectives* (4th Quarter), pp. 53–74.

Abbott, Andrew. 1988. *The System of Professions: An Essay on the Division of Expert Labor.* Chicago: University of Chicago Press.

Abraham, Katharine G., and Susan N. Houseman. 1995. "Earnings Inequality in Germany." In *Differences and Changes in Wage Structures,* edited by Richard B. Freeman and Lawrence F. Katz, 371–403. Chicago: University of Chicago and NBER.

Acemoglu, Daron. 1998. "Why Do New Technologies Complement Skills? Directed Technical Change and Wage Inequality." *Quarterly Journal of Economics* 113 (November), pp. 1055–1089.

———. 2002. "Technical Change, Inequality and the Labor Market." *Journal of Economic Literature* 40 (March), pp. 7–72.

———. 2003. "Cross Country Inequality Trends." *Economic Journal* 113 (February), pp. F121–F149.

Acemoglu, Daron, and Joshua Angrist. 2000. "How Large Are Human Capital Externalities? Evidence from Compulsory Schooling Laws." *NBER Macroeconomics Annual 2000* 15, pp. 9–59.

Acemoglu, Daron, Simon Johnson, and James A. Robinson. 2001. "Colonial Origins of Comparative Development: An Empirical Investigation." *American Economic Review* 91 (December), pp. 1369–1401.

———. 2002. "Reversal of Fortune: Geography and Institutions in the Making of the Modern World Income Distribution." *Quarterly Journal of Economics* 117 (November), pp. 1231–1294.

Acemoglu, Daron, and James A. Robinson. 2000. "Why Did the West Extend the Franchise? Democracy, Inequality, and Growth in Historical Perspective." *Quarterly Journal of Economics* 115 (November), pp. 1167–1199.

Acemoglu, Daron, and Fabrizio Zilbotti. 2001. "Productivity Differences." *Quarterly Journal of Economics* 116 (May), pp. 563–606.

Adams, Donald R. Jr. 1967. *Wage Rates in Philadelphia, 1790–1830.* New York: Arno Press.

Adams, Francis. 1969, orig. pub. 1875. *The Free School System of the United States.* New York: Arno Press.

Akerlof, George, and Rachel Kranton. 2002. "Identity and Schooling: Some Lessons for the Economics of Education." *Journal of Economic Literature* 40 (December), pp. 1167–1201.

Alesina, Alberto, Reza Baqir, and William Easterly. 1999. "Public Goods and Ethnic Divisions." *Quarterly Journal of Economics* 114 (November), pp. 1243–1284.

Allen, Steven G. 2001. "Technology and the Wage Structure." *Journal of Labor Economics* 19 (April), pp. 440–493.

American Association of University Professors. various years. *Bulletin of the American Association of University Professors* (to 1955, continued by *AAUP Bulletin*). Easton, PA: AAUP.

———. "Instructional Salaries in 42 Selected Colleges and Universities for the Academic Year 1948–49."

———. "Instructional Salaries in 41 Selected Colleges and Universities for the Academic Year 1949–50."

———. "Instructional Salaries in 40 Selected Colleges and Universities for the Academic Year 1951–52."

———. "Instructional Salaries in 41 Selected Colleges and Universities for the Academic Year 1953–54."

———. "Instructional Salaries in 41 Selected Colleges and Universities for the Academic Year 1955–56."

———. "Instructional Salaries in 39 Selected Colleges and Universities for the Academic Year 1957–58."

———. "Instructional Salaries in 39 Selected Colleges and Universities for the Academic Year 1959–60."

American Council on Education. 1934. *Report of Committee on Graduate Instruction.* Washington, DC: American Council on Education (April).

———. 1992. *1991 Statistical Report.* Washington, DC: American Council on Education.

———. 1995. *Who Took the GED? GED 1994 Statistical Report.* Washington, DC: American Council on Education.

————. 2005. *Who Passed the GED Tests? 2003 Statistical Report.* Washington, DC: American Council on Education.

Anderson, James D. 1988. *The Education of Blacks in the South, 1860–1935.* Chapel Hill: University of North Carolina Press.

Angrist, Joshua, Eric Bettinger, and Michael Kremer. 2006. "Long-Term Educational Consequences of Secondary School Vouchers: Evidence from Administrative Records in Columbia." *American Economic Review* 96 (June), pp. 847–862.

Angus, David L., and Jeffrey E. Mirel. 1999. *The Failed Promise of the American High School: 1890–1995.* New York: Teachers College Press.

Annuals. [various years to 1917]. *See* U.S. Office of Education, *Annual Report of the Commissioner of Education for [various years to 1917].* Washington, DC: G.P.O.

Atack, Jeremy. 1987. "Economies of Scale and Efficiency Gains in the Rise of the Factory in America, 1820–1900." In *Quantity and Quiddity: Essays in U.S. Economic History,* edited by Peter Kilby, 286–335. Middletown, CT: Wesleyan University Press.

Atack, Jeremy, Fred Bateman, and Robert A. Margo. 2004. "Skill Intensity and Rising Wage Dispersion in Nineteenth Century American Manufacturing." *Journal of Economic History* 64 (March), pp. 172–192.

————. 2005. "Capital Deepening and the Rise of the Factory: The American Experience during the Nineteenth Century." *Economic History Review* 58 (August), pp. 586–595.

Atkinson, A. B. 1983. *The Economics of Inequality.* 2nd ed. Oxford: Clarendon Press.

Attanasio, Orazio, Erich Battistin, and Hidehiko Ichimura. 2004. "What Really Happened to Consumption Inequality in the U.S.?" NBER Working Paper no. 10338 (March).

Aurner, Clarence R. 1914. *History of Education in Iowa.* Vol. 6. Iowa City: State Historical Society of Iowa.

Austen-Smith, David, and Roland G. Fryer. 2005. "An Economic Analysis of 'Acting White.'" *Quarterly Journal of Economics* 120 (May), pp. 551–583.

Autor, David H., Lawrence F. Katz, and Melissa S. Kearney. 2005a. "Trends in U.S. Wage Inequality: Re-Assessing the Revisionists." NBER Working Paper no. 11627 (September).

————. 2005b. "Rising Wage Inequality: The Role of Composition and Prices." NBER Working Paper no. 11628 (September).

————. 2006. "The Polarization of the U.S. Labor Market." *American Economic Review* 96 (May), pp. 189–194.

————. 2007. "Trends in U.S. Wage Inequality: Revising the Revisionists." *Review of Economics and Statistics* (forthcoming).

Autor, David, Lawrence F. Katz, and Alan B. Krueger. 1998. "Computing Inequality: Have Computers Changed the Labor Market?" *Quarterly Journal of Economics* 113 (November), pp. 1169–1213.

Autor, David H., Frank Levy, and Richard J. Murnane. 2002. "Upstairs, Downstairs: Computers and Skills on Two Floors of a Large Bank." *Industrial and Labor Relations Review* 55 (April), pp. 432–447.

————. 2003. "The Skill Content of Recent Technological Change: An Empirical Investigation." *Quarterly Journal of Economics* 118 (November), pp. 1279–1333.

Avery, Christopher, and Thomas J. Kane. 2004. "Student Perceptions of College Opportunities: The Boston COACH Program." In *College Choices*, edited by Caroline M. Hoxby, 355–391. Chicago: University of Chicago Press and NBER.

Barrett, David B., George T. Kurian, and Todd M. Johnson, eds. 2001. *World Christian Encyclopedia: A Comparative Survey of Churches and Religions in the Modern World*. 2nd ed. New York: Oxford University Press.

Barro, Robert J. 1991. "Economic Growth in a Cross Section of Countries." *Quarterly Journal of Economics* 116 (May), pp. 407–443.

Barro, Robert J., and Xavier Sala-i-Martin. 1991. "Convergence across States and Regions." *Brookings Papers on Economic Activity*, no. 1, pp. 107–182.

Bartel, Ann, Casey Ichniowski, and Kathryn Shaw. 2007. "How Does Information Technology Affect Productivity? Plant-Level Comparisons of Product Innovation, Process Improvement, and Worker Skills." *Quarterly Journal of Economics* 122 (November), pp. 1721–1758.

Bartelsman, Eric, and Wayne Gray. 1996. "The NBER Manufacturing Productivity Database." NBER Working Technical Paper no. 205 (October).

Bates, Ralph. 1965. *Scientific Societies in the United States*. Cambridge, MA: MIT Press.

Baumol, William J., Sue Anne Batey Blackman, and Edward N. Wolff. 1989. *Productivity and American Leadership: The Long View*. Cambridge, MA: MIT Press.

Becker, Gary S., and Kevin M. Murphy. 1988. "The Family and the State." *Journal of Law and Economics* 31 (January), pp. 1–18.

Bell, Philip W. 1951. "Cyclical Variations and Trend in Occupational Wage Differentials in American Industry since 1914." *Review of Economics and Statistics* 33 (November), pp. 329–337.

Bellamy, Edward. 1888. *Looking Backward: 2000–1887*. Boston: Ticknor and Company.

Berman, Eli, John Bound, and Zvi Griliches. 1994. "Changes in the Demand for Skilled Labor within U.S. Manufacturing Industries: Evidence from the Annual Survey of Manufacturing." *Quarterly Journal of Economics* 109 (May), pp. 367–397.

Berman, Eli, John Bound, and Stephen Machin. 1998. "Implications of Skill-Biased Technological Change: International Evidence." *Quarterly Journal of Economics* 113 (November), pp. 1245–1279.

Bettinger, Eric P., and Bridget Terry Long. 2005. "Addressing the Needs of Under-Prepared Students in Higher Education: Does College Remediation Work?" NBER Working Paper no. 11325 (May).

Biennial [various years from 1916–18 to 1956–58]. *See* U.S. Office [Bureau] of Education. *Biennial Survey of Education for* [*various years from 1916–18 to 1956–58*]. Washington, DC: G.P.O.

Bils, Mark, and Peter Klenow. 2000. "Does Schooling Cause Growth?" *American Economic Review* 90 (October), pp. 1160–1183.

Bishop, Eugene A. 1930. *The Development of a State School System: New Hampshire*. New York: Teachers College, Columbia University.

Bishop, John H. 1989. "Is the Test Score Decline Responsible for the Productivity Growth Decline?" *American Economic Review* 79 (March), pp. 178–197.

Blank, David M., and George J. Stigler. 1957. *The Demand and Supply of Scientific Personnel*. New York: National Bureau of Economic Research.

Blau, Francine D., and Lawrence M. Kahn. 2002. *At Home and Abroad.* New York: Russell Sage.

Board of Education for England and Wales. 1932. *Education in 1931.* London: His Majesty's Stationery Office.

Bolino, August C. 1973. *Career Education: Contributions to Economic Growth.* New York: Praeger Publishers.

Boothe, Viva. 1932. *Salaries and the Cost of Living in Twenty-Seven State Universities and Colleges, 1913–1932.* Columbus: Ohio State University Press.

Borjas, George J. 1994. "Long-Run Convergence of Ethnic Skill Differentials: The Children and Grandchildren of the Great Migration." *Industrial and Labor Relations Review* 47 (July), pp. 553–573.

———. 2003. "The Labor Demand Curve *Is* Downward Sloping: Reexamining the Impact of Immigration on the Labor Market." *Quarterly Journal of Economics* 118 (November), pp. 1335–1374.

Borjas, George J., Richard B. Freeman, and Lawrence F. Katz. 1997. "How Much Do Immigration and Trade Affect Labor Market Outcomes?" *Brookings Papers on Economic Activity*, no. 1, pp. 1–90.

Bound, John, and George Johnson. 1992. "Changes in the Structure of Wages in the 1980s: An Evaluation of Alternative Explanations." *American Economic Review* 82 (June), pp. 371–392.

Bound, John, and Sarah Turner. 2002. "Going to War and Going to College: Did World War II and the GI Bill Increase the Educational Attainment for Returning Veterans?" *Journal of Labor Economics* 20 (October), pp. 784–815.

———. 2007. "Cohort Crowding: How Resources Affect Collegiate Attainment." *Journal of Public Economics* 91 (June), pp. 877–899.

Bowles, Samuel, and Herbert Gintis. 1976. *Schooling in Capitalist America: Educational Reform and the Contradictions of Economic Life.* New York: Basic Books.

Braatz, Jay, and Robert D. Putnam. 1997. "Families, Communities, and Education in America: Exploring the Evidence." Unpublished paper, Harvard University.

Bradbury, Katharine, and Jane Katz. 2002. "Are Lifetime Incomes Growing More Unequal? Looking at New Evidence on Family Income Mobility?" *Regional Review*, no. 4, pp. 2–5.

Brandolini, Andrea, and Timothy M. Smeeding. 2006. "Inequality: International Evidence." In *The New Palgrave Dictionary of Economics*, edited by S. N. Durlauf and L. E. Blume. Basingstoke, England: Palgrave Macmillan.

Braverman, Harry. 1974. *Labor and Monopoly Capital: The Degradation of Work in the Twentieth Century.* New York: Monthly Review Press.

Bresnahan, Timothy F. 1999. "Computerization and Wage Dispersion: An Analytical Reinterpretation." *Economic Journal* 109 (June), pp. 390–415.

Bresnahan, Timothy F., Erik Brynjolfsson, and Lorin M. Hitt. 2002. "Information Technology, Workplace Organization and the Demand for Skilled Labor: Firm-level Evidence." *Quarterly Journal of Economics* 118 (February), pp. 339–376.

Bresnahan, Timothy, and Manuel Trajtenberg. 1995. "General Purpose Technologies: 'Engines of Growth'?" *Journal of Econometrics* 65 (January), pp. 83–108.

Brown, Elmer Ellsworth. 1899. *Secondary Education.* Monographs on Education in the United States. Department of Education for the United States Commission to the Paris Exposition of 1900. Albany, NY: J. B. Lyon.

Brubacher, John S., and Willis Rudy. 1958. *Higher Education in Transition: An American History, 1636–1956*. New York: Harper and Brothers.

Bryce, James. 1889. *The American Commonwealth*. Vol. 2. London: Macmillan.

Budd, Edward C. 1967. "Introduction." In *Inequality and Poverty*, edited by Edward C. Budd, vii–xxxiv. New York: W. W. Norton.

Burdge, Howard G. 1921. *Our Boys: A Study of the 245,000 Sixteen, Seventeen and Eighteen Year Old Employed Boys of the State of New York*. Albany, NY: J.B. Lyon.

Burtless, Gary. 1999. "Effect of Growing Wage Disparities and Family Composition Shifts in the Distribution of U.S. Income." *European Economic Review* 43 (April), pp. 853–865.

Burtless, Gary, and Christopher Jencks. 2003. "American Inequality and Its Consequences." In *Agenda for the Nation*, edited by H. Aaron, J. Lindsay, and P. Nivola, 61–108. Washington, DC: Brookings Institution.

Bush, George G. 1898. *History of Education in New Hampshire*. Washington, DC: G.P.O.

Cadena, Brian C., and Benjamin J. Keys. 2006. "Self-Control Induced Debt Aversion: Evidence from Interest-Free Student Loans." Unpublished paper, University of Michigan.

Cain, Louis P., and Donald G. Paterson. 1986. "Biased Technical Change, Scale, and Factor Substitution in American Industry, 1850–1919." *Journal of Economic History* 46 (March), pp. 153–164.

California Liaison Committee [of the Regents of the University of California]. 1960. *A Master Plan for Higher Education in California, 1960–1975*. Sacramento: California State Department of Education.

California, Office of Superintendent of Public Instruction. [various years] *(Number) Biennial Report of the Superintendent of Public Instruction*. Sacramento, CA. [e.g., 25th Biennial Report is 1911/1912]

Caliver, Ambrose. 1933a. "Secondary Education for Negroes." *U.S. Office of Education Bulletin*, no. 17, monograph 7. Washington, DC: G.P.O.

———. 1933b. "Education of Negro Teachers." *U.S. Office of Education Bulletin*, no. 10. Washington, DC: G.P.O.

Calvert, Monte A. 1967. *The Mechanical Engineer in America, 1830–1910*. Baltimore, MD: Johns Hopkins Press.

Cameron, Steven V., and James J. Heckman. 1993. "The Non-Equivalence of High School Equivalents." *Journal of Labor Economics* 11 (January), pp. 1–47.

Card, David. 1999. "The Causal Effect of Education on Earnings." In *Handbook of Labor Economics*, vol. 3A, edited by O. Ashenfelter and D. Card, 1801–1863. Amsterdam: North Holland.

———. 2001. "Estimating the Return to Schooling: Progress on Some Persistent Econometric Problems." *Econometrica* 69 (September), pp. 1127–1160.

———. 2005. "Is the New Immigration Really So Bad?" *Economic Journal* 115 (November), pp. F300–F323.

Card, David, John DiNardo, and Eugena Estes. 2000. "The More Things Change: Immigrants and the Children of Immigrants in the 1940s, the 1970s, and the 1990s." In *Issues in the Economics of Immigration*, edited by George J. Borjas, 227–269. Chicago: University of Chicago Press and NBER.

Card, David, and Alan B. Krueger. 1992a. "Does School Quality Matter? Returns to Education and Characteristics of Public Schools in the United States." *Journal of Political Economy* 100 (February), pp. 1–40.

————. 1992b. "School Quality and Black/White Relative Earnings: A Direct Assessment." *Quarterly Journal of Economics* 107 (February), pp. 151–200.

Card, David, and Thomas Lemieux. 2001. "Can Falling Supply Explain the Rising Return to College for Younger Men?" *Quarterly Journal of Economics* 116 (May), pp. 705–746.

Card, David, and Abigail Payne. 2002. "School Finance Reform, the Distribution of School Spending, and the Distribution of SAT Scores." *Journal of Public Economics* 83 (January), pp. 49–82.

Carlton, Frank Tracy. 1908. *Education and Industrial Evolution.* New York: Macmillan.

Carneiro, Pedro, and James J. Heckman. 2002. "The Evidence on Credit Constraints in Post-Secondary Schooling." *Economic Journal* 112 (October), pp. 705–734.

————. 2003. "Human Capital Policy." In *Inequality in America*, edited by James J. Heckman and Alan B. Krueger, 77–239. Cambridge, MA: MIT Press.

Carnoy, Martin. 2001. *School Vouchers: Examining the Evidence.* Washington, DC: Economic Policy Institute.

Carr-Saunders, A. M., D. Caradog Jones, and C. A. Moser. 1958. *A Survey of Social Conditions in England and Wales.* Oxford: Clarendon Press.

Carter, Susan, Scott S. Gartner, Michael Haines, Alan Olmstead, Richard Sutch, and Gavin Wright, eds. 2006. *Historical Statistics of the United States: Millennial Edition.* New York: Cambridge University Press. *Note:* We refer to this source as *Historical Statistics: Millennial Edition.* It is to be distinguished from the previous version, referred to here as *Historical Statistics.*

Cartter, Allan M. 1966. *An Assessment of Quality in Graduate Education: A Comparative Study of Graduate Departments in 29 Academic Disciplines.* Washington, DC: American Council on Education.

Cattell, James McKeen. 1938. *American Men of Science: A Biographical Directory.* 6th ed. New York: Science Press.

Cattell, Jaques. 1960. *American Men of Science: A Biographical Directory.* 10th ed. Tempe, AZ: Jaques Cattell Press.

Census of the United States. 1832. *Fifth Census or Enumeration of the Inhabitants of the United States, 1830.* Washington, DC: Duff Green.

Census of the United States. 1841. *Compendium of the Enumeration of the Inhabitants and Statistics of the United States . . . from the Returns of the Sixth Census.* Washington, DC: Thomas Allen.

Chadbourne, Ava H. 1928. *The Beginnings of Education in Maine.* New York: Teachers College, Columbia University.

Chandler, Alfred D., Jr. 1977. *The Visible Hand: The Managerial Revolution in American Business.* Cambridge, MA: Belknap Press.

Ciccone, Antonio, and Giovanni Peri. 2005. "Long-Run Substitutability between More and Less Educated Workers: Evidence from U.S. States, 1950–1990." *Review of Economics and Statistics* 87 (November), pp. 652–663.

Cline, William. 1997. *Trade and Income Distribution.* Washington, DC: Institute for International Economics.

Cohen, Miriam, and Michael Hanagan. 1991. "Work, School, and Reform: A Comparison of Birmingham, England, and Pittsburgh, USA, 1900–1950." *International Labor and Working-Class History* 40 (Fall), pp. 67–80.

College Blue Book. 1933. *The 1933 College Blue Book*, edited by Huber William Hurt and Harriett-Jeanne Hurt. Hollywood by-the-Sea, FL: The College Blue Book.

Conrad, Herbert S., and Ernest V. Hollis. 1955. "Trends in Tuition Charges and Fees." *Annals of the American Academy of Political and Social Science* 301 (September 1955), pp. 148–165.

Coombs, Whitney. 1926. *The Wages of Unskilled Labor in Manufacturing Industries in the United States, 1890–1924.* New York: Columbia University Press.

Corcoran, Sean, William N. Evans, and Robert S. Schwab. 2004. "The Changing Quality of Teachers over the Past Four Decades." *Journal of Policy Analysis and Management* 23 (Summer), pp. 449–470.

Costa, Dora. 1998. "The Unequal Work Day: A Long-Term View." *American Economic Review* 88 (May), pp. 330–334.

Counts, George S. 1926. *The Senior High School Curriculum.* Chicago: University of Chicago Press.

———. 1929. *Secondary Education and Industrialism.* The Inglis Lecture. Cambridge, MA: Harvard University Press.

Cremin, Lawrence A. 1951. *The American Common School: An Historic Conception.* New York: Bureau of Publications Teachers College, Columbia University.

Cubberley, Ellwood P. [1919] 1934, 1947. *Public Education in the United States: A Study and Interpretation of American Educational History.* Boston: Houghton Mifflin. (*Note:* The 1934 edition was mainly used.)

———. [1934] 1970. *Readings in Public Education in the United States: A Collection of Sources and Readings to Illustrate the History of Educational Practice and Progress in the United States.* Westport, CT: Greenwood Press.

Cubberley, Ellwood P., and Edward C. Elliott. 1915. *State and County School Administration.* New York: Macmillan.

Cullen, Donald E. 1956. "The Interindustry Wage Structure, 1899–1950." *American Economic Review* 46 (June), pp. 353–369.

Cullen, Julie Berry, Brian A. Jacob, and Steven D. Levitt. 2006. "The Effect of School Choice on Participants: Evidence from Randomized Lotteries." *Econometrica* 74 (September), pp. 1191–1230.

Cunha, Flavio, James J. Heckman, Lance Lochner, and Dimitriy V. Masterov. 2005. "Interpreting the Evidence on Lifetime Skill Formation." NBER Working Paper no. 11331 (May).

Currie, Janet. 2001. "Early Childhood Intervention Programs: What Do We Know?" *Journal of Economic Perspectives* 15 (Spring), pp. 212–238.

Currie, Janet, and Enrico Moretti. 2003. "Mother's Education and the Intergenerational Transmission of Human Capital: Evidence from College Openings." *Quarterly Journal of Economics* 118 (November), pp. 1495–1532.

Cutler, David M., and Lawrence F. Katz. 1991. "Macroeconomic Performance and the Disadvantaged." *Brookings Papers on Economic Activity,* no. 2, pp. 1–74.

———. 1992. "Rising Inequality? Changes in the Distribution of Income and Consumption in the 1980s." *American Economic Review* 82 (May), pp. 546–551.

D'Amico, Louis A., and W. Robert Bokelman. 1963. *Basic Student Charges, 1962–63: Tuition and Fees, Board and Room.* U.S. Department of Health, Education and Welfare. Office of Education. Washington, DC: G.P.O.

Davenport, Iowa. [year]. *Public School Directory of Davenport, Iowa for* [*1917 to 1934*].

Davis, Calvin Olin. 1927. *Our Evolving High School Curriculum.* Yonkers-on-Hudson, NY: World Book.

Dee, Thomas, and Brian Jacob. 2006. "Do High School Exit Exams Influence Educational Attainment or Labor Market Performance?" NBER Working Paper no. 12199 (April).

Deffenbaugh, Walter S., and Ward W. Keesecker. 1935. *Compulsory School Attendance Laws and Their Administration*. Bulletin No. 4, United States Department of the Interior, Office of Education. Washington, DC: G.P.O.

DeLong, J. Bradford, Goldin, Claudia, and Lawrence F. Katz. 2003. "Sustaining U.S. Economic Growth." In *Agenda for the Nation*, edited by H. Aaron, J. Lindsay, and P. Nivola, 17–60. Washington, DC: Brookings Institution Press.

Denison, Edward F. 1962. *The Sources of Economic Growth in the United States and the Alternatives before Us*. New York: Committee for Economic Development.

Devine, Warren. 1983. "From Shafts to Wires: Historical Perspective on Electrification." *Journal of Economic History* 43 (June), pp. 347–372.

Dew-Becker, Ian, and Robert J. Gordon. 2005. "Where Did the Productivity Growth Go? Inflation Dynamics and the Distribution of Income." *Brookings Papers on Economic Activity*, no. 2, pp. 67–150.

Dewhurst, J. Frederic, John O. Coppock, P. Lamartine Yates, and Associates. 1961. *Europe's Needs and Resources: Trends and Prospects in Eighteen Countries*. New York: Twentieth Century Fund.

Digest of Education Statistics. See U.S. Department of Education, NCES.

DiNardo, John, Nicole Fortin, and Thomas Lemieux. 1996. "Labor Market Institutions, and the Distribution of Wages, 1973–1992: A Semiparametric Approach." *Econometrica* 64 (September), pp. 1001–1044.

Doms, Mark, Timothy Dunne, and Kenneth R. Troske. 1997. "Workers, Wages, and Technology." *Quarterly Journal of Economics* 112 (February), pp. 253–290.

Donohue, John, and James J. Heckman. 1991. "Continuous Versus Episodic Change: The Impact of Civil Rights Policy on the Economic Status of Blacks." *Journal of Economic Literature* 29 (December), pp. 1603–1642.

Dos Passos, John. 1932. *1919*. New York: Harcourt Brace.

Douglas, Paul H. 1921. *The American Apprenticeship and Industrial Education*. New York: Columbia University Press.

———. 1926. "What Is Happening to the 'White-Collar-Job' Market?" *System: The Magazine of Business* 49 (December), pp. 719–721, 782, 784.

———. 1930. *Real Wages in the United States: 1890 to 1926*. Boston: Houghton Mifflin.

Du Boff, Richard B. 1979. *Electric Power in American Manufacturing 1889–1958*. New York: Arno Press.

Dunbar, Willis F. 1960. "The High School on Trial: The Kalamazoo Case." *Papers of the Michigan Academy of Science, Arts, and Letters*. Part 2, *Social Science* 45, pp. 187–200.

Dunne, Timothy, John Haltiwanger, and Kenneth R. Troske. 1996. "Technology and Jobs: Secular Changes and Cyclical Dynamics." NBER Working Paper no. 5656 (July).

Dustmann, Christian, Johannes Ludsteck, and Uta Schönberg. 2007. "Revisiting the German Wage Structure." Unpublished paper, University of Rochester.

Dynarski, Susan. 2002. "The Behavioral and Distributional Implication of Aid for College." *American Economic Review* 92 (May), pp. 279–285.

————. 2005. "Building the Stock of College-Educated Labor." Unpublished paper, Harvard University.

Dynarski, Susan, and Judith Scott-Clayton. 2006. "The Cost of Complexity in Federal Student Aid: Lessons from Optimal Tax Theory and Behavioral Economics." *National Tax Journal* 59 (June), pp. 319–356.

Easterlin, Richard A. 1981. "Why Isn't the Whole World Developed?" *Journal of Economic History* 61 (March), pp. 1–17.

Economic Report of the President, 2006. 2006. Washington, DC: G.P.O.

Edwards, Alba M. 1943. *Sixteenth Census of the United Sates: 1940. Population. Comparative Occupation Statistics for the United States, 1870 to 1940.* Washington, DC: G.P.O.

Eichengreen, Barry. 1992. "Can the Maastricht Treaty Be Saved?" *Princeton Studies in International Finance.* Princeton, NJ: Princeton University, International Economics Section.

Elbaum, Bernard. 1989. "Why Apprenticeship Persisted in Britain but Not in the United States." *Journal of Economic History* 49 (June), pp. 337–349.

Electrical Merchandising. 1922. *How to Retail Radio.* New York: McGraw-Hill.

Ellwood, David T., and Thomas Kane. 2000. "Who Is Getting a College Education? Family Background and the Growing Gaps in Enrollment." In *Securing the Future*, edited by Sheldon Danziger and Jane Waldfogel, 283–324. New York: Russell Sage Foundation.

Emmons, Frederick Earle. 1926. *City School Attendance Service.* Teachers College, Columbia University Contributions to Education, no. 200. New York: Teachers College, Columbia University.

Engerman, Stanley L., and Kenneth L. Sokoloff. 2005. "The Evolution of Suffrage Institutions in the New World." *Journal of Economic History* 65 (December), pp. 891–921.

Epple, Dennis, and Richard E. Romano. 1996. "Ends Against the Middle: Determining Public Service Provision When There Are Private Alternatives." *Journal of Public Economics* 62 (November), pp. 297–326.

Evans, Owen D. 1926. *Educational Opportunities for Young Workers.* New York: Macmillan.

Federal Board for Vocational Education. 1922. *Preliminary Report on the Senior Commercial Occupations Survey.* Mimeo. Washington, DC: N.p.

Fernandez, Raquel, and Richard Rogerson. 1995. "On the Political Economy of Education Subsidies." *Review of Economic Studies* 62 (April), pp. 249–262.

————. 2003. "Equity and Resources: An Analysis of Education Finance Systems." *Journal of Political Economy* 111 (August), pp. 858–897.

Ferrie, Joseph P. 2005. "History Lessons: The End of American Exceptionalism? Mobility in the U.S. Since 1850." *Journal of Economic Perspectives* 19 (Summer), pp. 199–215.

Feynman, Richard P. 1985. *"Surely You're Joking, Mr. Feynman!"* New York: W. W. Norton.

Field, Alexander. 1979. "Economic and Demographic Determinants of Educational Commitment: Massachusetts, 1855." *Journal of Economic History* 39 (June), pp. 439–459.

Fishlow, Albert. 1966a. "The American Common School Revival: Fact or Fancy?" In *Industrialization in Two Systems: Essays in Honor of Alexander Gerschenkron*, edited by Henry Rosovsky, 40–67. New York: Wiley Press.

————. 1966b. "Levels of Nineteenth-Century American Investment in Education." *Journal of Economic History* 26 (December), pp. 418–436.

Fiske, Edward B., and Helen F. Ladd. 2000. *When Schools Compete: A Cautionary Tale.* Washington, DC: The Brookings Institution.

Fogel, Robert William. 1964. *Railroads and American Economic Growth: Essays in Econometric History.* Baltimore, MD: Johns Hopkins Press.

Freeman, Richard B. 1975. "Overinvestment in College Training?" *Journal of Human Resources* 10 (Summer), pp. 287–311.

————. 2007. "Labor Market Institutions Around the World." NBER Working Paper no. 13243 (July).

Freeman, Richard B., and Lawrence F. Katz. 1994. "Rising Wage Inequality: The United States vs. Other Advanced Countries." In *Working Under Different Rules*, edited by Richard B. Freeman, 29–62. New York: Russell Sage Foundation.

Friedberg, Leora. 2003. "The Impact of Technological Change on Older Workers: Evidence from Data on Computers." *Industrial and Labor Relations Review* 56 (April), pp. 511–529.

Frydman, Carola. 2001. "Female Labor Force Participation and the Quality of the Teaching Force, Evidence form the High School Movement." Unpublished, Harvard University Department of Economics.

Fryer, Roland G., and Steven D. Levitt. 2004. "Understanding the Black-White Test Score Gap in the First Two Years of School." *Review of Economics and Statistics* 86 (May), pp. 447–464.

Fryer, Roland G., and Paul Torelli. 2006. "An Empirical Analysis of 'Acting White.'" Unpublished paper, Harvard University.

Furman, Jason, Lawrence H. Summers, and Jason Bordoff. 2007. "Achieving Progressive Tax Reform in an Increasingly Global Economy." Unpublished paper, The Brookings Institution, The Hamilton Project.

Gabler, Edwin. 1988. *The American Telegrapher: A Social History, 1860–1900.* New Brunswick, NJ: Rutgers University Press.

Galor, Oded, and Omer Moav. 2000. "Ability-Biased Technological Transition, Wage Inequality and Economic Growth." *Quarterly Journal of Economics* 115 (May), pp. 469–497.

Geiger, Roger L. 1990. "Organized Research Units—Their Role in the Development of University Research." *Journal of Higher Education* 61 (January/February), pp. 1–19.

General Electric Company. 1924. The Apprentice System." Pamphlet. West Lynn, MA: G.E.

Glenn, Charles L. 1988. *The Myth of the Common School.* Amherst: University of Massachusetts Press.

Goldberger, Marvin L., Brendan A. Maher, and Pamela Ebert Flattau, eds. 1995. *Research-Doctorate Programs in the United States: Continuity and Change.* Washington, DC: National Academy Press.

Goldin, Claudia. 1990. *Understanding the Gender Gap: An Economic History of American Women.* New York: Oxford University Press.

————. 1991. "Marriage Bars: Discrimination against Married Women Workers, 1920 to 1950." In *Favorites of Fortune: Technology, Growth, and Economic Development since the Industrial Revolution*, edited by Henry Rosovsky, David Landes, and Patrice Higonnet, 511–536. Cambridge, MA: Harvard University Press.

———. 1994. "The Political Economy of Immigration Restriction in the United States, 1890 to 1921." In *The Regulated Economy: A Historical Approach to Political Economy*, edited by Claudia Goldin and Gary D. Libecap, 223–257. Chicago: University of Chicago Press.

———. 1998. "America's Graduation from High School: The Evolution and Spread of Secondary Schooling in the Twentieth Century." *Journal of Economic History* 58 (June), pp. 345–374.

———. 1999. "Egalitarianism and the Returns to Education during the Great Transformation of American Education." *Journal of Political Economy* 107 (December), pp. S65–S94.

———. 2000. "Labor Markets in the Twentieth Century." In *The Cambridge Economic History of the United States*, vol. 3, *The Twentieth Century*, edited by Stanley L. Engerman and Robert E. Gallman, 549–624. Cambridge: Cambridge University Press.

Goldin, Claudia, and Lawrence F. Katz. 1995. "The Decline of 'Non-Competing Groups': Changes in the Premium to Education, 1890 to 1940." NBER Working Paper no. 5202 (August).

———. 1998. "The Origins of Technology-Skill Complementarity." *Quarterly Journal of Economics* 113 (June), pp. 693–732.

———. 1999a. "The Shaping of Higher Education: The Formative Years in the United States, 1890 to 1940." *Journal of Economic Perspectives* 13 (Winter), pp. 683–723. [Longer version is: NBER Working Paper no. 6537 (April 1998)]

———. 1999b. "Human Capital and Social Capital: The Rise of Secondary Schooling in America, 1910 to 1940." *Journal of Interdisciplinary History* 29 (Spring), pp. 683–723.

———. 2000. "Education and Income in the Early Twentieth Century: Evidence from the Prairies." *Journal of Economic History* 60 (September), pp. 782–818.

———. 2001a. "Decreasing (and then Increasing) Inequality in America: A Tale of Two Half Centuries." In *The Causes and Consequences of Increasing Inequality*, edited by Finis Welch, 37–82. Chicago: University of Chicago Press.

———. 2001b. "The Shaping of Higher Education in the United States and New England." *Regional Review* 11 (Q4), pp. 5–11.

———. 2003. "Mass Secondary Schooling and the State: The Role of State Compulsion in the High School Movement." NBER Working Paper no. 10075. (November).

———. 2005. "Why the United States Led in Education: Lessons from Secondary School Expansion, 1910 to 1940." Revised version of NBER Working Paper no. 6144 (August 1997).

Goldin, Claudia, Lawrence F. Katz, and Ilyana Kuziemko. 2006. "The Homecoming of American College Women: The Reversal of the College Gender Gap." *Journal of Economic Perspectives* 20 (Fall), pp. 133–156.

Goldin, Claudia, and Robert A. Margo. 1991. "Appendix to 'The Great Compression: The Wage Structure in the United States at Mid-Century': Skill Ratios and Wage Distributions: 1920s to 1950s." Unpublished manuscript.

———. 1992. "The Great Compression: The Wage Structure in the United States at Mid-Century." *Quarterly Journal of Economics* 107 (February), pp. 1–34.

Goldsmith, Selma. 1967. "Changes in the Size Distribution of Income." In *Inequality and Poverty*, edited by Edward C. Budd, 65–79. New York: W. W. Norton.

Goldsmith, Selma, George Jaszi, Hyman Kaitz, and Maurice Liebenberg. 1954. "Size Distribution of Income since the Mid-Thirties." *Review of Economics and Statistics* 36 (February), pp. 1–32.

Gordon, Robert J. 2000. "Interpreting the 'One Big Wave' in U.S. Long-Term Productivity Growth." NBER Working Paper no. 7752 (June).

———. 2004. "Why Europe Was Left at the Station When America's Productivity Locomotive Departed." NBER Working Paper no. 10661 (August).

Gottschalk, Peter, and Sheldon Danziger. 1998. "Family Income Mobility: How Much Is There, and Has it Changed?" In *The Inequality Paradox: Growth of Income Disparity*, edited by J. Auerbach and R. Belous, 92–111. Washington, DC: National Policy Association.

Gottschalk, Peter, and Mary Joyce. 1998. "Cross-National Differences in the Rise of Earnings Inequality: Market and Institutional Factors." *Review of Economics and Statistics* 80 (November), pp. 489–502.

Gottschalk, Peter, and Robert Moffitt. 1994. "The Growth of Earnings Instability in the U.S. Labor Market." *Brookings Papers on Economic Activity*, no. 2, pp. 217–272.

Griliches, Zvi. 1969. "Capital-Skill Complementarity." *Review of Economics and Statistics* 51 (November), pp. 465–468.

Grubb, W. Norton, and Marvin Lazerson. 2004. *The Education Gospel: The Economic Power of Schooling*. Cambridge: Harvard University Press.

Haider, Steven J. 2001. "Earnings Instability and Earnings Inequality in the United States, 1967–1991." *Journal of Labor Economics* 19 (October), pp. 799–836.

Hall, Clyde W. 1973. *Black Vocational, Technical, and Industrial Arts Education: Development and History*. Chicago: American Technical Society.

Handlin, Oscar, and Mary F. Handlin. 1970. *The American College and American Culture*. New York: McGraw-Hill.

Hanushek, Eric A. 2002. "The Failure of Input-Based Schooling Policies." *Economic Journal* 113 (February), pp. F64–F98.

Hanushek, Eric A., John F. Kain, Steven G. Rivkin, and Gregory F. Branch. 2006. "Charter School Quality and Parental Decision Making with School Choice." Unpublished paper, Hoover Institution, Stanford University.

Hanushek, Eric A., and Margaret E. Raymond. 2005. "Does School Accountability Lead to Improved School Performance?" *Journal of Policy Analysis and Management* 24 (Spring), pp. 297–327.

Heckman, James J., and Alan B. Krueger. 2003. *Inequality in America*. Cambridge, MA: MIT Press.

Heckman, James J., and Paul A. LaFontaine. 2006. "Bias-Corrected Estimates of GED Returns." *Journal of Labor Economics* 24 (July), pp. 661–700.

Heckman, James J., and Lance Lochner. 2000. "Rethinking Education and Training Policy: Understanding the Sources of Skill Formation in a Modern Economy." In *Securing the Future*, edited by Sheldon Danziger and Jane Waldfogel, 47–83. New York: Russell Sage Foundation.

Heckman, James J., Lance Lochner, and Christopher Taber. 1998. "Explaining Rising Wage Inequality: Explorations with a Dynamic General Equilibrium Model of Labor Earnings with Heterogeneous Agents." *Review of Economic Dynamics* 1 (January), pp. 11–58.

Heckman, James J., and Dimitriy V. Masterov. 2007. "The Productivity Argument for Investing in Young Children." NBER Working Paper no. 13016 (April).

Heckman, James J., and Yona Rubinstein. 2001. "The Importance of Noncognitive Skills: Lessons from the GED Testing Program." *American Economic Review* 91 (May), pp. 145–149.

Herbst, Jurgen. 1996. *The Once and Future School: Three Hundred and Fifty Years of American Secondary Education.* New York: Routledge.

Heston, Alan, Robert Summers, and Bettina Aten. 2002. Penn World Table, Version 6.1, Center for International Comparisons at the University of Pennsylvania (CICUP), accessed at http://pwt.econ.upenn.edu/.

Higher Education Publications. 1992. *Higher Education Directory.* Washington, DC: Higher Education Publications.

Historical Statistics. See U.S. Bureau of the Census (1975).

Historical Statistics: Millennial Edition. See Carter et al. (2006).

Hofstadter, Richard, and C. DeWitt Hardy. 1952. *The Development and Scope of Higher Education in the United States.* New York: Columbia University Press.

Hogan, David John. 1985. *Class and Reform: School and Society in Chicago, 1880–1930.* Philadelphia: University of Pennsylvania Press.

Hounshell, David. 1984. *From the American System to Mass Production.* Baltimore, MD: Johns Hopkins Press.

Howell, William G., and Paul E. Peterson. 2004. "Uses of Theory in Randomized Field Trials: Lessons from School Voucher Research on Disaggregation, Missing Data, and the Generalization of Findings." *American Behavioral Scientist* 47 (January), pp. 634–657.

Hoxby, Caroline M. 1996. "Are Efficiency and Equity in School Finance Substitutes or Complements?" *Journal of Economic Perspectives* 10 (Fall), pp. 51–72.

———. 1997. "The Changing Market Structure of U.S. Higher Education." Working Paper, Harvard University.

———. 1998. "How Much Does School Spending Depend on Family Income? The Historical Origins of the Current School Finance Dilemma." *American Economic Review, Papers and Proceedings* 88 (May), pp. 309–314.

———. 1999. "The Productivity of Schools and Other Local Public Goods Providers." *Journal of Public Economics* 74 (October), pp. 1–30.

———. 2001. "All School Finance Equalizations Are Not Created Equal." *Quarterly Journal of Economics* 116 (November), pp. 1189–1231.

———. 2003. "School Choice and School Productivity: Could School Choice Be a Tide that Lifts All Boats?" In *The Economics of School Choice,* edited by Caroline M. Hoxby, 287–341. Chicago: University of Chicago Press and NBER.

Hoxby, Caroline M., and Ilyana Kuziemko. 2004. "Robin Hood and His Not So Merry Plan." NBER Working Paper no. 10722 (September).

Hsieh, Chang-Tan, and Miguel Urquiola. 2006. "The Effects of Generalized School Choice on Achievement and Stratification: Evidence from Chile's Voucher Program." *Journal of Public Economics* 90 (September), pp. 1477–1503.

Hughes, Gordon, and Barry McCormick. 1987. "Housing Markets, Unemployment and Labour Market Flexibility in the U.K." *European Economic Review* 31 (April), pp. 615–641.

Inter-university Consortium for Political and Social Research (ICPSR). 1984. *Census of Population, 1940* [United States]: *Public Use Microdata Sample* (ICPSR 8236). Ann Arbor, MI: ICPSR.

Iowa. [various years]. *Report of the Department of Public Instruction for* [*year*]. *Iowa School Report.* Des Moines, IA: State Printer.

Iowa. 1925. *Iowa Education Directory for the School Year Commencing July 1, 1924.* Des Moines, IA: State Printer.

Iowa Department of Public Instruction. [year] *Biennial Report of the Department of Public Instruction to the Governor of Iowa.* Des Moines, IA: State Printer.

Jackson, C. Kirabo. 2007. "A Little Now for a Lot Later: A Look at a Texas Advanced Placement Incentive Program." Unpublished paper, Harvard University.

Jacob, Brian A. 2005. "Accountability, Incentives and Behavior: The Impact of High-Stakes Testing in Chicago." *Journal of Public Economics* 89 (September), pp. 761–796.

Jaffe, Adam B. 1989. "Real Effects of Academic Research." *American Economic Review* 79 (December), pp. 957–970.

James, John. 1983. "Structural Change in American Manufacturing, 1850–1890." *Journal of Economic History* 43 (June), pp. 433–459.

James, John A., and Jonathan S. Skinner. 1985. "The Resolution of the Labor Scarcity Paradox." *Journal of Economic History* 45 (September), pp. 513–540.

Jencks, Christopher, and Susan Mayer. 1990. "The Social Consequences of Growing Up in a Poor Neighborhood." In *Inner-City Poverty in the United States,* edited by Laurence E. Lynn, Jr., and Michael G.H. McGeary, 111–186. Washington, DC: National Academy Press.

Jencks, Christopher, and Meredith Phillips, eds. 1998. *The Black-White Test Score Gap.* Washington, DC: Brookings Institution Press.

Jerome, Harry. 1934. *Mechanization in Industry.* New York: National Bureau of Economic Research.

Jessen, Carl A. 1928. *Requirements for High-School Graduation.* Bureau of Education Bulletin 1928, no. 21. Washington, DC: G.P.O.

Jones, Charles I. 1995. "R&D-Based Models of Economic Growth." *Journal of Political Economy* 103 (August), pp. 759–784.

———. 2002. "Sources of U.S. Growth in a World of Ideas." *American Economic Review* 92 (March), pp. 220–239.

Jorgenson, Dale, and Mun S. Ho 1999. "The Quality of the U.S. Workforce, 1948–95." Department of Economics, Harvard University.

Judd, Charles Hubbard. 1928. *The Unique Character of American Secondary Education.* The Inglis Lecture. Cambridge: Harvard University Press.

Kaestle, Carl. 1973. *The Evolution of an Urban School System: New York City, 1750 to 1850.* Cambridge: Harvard University Press.

———. 1983. *Pillars of the Republic: Common Schools and American Society, 1780–1860.* New York: Hill and Wang.

Kaestle, Carl, and Maris Vinovskis. 1980. *Education and Social Change in Nineteenth-Century Massachusetts.* Cambridge: Harvard University Press.

Kahn, Lawrence M. 2000. "Wage Inequality, Collective Bargaining, and Relative Employment 1985–94: Evidence from Fifteen OECD Countries." *Review of Economics and Statistics* 82 (November), pp. 564–579.

Kandel, I. L. 1930. *History of Secondary Education: A Study in the Development of Liberal Education.* Boston: Houghton Mifflin.

———. 1934. *The Dilemma of Democracy.* The Inglis Lecture. Cambridge: Harvard University Press.

————. 1955. *A New Era in Education: A Comparative Study.* Boston: Houghton Mifflin.

Kane, Thomas J. 1999. *The Price of Admission.* Washington, DC: Brookings Institution Press and Russell Sage.

Kane, Thomas J., Jonah E. Rockoff, and Douglas Staiger. 2006. "What Does Certification Tell Us About Teacher Effectiveness? Evidence from New York City." Unpublished paper, Harvard Graduate School of Education.

Kane, Thomas J., and Cecilia E. Rouse. 1999. "The Community College: Educating Students at the Margin between College and Work." *Journal of Economic Perspectives* 13 (Winter), pp. 63–84.

Kansas, State Superintendent of Public Instruction. [year]. *Biennial Report of the State Superintendent of Public Instruction for the Years Ending* [years]. Topeka, KS: State Printing Office.

Kaplan, David L., and M. Claire Casey. 1958. *Occupational Trends in the United States, 1900 to 1950.* Bureau of the Census Working Paper No. 5. Washington, DC: G.P.O.

Katz, Lawrence F., and David H. Autor. 1999. "Changes in the Wage Structure and Earnings Inequality." In *Handbook of Labor Economics,* vol. 3A, edited by Orley Ashenfelter and David Card, 1463–1555. Amsterdam: Elsevier Science.

Katz, Lawrence F., Gary W. Loveman, and David G. Blanchflower. 1995. "A Comparison of Changes in the Structure of Wages in Four OECD Countries." In *Differences and Changes in Wage Structures,* edited by Richard B. Freeman and Lawrence F. Katz, 25–65. Chicago: University of Chicago Press and NBER.

Katz, Lawrence F., and Kevin M. Murphy. 1992. "Changes in Relative Wages, 1963–87: Supply and Demand Factors." *Quarterly Journal of Economics* 107 (February), pp. 35–78.

Katz, Michael B. 1968. *The Irony of Early School Reform: Educational Innovation in Mid-Nineteenth Century Massachusetts.* Cambridge: Harvard University Press.

Keat, Paul G. 1960. "Long-Run Changes in Occupational Wage Structure, 1900–1956." *Journal of Political Economy* 68 (December), pp. 584–600.

Kelly, Frederick J., and Betty A. Patterson. 1934. *Residence and Migration of College Students,* Pamphlet no. 48, U.S. Bureau of Education. Washington, DC: G.P.O.

Kemple, James J., and Judith Scott-Clayton. 2004. *Career Academies: Impacts on Labor Market Outcomes and Educational Attainment.* New York: MDRC.

Kendrick, John W. 1961. *Productivity Trends in the United States.* Princeton, NJ: Princeton University Press (for NBER).

Kessler, Judd. 2007. "Crowding Out High School? The Effect of an Increase in GED Passing Standards on High School Graduation." Unpublished paper, Harvard University.

Kevles, Daniel. 1979. "The Physics, Mathematics, and Chemistry Communities: A Comparative Analysis." In *The Organization of Knowledge in Modern America, 1860–1920,* edited by Alexandra Oleson and John Voss. Baltimore, MD: Johns Hopkins University Press.

Kiger, Joseph C. 1982. *Research Institutions and Learned Societies.* Westport, CT: Greenwood Press.

Kimball, Bruce A. 1992. *The "True Professional Ideal" in America: A History.* Cambridge MA and Oxford UK: Blackwell Publishers.

Kling, Jeffrey R., Jeffrey B. Liebman, and Lawrence F. Katz. 2007. "Experimental Analysis of Neighborhood Effects." *Econometrica* 75 (January), pp. 83–119.

Kocka, Jürgen. 1980. *White Collar Workers in America, 1890–1940: A Social-Political History in International Perspective*. Translated by Maura Kealey. Beverly Hills, CA: Sage Publications.

Kopczuk, Wojciech, Emmanuel Saez, and Jae Song. 2007. "Uncovering the American Dream: Inequality and Mobility in Social Security Earnings Data since 1937." NBER Working Paper no. 13345 (August).

Krueger, Alan B. 1999. "Experimental Estimates of Educational Production Functions." *Quarterly Journal of Economics* 114 (May), pp. 497–532.

———. 2003. "Inequality, Too Much of a Good Thing." In *Inequality in America*, edited by James J. Heckman and Alan B. Krueger, 1–75. Cambridge, MA: MIT Press.

Krueger, Alan B., and Mikael Lindahl. 2001. "Education for Growth: Why and for Whom?" *Journal of Economic Literature* 39 (December), pp. 1101–1136.

Krueger, Alan B., and Pei Zhu. 2004. "Another Look at the New York City Voucher Experiment." *American Behavioral Scientist* 47 (January), pp. 658–698.

Krug, Edward A. 1962. "Graduates of Secondary Schools in and Around 1900: Did Most of Them Go to College?" *The School Review* 70 (Autumn), pp. 266–272.

———. 1964. *The Shaping of the American High School: 1880–1920*. Madison: University of Wisconsin Press.

Krugman, Paul R. 1990. *The Age of Diminished Expectations: U.S. Economic Policies in the 1990s*. Cambridge, MA: MIT Press.

Kuznets, Simon. 1953. *Shares of Upper Income Groups in Income and Savings*. New York: National Bureau of Economic Research.

Kuznets, Simon, Ann Ratner Miller, and Richard A. Easterlin. 1960. *Population Redistribution and Economic Growth: United States, 1870–1950*. Vol. 2, *Analyses of Economic Change*. Philadelphia: American Philosophical Society.

Labaree, David F. 1988. *The Making of an American High School: The Credentials Market and the Central High School of Philadelphia, 1838–1939*. New Haven, CT: Yale University Press.

Landes, David. 1972. *The Unbound Prometheus: Technological Change and Industrial Development in Western Europe from 1750 to the Present*. New York: Cambridge University Press.

Landes, William M., and Lewis C. Solmon. 1972. "Compulsory Schooling Legislation: An Economic Analysis of Law and Social Change in the Nineteenth Century." *Journal of Economic History* 32 (March), pp. 54–91.

Lebergott, Stanley. 1947. "Wage Structures." *The Review of Economic Statistics* 29 (November), pp. 274–285.

———. 1984. *The Americans: An Economic Record*. New York: W. W. Norton.

Lee, Everett S. 1961. "The Turner Thesis Reexamined." *American Quarterly* 13 (Spring), pp. 77–83.

Lee, Everett S., Ann Ratner Miller, Carol P. Brainerd, and Richard A. Easterlin. 1957. *Population Redistribution and Economic Growth, United States, 1870–1950*. Vol. 1, *Methodological Considerations and Reference Tables*. Philadelphia: American Philosophical Society.

Lemieux, Thomas. 2006a. "Postsecondary Education and Increased Wage Inequality." *American Economic Review* 96 (May), pp. 195–199.

———. 2006b. "Increased Residual Wage Inequality: Composition Effects, Noisy Data, or Rising Demand for Skill." *American Economic Review* 96 (June), pp. 461–498.

Lemon, James T. 1972. *The Best Poor Man's Country*. Baltimore, MD: Johns Hopkins Press.

Levy, Frank, Amy Beamish, Richard Murnane, and David Autor. 1999. "Computerization and Skills: Examples from a Car Dealership." Unpublished paper, MIT.

Levy, Frank, and Richard J. Murnane. 1996. "With What Skills Are Computers a Complement?" *American Economic Review* 86 (May), pp. 258–262.

———. 2004. *The New Division of Labor*. New York: Russell Sage.

Levy, Frank, and Peter Temin. 2007. "Inequality and Institutions in Twentieth Century America." NBER Working Paper no. 13106 (May).

Lewis, Sinclair. 1917. *The Job: An American Novel*. New York: Harper Brothers.

———. 1920. *Main Street: The Story of Carol Kennicott*. New York: Harcourt, Brace, and Howe.

Lindert, Peter H. 1994. "The Rise of Social Spending: 1880–1930." *Explorations in Economic History* 31 (January), pp. 1–37.

———. 1996. "What Limits Social Spending?" *Explorations in Economic History* 33 (January), pp. 1–34.

———. 2000. "The Comparative Political Economy of Mass Schooling before 1914." Working Paper, Department of Economics, University of California at Davis.

———. 2004. *Growing Public: Social Spending and Economic Growth since the Eighteenth Century*. Vol. 1, *The Story*. Vol. 2, *The Evidence*. New York: Cambridge University Press.

Lleras-Muney, Adriana. 2002. "Were Compulsory Attendance and Child Labor Laws Effective? An Analysis from 1915 to 1939." *Journal of Law and Economics* 45 (October), pp. 401–435.

———. 2005. "The Relationship between Education and Adult Mortality in the United States." *Review of Economic Studies* 72 (January), pp. 189–221.

Lochner, Lance, and Enrico Moretti. 2004. "The Effect of Education on Crime: Evidence from Prison Inmates, Arrests, and Self-Reports." *American Economic Review* 94 (March), pp. 155–189.

Long, Clarence D. 1960. *Wages and Earnings in the United States, 1860–1890*. Princeton, NJ: Princeton University Press.

Long, Jason, and Ferrie, Joseph P. 2007. "The Path to Convergence: Intergenerational Occupational Mobility in Britain and the US in Three Eras." *Economic Journal* 117 (March), pp. C61–C71.

Ludwig, Jens, and Douglas L. Miller. 2007. "Does Head Start Improve Children's Life Chances? Evidence from a Regression Discontinuity Design." *Quarterly Journal of Economics* 122 (February), pp. 159–208.

Machin, Stephen, and John Van Reenen. 1998. "Technology and Changes in Skill Structure: Evidence from Seven OECD Countries." *Quarterly Journal of Economics* 113 (November), pp. 1215–1244.

Maddison, Angus. 1987. "Growth and Slowdown in Advanced Capitalist Economies: Techniques of Quantitative Assessment." *Journal of Economic Literature* 25 (June), pp. 649–698.

Mann, Horace. 1841. *Fifth Annual Report of the Secretary of the Board of Education of Massachusetts*. [The *Fifth Annual Report* is included in Mann (1891)]
———. 1891. *Life and Works of Horace Mann*. Vols. 3 and 4, *Annual Reports of the Secretary of the Board of Education of Massachusetts for the Years 1839–44* [vol. 3], *1845–48* [vol. 4]. Boston: Lee and Shepard Publishers.
Margo, Robert A. 1990. *Race and Schooling in the South, 1880–1950: An Economic History*. Chicago: University of Chicago Press.
———. 2000. *Wages and Labor Markets in the United States, 1820–1860*. Chicago: University of Chicago Press.
Margo, Robert A., and T. Aldrich Finegan. 1996. "Compulsory Schooling Legislation and School Attendance in Turn-of-the Century America: A 'Natural Experiment' Approach." *Economics Letters* 53 (October), pp. 103–110.
Marks, Daniel. 1989. "Statewide Achievement Testing: A Brief History." *Educational Research Quarterly* 13 (March), pp. 36–43.
Matthews, Roderick Donald. 1932. *Post-Primary Education in England*. Ph.D. diss., University of Pennsylvania, Philadelphia.
McClure, Arthur F., James Riley Chrisman, and Perry Mock. 1985. *Education for Work: The Historical Evolution of Vocational and Distributive Education in America*. Rutherford, NJ: Fairleigh Dickinson University Press.
McKenzie, Fred A. c.1901. *The American Invaders: Their Plans, Tactics, and Progress*. New York: Street and Smith, Publishers.
Meyer, John W., David Tyack, Joane Nagel, and Audri Gordon. 1979. "Public Education as Nation-Building in America: Enrollments and Bureaucratization in the American States, 1870–1930." *American Journal of Sociology* 85 (November), pp. 591–613.
Michaels, Guy. 2007. "The Division of Labor, Coordination and the Demand for Information Processing." Unpublished paper, London School of Economics.
Michaelsen, Robert. 1970. *Piety in the Public School*. London: Macmillan.
Miller, Nelson A. 1947. *State, Regional, and Local Market Indicators, 1939–45*. Office of Domestic Commerce, Economic Studies No. 60. Washington, DC: G.P.O.
Milligan, Kevin, Enrico Moretti, and Philip Oreopoulos. 2004. "Does Education Improve Citizenship? Evidence from the U.S. and the U.K." *Journal of Public Economics* 88 (August), pp. 1667–1695.
Mincer, Jacob. 1974. *Schooling, Experience, and Earnings*. New York: Columbia University Press for the National Bureau of Economic Research.
Mishel, Lawrence, Jared Bernstein, and Sylvia Allegretto. 2005. *The State of Working America, 2004–5*. Ithaca, NY: I.L.R. Press.
———. 2007. *The State of Working America, 2006–2007*. Ithaca, NY: I.L.R. Press.
Mishel, Lawrence, and Joydeep Roy. 2006. *Rethinking High School Graduation Rates and Trends*. Washington, DC: Economic Policy Institute.
Mitchell, Daniel J. B. 1980. *Unions, Wages, and Inflation*. Washington, DC: The Brookings Institution.
———. 1985. "Shifting Norms in Wage Determination." *Brookings Papers on Economic Activity*, no. 2, pp. 575–599.
Morse, Perley. 1932. *Business Machines: Their Practical Application and Educational Requirements*. New York: Longmans, Green and Co.

Mroz, Thomas, Paul Rhode, and Koleman Strumpf. 2006. "Local Educational Investments and Migration: Evidence from 1940." Unpublished paper, University of North Carolina.

Murphy, Kevin M., and Finis Welch. 1993. "Occupational Change and the Demand for Skill, 1940–1990." *American Economic Review, Papers and Proceedings* 83 (May), pp. 122–126.

Murray, Sheila E., William N. Evans, and Robert M. Schwab. 1998. "Education-Finance Reform and the Distribution of Education Resources." *American Economic Review* 88 (September), pp. 789–812.

Nasar, Sylvia. 1992, August 16. "The Rich Get Richer, but Never the Same Way Twice." *New York Times.*

National Cash Register Company. 1904. *National Cash Register Factory, Dayton, Ohio, U.S.A., as Seen by English Experts of the Mostly Industrial and Educational Commissions,* compiled by Alfred A. Thomas. Dayton, OH: National Cash Register Company.

————. 1919. *Outline of Plan for the Education of Apprentices.* Dayton, OH: National Cash Register Company.

National Catholic Welfare Conference. [Various years] *Summary of Catholic Education.* Washington, DC: National Catholic Welfare Conference.

National Education Association. 1918. *Cardinal Principles of Secondary Education.* A Report of the Commission on the Reorganization of Secondary Education. Washington, DC: G.P.O.

Neal, Derek. 2006. "Why Has Black-White Skill Convergence Stopped?" In *Handbook of Economics of Education,* vol. 1, edited by E. Hanushek and F. Welch, 512–576. Amsterdam: North Holland.

Nelson, Daniel. 1987. "Mass Production in the U.S. Tire Industry." *Journal of Economic History* 47 (June), pp. 329–339.

Nelson, Richard R., and Edmund S. Phelps. 1966. "Investment in Humans, Technological Diffusion, and Economic Growth." *American Economic Review* 56 (May), pp. 69–75.

Nelson, Richard R., and Gavin Wright. 1992. "The Rise and Fall of American Technological Leadership: The Postwar Era in Historical Perspective." *Journal of Economic Literature* 30 (December), pp. 1931–1964.

New York City, Department of Education. 1911. *Thirteenth Annual Report of the City Superintendent of Schools for the Year Ending July 31, 1911.* New York.

New York Daily Times. 1852 to 1857.

New York State Regents. [Various years]. *Annual Report of the Regents of the University of the State of New York.*

Nickerson, Kermit. 1970. *150 Years of Education in Maine.* Augusta: State of Maine Department of Education.

Nord, Warren A. 1995. *Religion and American Education: Rethinking a National Dilemma.* Chapel Hill: University of North Carolina Press.

North Carolina State. 1910. *The North Carolina High School Bulletin,* vol. 1., edited by N. W. Walker. Chapel Hill: University of North Carolina.

Nye, David E. 1990. *Electrifying America: Social Meanings of a New Technology.* Cambridge, MA: MIT Press.

O. Henry. 1906. *The 4 Million.* New York: A.L. Burt Company.

Ober, Harry. 1948. "Occupational Wage Differentials, 1907–1947." *Monthly Labor Review* (August), pp. 27–134.

————. 1953. "Occupational Wage Differentials in Industry." In W.S. Woytinsky (and associates), *Employment and Wages in the United States*, chapter 40, pp. 466–474, 758–762 (appendix tables). New York: The Twentieth Century Fund.

Office Equipment Catalogue. 1924. *A Compilation of Condensed and Standardized Catalogue Data*. First Annual Edition. New York: Office Equipment Catalogue.

Oleson, Alexandra, and John Voss. 1979. *The Organization of Knowledge in Modern America, 1860–1920*. Baltimore, MD: Johns Hopkins University Press.

O'Neill, June. 1971. *Resource Use in Higher Education: Trends in Output and Inputs, 1930 to 1967*. Berkeley, CA: Carnegie Commission on Higher Education.

O'Neill, William L. 1972. *Women at Work, Including "The Long Day: The Story of a New York Working Girl" by Dorothy Richardson*. Chicago: Quadrangle Books.

Oregon. 1923. *Official Directory of Superintendents, Supervisors, Principals, High School Teachers, and Standard High School of the State of Oregon, 1923–1924*. Salem, OR: State Printing Department.

Oreopoulos, Philip. 2003. "Do Dropouts Drop Out Too Soon? Evidence from Changes in School-Leaving Laws." Unpublished paper, University of Toronto.

————. 2007. "Would More Compulsory Schooling Help Disadvantaged Youth? Evidence from Recent Changes to School-Leaving Laws." Unpublished paper, University of Toronto.

Organization for Economic Cooperation and Development [OECD]. [year]. *Education at a Glance: OECD Indicators, [Year]*. Paris: OECD.

Ottaviano, Gianmarco I. P., and Giovanni Peri. 2006. "Rethinking the Effects of Immigration on Wages." NBER Working Paper no. 12497 (August).

Ottumwa, Iowa. [year]. *Annual Directory of the Public Schools for the School Year Beginning September [1900 to 1929]*.

Perlmann, Joel, and Robert A. Margo. 2001. *Women's Work? American Schoolteachers, 1650–1920*. Chicago: University of Chicago Press.

Piketty, Thomas, and Emmanuel Saez. 2003. "Income Inequality in the United States, 1913 to 1998." *Quarterly Journal of Economics* 118 (February), pp. 1–39.

————. 2006. "The Evolution of Top Incomes: A Historical and International Perspective." *American Economic Review* 96 (May), pp. 200–205.

————. 2007a. "Updated Tables and Figures to Income Inequality in the United States: 1913–2002." Unpublished paper, University of California, Berkeley, available at http://elsa.berkeley.edu/~saez/TabFig2005prel.xls.

————. 2007b. "How Progressive is the U.S. Federal Tax System? A Historical and International Perspective." *Journal of Economic Perspectives* 21 (Winter), pp. 3–24.

Piore, Michael J., and Charles F. Sabel. 1984. *The Second Industrial Divide*. New York: Basic Books.

Portland, Oregon. 1920. *Report of the Superintendent of Schools. Portland School Report, 1920*. Portland, OR:

Poterba, James. 1997. "Demographic Structure and the Political Economy of Public Education." *Journal of Policy Analysis and Management* 16 (Winter), pp. 48–66.

Prestowitz, Clyde V. 1988. *Trading Places: How We Allowed Japan to Take the Lead*. New York: Basic Books.

Proffitt, Maris M. 1930. "Statistics of Private Commercial and Business Schools, 1928–29." U.S. Office of Education, Bulletin no. 25. Washington, DC: G.P.O.

Quigley, John M., and Daniel L. Rubinfeld. 1993. "Public Choices in Public Higher Education." In *Studies of Supply and Demand for Higher Education*, edited by Charles T. Clotfelter and Michael Rothschild, 243–273. Chicago: University of Chicago Press.

Quinn, Lois. 1993. "The Test That Became an Institution: A History of the GED." Xerox.

Randall, Samuel. 1844. *A Digest of the Common School System of the State of New York.* Albany, NY: C. Van Benthuysen & Co.

Ravitch, Diane. 1974. *The Great School Wars: New York City, 1805–1973: A History of the Public Schools as Battlefield of Social Change.* New York: Basic Books.

———. 2000. *Left Back: A Century of Failed School Reforms.* New York: Simon and Schuster.

Reese, William J. 1995. *The Origins of the American High School.* New Haven, CT: Yale University Press.

Reich, Robert B. 1991. *The Work of Nations.* New York: Knopf.

Ringer, Fritz K. 1979. *Education and Society in Modern Europe.* Bloomington, IN: Indiana University Press.

Riordan, Cornelius. 1990. *Girls and Boys in School: Together or Separate?* New York: Teachers College Press.

Romer, Paul H. 1990. "Endogenous Technological Change." *Journal of Political Economy* 89 (October), pp. S71–S102.

Roose, Kenneth D., and Charles J. Andersen. 1970. *A Rating of Graduate Programs.* Washington, DC: American Council on Education.

Rosenbaum, James E. 1995. "Changing the Geography of Opportunity by Expanding Residential Choice: Lessons from the Gautreaux Program." *Housing Policy Debate* 6 (1), pp. 231–269.

Rosenberg, Nathan, and Richard R. Nelson. 1996. "American Universities and Technical Advance in Industry." In *The Sources of Economic Growth*, edited by Richard R. Nelson, 189–229. Cambridge: Harvard University Press.

Ross, Dorothy. 1979, "The Development of the Social Sciences." In *The Organization of Knowledge in Modern America, 1860–1920*, edited by Alexandra Oleson and John Voss. Baltimore, MD: Johns Hopkins University Press.

Rossiter, Margaret W. 1979. "The Organization of the Agricultural Sciences." In *The Organization of Knowledge in Modern America, 1860–1920*, edited by Alexandra Oleson and John Voss. Baltimore, MD: Johns Hopkins University Press.

Rostow, Walt Whitman. 1960. *The Stages of Economic Growth: A Non-Communist Manifesto.* New York: Cambridge University Press.

Rotella, Elyce. 1981. *From Home to Office: U.S. Women at Work, 1870–1930.* Ann Arbor, MI: UMI Research Press.

Rouse, Cecilia E. 1998. "Private School Vouchers and Student Achievement: An Evaluation of the Milwaukee Parental Choice Program." *Quarterly Journal of Economics* 113 (May), pp. 553–602.

Rudolph, Frederick. 1965. *Essays on Education in the Early Republic.* Cambridge, MA: Belknap Press of Harvard University Press.

Ryan, Bryce, and Neal Gross. 1950. *Acceptance and Diffusion of Hybrid Corn Seed in Two Iowa Communities.* Research Bulletin no. 372, January. Agricultural Experiment Station, Sociology Subsection. Ames: Iowa State College of Agriculture and Mechanic Arts.

Schmidt, Stefanie. 1996. "School Quality, Compulsory Education Laws, and the Growth of American High School Attendance, 1915–1935." Ph.D. diss., Department of Economics, MIT.

Schultz, Theodore W. 1960. "Capital Formation by Education." *Journal of Political Economy* 68 (December), pp. 571–583.

———. 1964. *Transforming Traditional Agriculture*. New Haven, CT: Yale University Press.

Scott-Clayton, Judith. 2007. "What Explains Rising Labor Supply Among U.S. Undergraduates, 1970–2003?" Unpublished paper, Harvard University.

Sizer, Theodore R., ed. 1964a. *The Age of the Academies*. New York: Bureau of Publications Teachers College, Columbia University.

———. 1964b. *Secondary Schools at the Turn of the Century*. New Haven, CT: Yale University Press.

Slichter, Sumner H. 1950. "Notes on the Structure of Wages." *Review of Economics and Statistics* 32 (February), pp. 80–91.

Smith, James P., and Michael P. Ward. 1984. *Women's Wages and Work in the Twentieth Century*. Santa Monica, CA: The Rand Corporation.

Sokoloff, Kenneth L. 1984. "Was the Transition from the Artisanal Shop to the Non-Mechanized Factory Associated with Gains in Efficiency? Evidence from the U.S. Manufacturing Censuses of 1820 and 1850." *Explorations in Economic History* 21 (October), pp. 351–382.

———. 1986. "Productivity Growth in Manufacturing during Early Industrialization." In *Long-Term Factors in American Economic Growth*, Studies in Income and Wealth, NBER, vol. 51, edited by Stanley L. Engerman and Robert E. Gallman, 679–725. Chicago: University of Chicago Press.

Sokoloff, Kenneth L., and Stanley L. Engerman. 2000. "Institutions, Factor Endowments, and Paths of Development in the New World." *Journal of Economic Perspectives* 14 (Summer), pp. 217–232.

Solow, Robert M. 1956. "A Contribution to the Theory of Economic Growth." *Quarterly Journal of Economics* 70 (February), pp. 65–94.

———. 1957. "Technical Change and the Aggregate Production Function." *Review of Economics and Statistics* 39 (August), pp. 312–320.

Stanley, Marcus. 2003. "College Education and the Midcentury GI Bills." *Quarterly Journal of Economics* 118 (May), pp. 671–708.

Starr, Paul. 1982. *The Social Transformation of American Medicine*. New York: Basic Books.

Steinhilber, August W., and Carl J. Sokolowsi. 1966. *State Law on Compulsory Attendance*. U.S. Department of Health, Education and Welfare, Circular no. 793. Washington, DC: G.P.O.

Stewart, Rolland M. 1914. *Co-operative Methods in the Development of School Support in the United States*. Iowa City, IA: Chestnut Printing Co.

Stigler, George J. 1950. *Employment and Compensation in Education*. Occasional paper no. 33. New York: National Bureau of Economic Research.

———. 1956. *Trends in Employment in the Service Industries*. Princeton, NJ: Princeton University Press.

Stinebrickner, Ralph, and Todd R. Stinebrickner. 2003. "Working during School and Academic Performance." *Journal of Labor Economics* 21 (April), pp. 473–491.

Stokes, Anson Phelps, and Leo Pfeffer. 1964. *Church and State in the United States.* New York: Harper and Row.

Strom, Sharon Hartman. 1992. *Beyond the Typewriter: Gender, Class, and the Origins of Modern American Office Work, 1900–1930.* Urbana: University of Illinois Press.

Summers, Lawrence H. 1994. *Investing in All the People: Educating Women in Developing Countries.* Economic Development Institute Seminar Paper No. 45. Washington, DC: The World Bank.

Swift, Fletcher Harper. 1933. *European Policies of Financing Public Educational Institutions: France, Czechoslovakia, Austria, Germany, England and Wales.* Vol. 1, *France.* Berkeley: University of California Press.

Taubman, Paul, and Terence Wales. 1972. *Mental Ability and Higher Educational Attainment in the 20th Century.* NBER Occasional Paper 118. New York: National Bureau of Economic Research.

Thorndike, Edward L. 1907. "A Neglected Aspect of the American High School." *Educational Review* 33 (March), pp. 245–255.

Tinbergen, Jan. 1974. "Substitution of Graduate by Other Labor." *Kyklos* 27 (2), pp. 217–226.

———. 1975. *Income Distribution: Analysis and Policies.* Amsterdam: North-Holland.

Tocqueville, Alexis de. [1832] 1981. *Democracy in America.* New York: Modern Library.

Torpey, William George. 1948. *Judicial Doctrines of Religious Rights in America.* Chapel Hill: The University of North Carolina Press.

Troen, Selwyn K. 1975. *The Public and the Schools: Shaping the St. Louis System, 1838–1920.* Columbia: University of Missouri Press.

Trow, Martin. 1961. "The Second Transformation of American Secondary Education." *International Journal of Comparative Sociology* 2 (September), pp. 144–166.

Turner, Sarah E. 2004. "Going to College and Finishing College." In *College Choices*, edited by Caroline M. Hoxby, 13–56. Chicago: University of Chicago Press and NBER.

Tyack, David B. 1967. *Turning Points in American Educational History.* Waltham, MA: Blaisdell Publishing Co.

———. 1974. *The One Best System: A History of American Urban Education.* Cambridge: Harvard University Press.

Tyack, David B., and Elisabeth Hansot. 1990. *Learning Together: A History of Coeducation in American Schools.* New Haven, CT: Yale University Press.

Tyack, David B., Robert Lowe, and Elisabeth Hansot. 1990. *Public Schools in Hard Times: The Great Depression and Recent Years.* Cambridge: Harvard University Press.

Tyler, John H. 2004. "Does the GED Improve Earnings? Estimates from a Sample of Both Successful and Unsuccessful GED Candidates." *Industrial and Labor Relations Review* 57 (July), pp. 579–598.

Ueda, Reed. 1987. *Avenues to Adulthood: The Origins of the High School and Social Mobility in an American Suburb.* New York: Cambridge University Press.

U.S. Bureau of the Census. 1912. *Thirteenth Census of the United States: 1910. Population.* Washington, D.C.: G.P.O.

———. 1913. *Thirteenth Census of the United States, 1910.* Vol. 8, *Manufactures, 1909, General Report and Analysis.* Washington, DC: G.P.O.

———. 1923a. *Fourteenth Census of the United States, 1920.* Vol. 8, *Manufactures, 1919, General Report and Analytical Tables.* Washington, DC: G.P.O.

———. 1923b. *Fourteenth Census of the United States, 1920. Population.* Washington, DC: G.P.O.

———. 1927. *Financial Statistics of Cities, 1925.* Washington, DC: G.P.O.

———. 1932. *Fifteenth Census of the United States, 1930.* Vol. 3, *Population.* Washington, DC: G.P.O.

———. 1933. *Fifteenth Census of the United States, 1930.* Vol. 1, *Manufactures, 1929, General Report.* Washington, DC: G.P.O.

———. 1942. *Sixteenth Census of the United States, 1940. Manufactures 1939.* Vol. 1, *Statistics by Subject.* Washington, DC: G.P.O.

———. 1975. *Historical Statistics of the United States from Colonial Times to 1970.* Washington, DC: G.P.O. (*Note:* We refer to this source as *Historical Statistics.* It is to be distinguished from the more recent *Historical Statistics, Millennial Edition.*)

U.S. Bureau of Education. [Various years]. *Statistics of State Universities and Other Institutions of Higher Education Partially Supported by the State.* Bulletins 1908 no. 8; 1909 no. 11; 1910 no. 6; 1911 no. 19; 1913 no. 60. Washington, DC: G.P.O.

U.S. Bureau of Labor. 1897. *Eleventh Annual Report of the Commissioner of Labor, 1895–96: Work and Wages of Men, Women, and Children.* Washington, DC: G.P.O.

U.S. Census Bureau. 2005a. "Historical Income Tables—Experimental Measures," available at http://www.census.gov/hhes/www/income/histinc/incexper.html. Updated December 20, 2005.

———. 2005b. *Annual Survey of Manufactures. Statistics for Industry Groups and Industries: 2004.* Washington, DC: G.P.O.

———. 2006. "Historical Poverty Tables—Table 3," available at http://www.census .gov/hhes/www/poverty/histpov/hstpov3.html. Updated September 6, 2006.

———. 2007. "The Living Arrangement of Children in 2005," available at http:// www.census.gov/population/pop-profile/dynamic/LivArrChildren.pdf. Updated May 2007.

U.S. Census Office. 1853. *Seventh Census of the United States.* Washington, DC: Robert Armstrong.

———. 1864. *Eighth Census of the United States. Population of the United States in 1860.* Washington, DC: G.P.O.

———. 1872. *Ninth Census of the United States.* Vol. 1, *Population of the United States, 1870.* Washington, DC: G.P.O.

———. 1895a. *Report on Manufacturing Industries in the United States at the Eleventh Census: 1890.* Part 1, *Totals for States and Industries.* Washington, DC: G.P.O.

———. 1895b. *Report on Manufacturing Industries in the United States at the Eleventh Census: 1890.* Part 2, *Statistics of Cities.* Washington, DC: G.P.O.

———. 1895c. *Report on Manufacturing Industries in the United States at the Eleventh Census: 1890.* Part 3, *Selected Industries.* Washington, DC: G.P.O.

———. 1897. *Report on Population of the United States at the Eleventh Census: 1890.* Part 2. Washington, DC: G.P.O.

U.S. Commissioner of Education. 1895. *Report of the U.S. Commissioner of Education, 1891/92.* Vol. 5, part 2, chapter 26, "Coeducation of the Sexes in the United States," by A. Tolman Smith. Washington, DC: G.P.O.

———. 1906. *Report of the U.S. Commissioner of Education, 1904.* Vol. 2. Washington, DC: G.P.O.

U.S. Department of Commerce. [year]. *Statistical Abstract of the United States* [year]. Washington, DC: G.P.O.

———. 1930. *Religious Bodies: 1926.* Vol. 1, *Summary and Detailed Tables.* Washington, DC: G.P.O.

U.S. Department of Education, National Center for Education Statistics. [year]. *Digest of Education Statistics,* [year]. Washington, DC: G.P.O. (*Note:* We refer to this source as *Digest of Education Statistics* [year].)

———. 1993. *120 Years of American Education: A Statistical Portrait.* Washington, DC: G.P.O.

———. 1998. *Pursuing Excellence: A Study of U.S. Twelfth-Grade Mathematics and Science Achievement in International Context.* Washington, D.C.: G.P.O. http://nces.ed.gov/timss

———. 2004. *International Outcomes of Learning in Mathematics Literacy and Problem Solving: PISA 2003 Results from the U.S. Perspective.* Washington, D.C.: G.P.O. http://nces.ed.gov/pubs2005/2005003.pdf

———. 2007. *Condition of Education 2007.* Washington, DC: G.P.O.

U.S. Department of Health, Education and Welfare. [year]. *Statistics of State School Systems,* [year] *Final Report.* Washington, DC: G.P.O.

U.S. Department of Labor, Bureau of Labor Statistics. 1918–21. *Descriptions of Occupations: Coal and Water Gas, Paint and Varnish, Paper, Printing Trades, Rubber Goods; Electrical Manufacturing Distribution and Maintenance; Glass; Medicinal Manufacturing.* Prepared for the U.S. Employment Service. Washington, DC: G.P.O.

———. 1934. *History of Wages in the United States from Colonial Times to 1928.* B.L.S. Bulletin no. 604. Washington, DC: G.P.O.

———. 1938. "Earnings and Hours of Labor in Private Shipyards, 1936 and 1937." *Monthly Labor Review* (September).

———. 1938a. "Hourly Earnings in Furniture Manufacturing October 1937." *Monthly Labor Review* (November).

———. 1938b. "Average Hourly Earnings in Cotton-Goods Industry, 1937." *Monthly Labor Review* (April).

———. 1938c. "Earnings and Hours in the Soap Industry, January 1938." *Monthly Labor Review* (June).

———. 1940a. "Hourly Earnings in Dyeing and Finishing of Cotton, Rayon, and Silk." *Monthly Labor Review* (January).

———. 1940b. "Earnings and Hours in the Iron and Steel Industry, April 1938." *Monthly Labor Review* (August).

———. 1940c. "Earnings in Gray-Iron and Malleable-Iron Foundries, 1938–39." *Monthly Labor Review* (November).

———. 1941a. "Hourly Earnings in the Lumber and Timber Products Industry." *Monthly Labor Review* (July).

———. 1941b. "Earnings and Hours in the Rayon and Silk Industry, 1940." *Monthly Labor Review* (August).

———. 1941c. "Hours and Earnings in the Cigar Industry, 1940." *Monthly Labor Review* (December).

———. 1942a. "Earnings and Hours in Manufacture of Cigarettes, Chewing and Smoking Tobacco, and Snuff, December 1940." *Monthly Labor Review* (January).

———. 1942b. "Earnings in the Grain-Mill Products Industries, 1941." *Monthly Labor Review* (April).

U.S. Office of Education. [various years to 1917]. *Annual Report of the Commissioner of Education for [various years to 1917]*. Washington, DC: G.P.O. (*Note:* We refer to this source as the *Annuals.*)

———. [various years from 1916–18 to 1956–58]. *Biennial Survey of Education for [various years from 1916–18 to 1956–58]*. Washington, DC: G.P.O. (*Note:* We refer to this source as the *Biennials.* After 1953, the Office of Education was housed in the Department of Health, Education and Welfare.)

———. 1906. *Report of the Commissioner of Education for the Year Ending June 30, 1904.* Vol. 2. Washington, DC: G.P.O.

———. 1920. "Private Commercial and Business Schools, 1917–1918." Bulletin no. 47. Washington, DC: G.P.O.

———. 1961. *Higher Education Planning and Management Data.* Washington, DC: G.P.O.

Valletta, Robert G. 2006. "Computer Use and the U.S. Wage Distribution, 1984–2003." Federal Reserve Bank of San Francisco Working Paper 2006–34 (October).

Vermont Superintendent of Education. 1900. *Annual Report, 1900.* Montpelier, VT.

Veysey, Laurence R. 1965. *The Emergence of the American University.* Chicago: University of Chicago Press.

Vinovskis, Maris A. 1972. "Trends in Massachusetts Education, 1826–1860." *History of Education Quarterly* 12 (Winter), pp. 501–529.

Vinovskis, Maris. 1985. *The Origins of Public High Schools: A Reexamination of the Beverly High School Controversy.* Madison: University of Wisconsin Press.

———. 1995. *Education, Society, and Economic Opportunity.* New Haven, CT: Yale University Press.

Vinovskis, Maris A., and Richard M. Bernard. 1978. "Beyond Catharine Beecher: Female Education in the Antebellum Period." *Signs: Journal of Women in Culture and Society* 3 (Summer), pp. 856–869.

Wagoner, Harless D. 1966. *The U.S. Machine Tool Industry from 1900 to 1950.* Cambridge, MA: MIT Press.

Washington Higher Education Coordinating Board. [year] *Tuition and Fee Rates: A National Comparison.* Available on the HECB website at www.hecb.wa.gov.

Washington State. 1922. *Twenty-sixth Biennial Report of the Superintendent of Public Instruction. Report of High School Inspector.* Olympia, WA.

Watson, Tara. 2006. "Metropolitan Growth, Inequality and Neighborhood Segregation by Income." *Brookings-Wharton Papers on Urban Affairs*, no. 7, pp. 1–52.

WebCASPAR. *Integrated Science and Engineering Resource System of the National Science Foundation.* Available at http://caspar.nsf.gov/

Weiss, Janice Harriet. 1978. *Educating for Clerical Work: A History of Commercial Education in the United States since 1850.* Ann Arbor, MI: UMI Press.

Welch, Finis. 1970. "Education in Production." *Journal of Political Economy* 78 (February), pp. 35–59.

Whaples, Robert. 1990. "The Shortening of the American Workweek: An Economic and Historical Analysis." Ph.D. diss., Department of Economics, University of Pennsylvania.

Williamson, Jeffrey G. 1975. "The Relative Costs of American Men, Skills, and Machines: A Long View." Institute for Research on Poverty Discussion Paper 289–75. University of Wisconsin, Madison.

Williamson, Jeffrey G., and Peter H. Lindert. 1980. *American Inequality: A Macroeconomic History.* New York: Academic Press.

Wolff, Edward N. 1995. *Top Heavy: A Study of the Increasing Inequality of Wealth in America.* New York: Twentieth Century Fund Press.

Woolf, Arthur George. 1980. "Energy and Technology in American Manufacturing: 1900–1929." Ph.D. diss., Department of Economics, University of Wisconsin.

Woytinsky, W. S. (and Associates). 1953. *Employment and Wages in the United States.* New York: The Twentieth Century Fund.

Yates, Joanne. 1989. *Control through Communication: The Rise of System in American Management.* Baltimore, MD: Johns Hopkins University Press.

Zook, George F. 1926. *Residence and Migration of University and College Students.* Bulletin 1926, no. 11, U.S. Bureau of Education. Washington, DC: G.P.O.

Acknowledgments

This volume is a melding of our intellectual passions. Lawrence Katz has measured, tracked, and dissected the wage structure and economic inequality for more than two decades; Claudia Goldin has investigated the history of education and human capital for almost as long. Together we have worked on the labor market impact of technological change, the returns to education, and the long-run evolution of the U.S. wage structure.

We do not remember exactly when we embarked on the project that has culminated in this book. One part of the project was begun when Goldin was a fellow at the Brookings Institution and investigated the high school movement and its impact on the wage structure. Several years later, when we were fellows at the Russell Sage Foundation, we worked on the history of higher education. At other times we worked on skill-biased technological change and the wage structure. We began the actual writing of the book about three years ago, and chapters were written during the summer months when we were free of teaching obligations.

A National Science Foundation grant (SBR-951521) supported our work on skill-biased technological change and long-run changes in wage ratios by skill. A Spencer Foundation grant (199600128) funded work on the history of education, in particular the collection of the 1915 Iowa State Census sample. The Brookings Institution funded a leave for Goldin; the Russell Sage Foundation did so for both Goldin and Katz in 1997–98 as did the Radcliffe Institute for Advanced Study in 2005–6. We gratefully acknowledge funding and support from all of these institutions and agencies.

We have many individuals to acknowledge and thank. Colleagues have been enormously generous in providing comments, information, and data. We will never be able to recall all of them, but they include the following.

Stanley Engerman, Alan Krueger, Ilyana Kuziemko, Robert A. Margo, and Sarah Turner read the entire draft and gave us comments and suggestions that went far beyond usual collegial help. In her final weeks as a Harvard graduate student, Ilyana became our indefatigable advisor and we will forever be grateful to her for companionship and advice. Michael Aronson, our Harvard University Press editor, encouraged us, offered suggestions, and asked probing questions. We are deeply in the debt of each of these incredibly supportive and giving individuals.

Among those who provided information and data are the following (in alphabetical order). David Autor helped construct measures of wage inequality, educational wage differentials, and relative skill supplies from U.S. Census of Population and Current Population Surveys; Robert Barro supplied world religion data that he and his collaborators had compiled; Eric Hilt provided information about mid-nineteenth-century New York State academies; Adriana Lleras-Muney shared her coding of twentieth-century state compulsory education and child labor laws; Robert Margo lent us microfilms of the U.S. Census, Social Statistics for 1850 to 1870; Paul Rhode provided data on defense contracts; Alicia Sasser gave us state-level K-12 expenditure data; Judith Scott-Clayton provided school enrollment and educational attainment data from the October Current Population Surveys; Sarah Turner sent us many series for Chapters 7 and 9 and, with unceasing support, guided us through educational data mazes; Robert Whaples provided city-level data; and Arthur Woolf gave us his industry-level data from the census of manufactures. We are grateful to them all.

Many individuals provided comments along the way. They include (in alphabetical order): Daron Acemoglu, David Autor, Howard Bodenhorn, Susan Dynarski, Roland Fryer, Edward Glaeser, James J. Heckman, Caroline Hoxby, Brian Jacob, Peter Lindert, Lawrence

Mishel, Paul Rhode, Andrei Shleifer, Marcus Stanley, Lawrence Summers, John Wallis, and David Wessel. They have each made the volume better with their pointed questions and remarks.

We have been fortunate to have had many collaborators from whom we have learned a great deal. We are grateful to David Autor, George Borjas, Brad DeLong, Richard Freeman, Melissa S. Kearney, Alan Krueger, Ilyana Kuziemko, Robert A. Margo, and Kevin M. Murphy for their past work with us on topics closely related to those of this book.

Research assistants have helped at every stage of the project beginning with Linda Tuch, who worked for Goldin at the Brookings Institution. Cheryl Seleski and Kerry Woodward helped us at the Russell Sage Foundation. Many assistants helped collect 60,000 observations from the 1915 Iowa State Census. They include Todd Braunstein, Brigit Chen, Michael Cress, Misha Dewan, Serena Mayeri, Arvind Kirshnamurthy, and especially Allegra Ivey, who collected most of the rural sample and much of the urban one.

A group of extremely talented assistants worked with us every summer including Anne Berry, Maya Federman, Katherine Kaplan, Fabiana Silva, Abigail Waggoner Wozniak, and most recently Crystal Yang, who has been our very able assistant during most of the writing phase of the project. Leah Platt Boustan was of enormous assistance in the crafting of Chapter 4. Carola Frydman worked on parts of Chapter 6, especially that on teachers on which she wrote an excellent paper for a course she took from Goldin. Both Leah and Carola labored joyously at various archives, demonstrating their penchant for history as well as their scholarship.

The National Bureau of Economic Research provided the very best intellectual environment for us during our summers of writing and gave our research assistants desks nearby and ample resources. We thank the fine staff of the NBER for their support.

We have received an enormous number of helpful comments from people who attended our seminars and public lectures at various universities, colleges, and other forums. We cannot list them here, but we thank those in attendance for their support and suggestions. Finally, we thank Meg Fergusson who carefully copy-edited the manuscript.

The title of this book was inspired by Jan Tinbergen, who used a similar metaphor in describing the forces affecting the historical evolution of the wage structure.

This book differs, sometimes markedly, from the various papers we have written and published on inequality, education, and technology. Nevertheless, there are several places where tables, figures, and even some text have been borrowed from one of our published papers. The papers and the chapters are the following:

Chapter 1: DeLong, J. Bradford, Claudia Goldin, and Lawrence F. Katz. 2003. "Sustaining U.S. Economic Growth." In H. Aaron, J. Lindsay, and P. Nivola, eds., *Agenda for the Nation*. Washington, DC: Brookings Institution Press.

Chapter 2: Goldin, Claudia, and Lawrence F. Katz. 2000. "Education and Income in the Early Twentieth Century: Evidence from the Prairies." *Journal of Economic History* 60 (September), pp. 782–818.

Chapter 2: Goldin, Claudia, and Lawrence F. Katz. 2001. "Decreasing (and then Increasing) Inequality in America: A Tale of Two Half Centuries." In Finis Welch, ed., *The Causes and Consequences of Increasing Inequality*. Chicago: University of Chicago Press, pp. 37–82.

Chapter 3: Goldin, Claudia, and Lawrence F. Katz. 1998. "The Origins of Technology-Skill Complementarity." *Quarterly Journal of Economics* 113 (June), pp. 693–732.

Chapter 6: Goldin, Claudia, and Lawrence F. Katz. 1999. "Human Capital and Social Capital: The Rise of Secondary Schooling in America, 1910 to 1940." *Journal of Interdisciplinary History* 29 (Spring), pp. 683–723.

Chapter 7: Goldin, Claudia, and Lawrence F. Katz. 1999. "The Shaping of Higher Education: The Formative Years in the United States, 1890 to 1940." *Journal of Economic Perspectives* 13 (Winter), pp. 683–723.

We thank one another for being the essential part that completes the whole. We are each imperfect, and neither of us could have written the volume without the other. Finally, we thank Prairie, who does not know about chapters, bibliographies, tables, and figures. She maintained our sanity throughout the years with licks and tail wagging, and she continues to produce volumes of love and affection.

Index